Energy Storage Devices for Electronic Systems

Energy Storage Devices for Electronic Systems

Rechargeable Batteries and Supercapacitors

Nihal Kularatna
School of Engineering, The University of Waikato
Hamilton, New Zealand

ELSEVIER

AMSTERDAM • BOSTON • HEIDELBERG • LONDON
NEW YORK • OXFORD • PARIS • SAN DIEGO
SAN FRANCISCO • SINGAPORE • SYDNEY • TOKYO
Academic Press is an imprint of Elsevier

AP

Acquiring Editor: Joe Hayton
Editorial Project Manager: Chelsea Johnston
Project Manager: Preethy Simon
Designer: Maria Inês Cruz

Academic Press is an imprint of Elsevier
32 Jamestown Road, London NW1 7BY, UK
525 B Street, Suite 1800, San Diego, CA 92101-4495, USA
225 Wyman Street, Waltham, MA 02451, USA
The Boulevard, Langford Lane, Kidlington, Oxford OX5 1GB, UK

Library of Congress Cataloging-in-Publication Data
A catalog record for this book is available from the Library of Congress

British Library Cataloguing in Publication Data
A catalogue record for this book is available from the British Library

For information on all Academic Press publications
visit our web site at store.elsevier.com

ISBN: 978-0-12-407947-2

Dedication

This book is dedicated to Professor W. P Jayasekara,
who honourably dedicated his long academic career in Sri Lanka
to teach fundamentals of electrical
engineering to many engineering students,
supported by his capable team of academic staff at the Faculty of
Engineering,
and giving me,
the wisdom of applying simple fundamentals to practical circuits.

Contents

Preface

Electronic engineering advances at a rapid rate where designers and researchers are expected to maintain their *knowledge half-life* at 3-5 year levels. New subjects enter the field of electronics, and we have to keep up-to-date by learning these new subjects and still maintain our historic and practical experience-based know-how in the background to mix with the new subject areas. In this process one simple thing has never changed—that is the simple fundamentals we apply in circuits on a daily basis.

During my last 13 years of the full-time academic career, I have learned something important. If the simple fundamentals can be practically and effectively applied to new circuit topologies and state-of-the-art devices, we can come up with novel techniques, and develop them into very useful industrial applications, provided we also learn from our globally distributed subject experts and mix those knowledge elements with our own work. In this exercise, attributes such as perseverance, creativity, and life-long learning help us immensely.

This book is my eighth major work, and this is mostly based on my last 10 years of the academic and industrial experience in working with energy storage devices (ESDs), particularly in the subject of nontraditional supercapacitor applications. In this subject area, I was once again convincing myself of the most important aspect of learning—how to apply simple circuit fundamentals with new (energy storage or other) devices, with persevering efforts to understand the underlying physics and electrochemistry. Particularly in developing the Chapters 2, 3, and 5 of the work, I practically learnt that there is a massive ongoing concerted effort by the electrochemists, physicists, and technologists to enhance the performance specifications of ESDs. I sincerely thank all of them for their concerted efforts to give the electronic circuit and system designers new packaged devices where we could start with simplified equivalent circuits, and industrial device data sheets to develop new applications.

This work reflects a summary of the applications knowledge gained by a team of researchers lead by me in New Zealand, directly and indirectly supported by the concerted efforts of the global team. In this work, most of research projects we work on could help the nature and the environment too, with an effort to understand that the energy resources are limited and renewable energy area should be well supported by new research efforts on ESDs and applications.

Thank you,

Nihal Kularatna
School of Engineering
The University of Waikato
Hamilton
New Zealand
8th September 2014

Acknowledgments

From my university graduation with a degree in electrical engineering, I have spent over 38 years in the industry and academia, and completing almost 13 years in a full-time academic career in New Zealand. All my subject learning mixed with life experience tells me one thing... learning a subject is a life-long exercise, and it is a moving target. In this journey, all my school teachers, university teachers, senior industry engineers, and the mentors/advisors have helped me immensely. I thank all of them who helped me come a long-way in an electronic-engineering career.

For my engineering expertise I thank Professor W.P. Jaysekara and his team at University of Peradeniya, Sri Lanka who gave me the courage and the perseverance to think of fundamentals first in any EE project. My previous work places such as the Area Control Centre of Department of Civil Aviation, Sri Lanka; Saudi Telephone; Arthur C Clarke Institute for Modern Technologies (ACCIMT), Sri Lanka provided me all the opportunities to gain real-world on-the job engineering experience, for which I am very grateful. At ACCIMT I was inspired by Sir Arthur C Clarke, and also he introduced me top US engineers such as John Robinson Pierce who named the transistor. I strongly believe that my mid-career was well supported by the opportunities, facilities, and the team at the ACCIMT and I am very thankful to Sir Arthur and the ACCIMT team with whom I worked for 16 years. I am very grateful to a mentor such as Prof Pierce, who guided me in 1992 on how to complete a book project successfully.

I am very thankful to Prof. John T. Boys of University of Auckland, who invited me to come into the academia in 2002. After 25 years in the industry, this move made me develop a new research area with an academic inclination into the depth of a selected subject utilizing my breadth of the subject area and a strong team work.

I thank all my colleagues and the senior management at the University of Waikato who continually encourage me to contribute. During the last 7 years with various commercial type grants from the university's commercial arm, WaikatoLink Limited (WLL), we were very much encouraged to develop new commercially useful supercapacitor-based techniques, and I am very thankful to the whole team at WLL. I am very thankful to Professors Janis Swan, Ilanko Sinniah, Brian Gabbitas who were my line managers who encouraged me to develop my academic research programs with a strong industrial flavor. In my post graduate research work, for cosupervision, I am very grateful to my colleagues such as Alistair Steyn-Ross, Rainer Kunnemeyer, Howell Round, and Sadhana Talele. I am also very thankful to my school administrators Janine Williams and Mary Dalbeth for assisting me on a daily basis to perform my teaching and research tasks effectively.

My former PhD student Kosala Gunawardane and the current student Jayathu Fernando are gratefully acknowledged for coauthoring three chapters of the work. Many other postgraduate students helped me developed lot of details in various research projects and I am thankful to all of them for their efforts, which are directly or indirectly used in chapter contents. Jayathu Fernando, Tanya Jayasuriya, and Dhanya Herath were helping me immensely to generate a significant number of figures required for the book and I am very thankful to him for that work.

From establishing the initial contract into the final production of the printed work I gratefully acknowledge the editorial and the production team of Elsevier with special thank you notes to Joe Hayton, Tiffany Gasbarriny, Kattie Washington, Preethy Mampally, and Chelsea Johnston. I also thank IEEE, PET Magazine, EDN, CRC Press, and Elsevier Publications for me to reuse various figures required in developing the manuscript. Natalie Guest, Nicoloy Gurusinghe, and Jayathu Fernando are gratefully acknowledged for the cover page image of the book.

In my home environment, I am fully supported and always encouraged by my loving wife Priyani, daughters Dulsha and Malsha, son-in-laws Rajith and Kasun to do my technical writing. I am most grateful to them for this continuous encouragement. Two little grand-daughters Nethuli and Mineli are keeping me and Priyani entertained and it provides me a reason to think of future and the environment.

I thank all my friends and extended families who have always appreciated my technical work and the authorship of technical books.

I trust that my current work on ESD applications will help the environment in the longer run, and provides us creative means apply these devices in unique ways.

Last, I thank all my past and future students who will use this new book and encourage them to point any errors, and suggestive improvements.

Nihal Kularatna
29, Langdale Court
Huntington
Hamilton 3210
New Zealand
9th September 2014

Energy storage devices—a general overview

1

1.1 Introduction

From the invention of the transistor in late 1940s, electronic products and systems have become an essential part of the modern world with interconnected global communication systems, intelligent white goods, electronic subsystem dominated automotive and a plethora of portable devices. With the world population growing beyond 7 billion in 2012, and with the portable electronic devices getting popular even in the poorest parts of the world, worldwide energy consumption keeps growing at a steady rate. In 2008, world's energy consumption was estimated to be around 144,000 TWh. Energy supply usually comes from several dominant sources, namely, fossil fuel and nuclear (as nondominant major sources), and from many renewable sources such as hydroelectric, solar, wind, geothermal, biomass, and biofuels.

In the world of energy usage, energy storage requirement comes in different forms. In our home and work environments, sometimes we need to store energy in systems such as the uninterruptible power supplies (UPSs) used for information systems and other critical facilities such as hospitals, airports, and factories in order to safeguard against power outages. In these situations, different types of energy storage systems (ESS) such as batteries, supercapacitor (SC) banks, flywheels, and compressed air are used together with suitable electromechanical energy conversion systems. In these systems, we usually deal with few 100 W to few megawatts order requirements with the outages expected in the range of fractional seconds to few hours.

In our modern hybrid electric vehicles (HEV) and electric vehicles (EV), battery banks and fuel cells are used to deal with kilowatt order requirements lasting for few minutes to few hours of driving, where the ESS comes into play. With the inefficiencies of internal combustion engine together with mechanical drive systems in the cars, overall energy efficiency is in the range of 15% to over 20%, encouraging the development of more energy efficient HEVs and EVs. In these applications, where tens of kilowatts of power is required, the capacity of the battery banks in terms of energy storage should be in the range of kilowatt-hours.

Compared to the above scenario, if we consider our portable electronic devices such as cell phones, PDAs, and laptops where battery packs are used, we commonly come across the capacity in terms of milliampere-hours (mAh) or ampere-hours (Ah), with battery packs with nominal voltage values varying from about 2.5 to about 16 V. Given these nominal voltage values, assumed to be constant over the operational range, will give us the energy storage capacity in mWh to Wh. Compared to the two above application areas, if we take much smaller power and energy requirements such as wrist watches, hearing aids and implants, the battery pack capacity may be specified in microwatt-hours to few milliwatt-hours only. Terminal voltages of these battery packs, mostly coming in the form of one or few single cells, may be in the range of 0.5 V to few Volts per cell.

Given a few common day-to-day examples of battery-based systems, we could practically appreciate that a few commonly identified electrical parameters such as terminal voltage, and the energy storage capacity in terms of Watt-hours or milliwatt-hours allow us to compare the capabilities of the battery packs. If a load consumes power at a steady rate of 1 W for 1 h, the device has consumed a total energy of 1 Wh during that period. In standard SI units, this translates to a value of 3600 J, since 1 W dissipated during 1 s is equal to 1 J.

So far we have considered only batteries as an energy storage device (ESD) family, but in general many different methods and devices can be used for energy storage suitable for electrical and electronic systems. This chapter provides an overview of energy storage and delivery devices, methods, and the essential fundamentals applicable to energy transfer into electronic systems and devices. With some quantification approach, the contents will assist comparing the details of energy storage capabilities, and energy delivery capabilities and limitations of ESDs in electronic environments in a practically useful way. The contents of this and the overall work are aimed as a practical guideline to engineers, designers, and researchers working in these areas. Reader is expected to have some essential prior knowledge on circuit theory applicable to simple circuits.

1.2 Simple fundamentals

Let us start with the simple analogy of a water tank at a certain height above the ground level. If we have a hose connected from the bottom of the tank, with a tap connected at the end of the hose, water flow starts when the tap is opened. We can measure the rate of water flow in terms of liters/second. If you consider the tank as the storage device, at this rate we can estimate the time to release the total quantity of water. If we make the tank bigger, we get more quantity of water stored in the tank. If the bottom of the tank is at the same level as before, we will have the same rate of flow at the end of the pipe. We will take a longer time to release that water at that same rate. Now we can raise the bottom level of the tank which will increase the rate of flow. This means the tank will empty faster. We can also increase the diameter of the water hose connected to achieve a higher flow rate of water. This happens since the resistance to water flow will reduce due to higher diameter of the hose. If you increase the length of the hose, the flow rate will be slower.

The above simple analogy helps us understand ESDs. The level of the water tank (above the reference level of ground) is analogous to the (positive) terminal voltage of the device of the with respect to the negative terminal (which is the reference voltage level). Volume of the tank is analogous to the amount of total charge (or energy) available for release into the external circuit. Water flow in the hose is analogous to the current (in Amperes or coulombs/second) in the external circuit. By changing length or diameter of the hose, we create the analogy of a resistance of a conductor (which decreases with the diameter and increases with the length). Sometimes changing the material of the hose, we can notice the different rate of flow for the same diameter/length, which indicates us the analogy of the resistivity of an electrical material. If we extend this to the behavior of the tap, opening and closing is analogous to a short circuit (with ideally zero resistance) and an open circuit (of infinite resistance) condition, respectively.

1.2.1 Work, power, and energy

We all have an intuitive sense of the term *work*. Precise definition of work reflects this intuition and we can define work done, W, by an applied force $\vec{F}$ (a vector quantity, which has an amplitude and a direction) moving an object by a distance Δx,

$$W = \vec{F}\Delta x \tag{1.1}$$

In SI units, force is measured in Newtons and distance moved is in meters, giving units of Newton-meter. Newton-meter is given the name Joule. It is important to note that to have distance moved and force applied should be in the same direction. Otherwise, respective directional components need to be used. Work is a scalar quantity and has no direction, but the force and distance both are vector quantities.

Closely related to work, *energy* is the ability to do work. Work and energy have the same unit, Joule. Energy, which is one of the most important concepts in all of physics, comes in different forms. For example, a moving object of mass m moving at velocity of v m/s carries a kinetic energy, K, given by $(1/2)mv^2$. Work-energy theorem indicates us that the change in an object's kinetic energy is equal to the net work done on the object:

$$\Delta k = W_{\text{net}} \tag{1.2}$$

In practice, we come across many different units that are used for energy, particularly applicable to different subject areas. Examples are *electron-volt* used in nuclear, atomic, and molecular physics; and the *calorie* used in thermodynamics and chemistry. Older English units include *foot-pound* and the *British thermal unit (BTU)* and the latter is used in heating and cooling systems. Electricity companies use *kilowatt-hours* (kWh). For all these commonly used practical units, one can use conversion factors as in Table 1.1 to convert to Joules.

Next important item in energy conversion and ESDs is the *power*, which is the rate of doing work or rate of change of energy, which is measured in *watts*. Power is defined as the rate of doing work. *Average power*, $\overline{P}$, can be expressed as

$$\overline{P} = \frac{\Delta W}{\Delta t} \tag{1.3}$$

In practice, as the rate of doing work varies with time it is useful to define the *instantaneous power*, P, as

Table 1.1 Conversion coefficients for different units for energy and power

	To convert from	To	Multiplication factor
Energy	Barrel of oil	Gigajoules (GJ)	≈ 6
	Cubic foot of methane (at standard temperature and pressure)	Megajoules (MJ)	≈ 1
	KiloWatt-hour	Joule	3.6 x 10^{6}
	Calorie (mean)[a]	Joule	4.19002
	British Thermal unit (BTU)[b]	Joule	1055.87
	Electron-volt	Joule	1.60206 x 10^{-19}
Power	Foot lbf/hour	Watt	3.7662 x 10^{-4}
	Horse power (electrical)	Watt	746
	Horse power (mechanical)	Watt	735

[a]There are slightly different conversion factors for calorie used in different cases (see da Rosa, 2013 for details).
[b]Mean value is indicated (see da Rosa, 2013 for details).
Adapted from da Rosa, 2013.

$$P = \lim_{\Delta t \to 0} \frac{\Delta W}{\Delta t} = \frac{dW}{dt} \tag{1.4}$$

Since our subject of the book is ESDs for electrical or electronic engineering environments, we deal with power and energy associated with electrical, electrochemical, or electromechanical systems, we can measure electrical power consumed or generated based on the terminal voltage and the current passing through an electrical component as

$$P = VI \tag{1.5}$$

Unit of power is Joules/second and this unit is given the name *watt* (W) in honor of James Watt, a Scottish engineer who developed the steam engine. Watt himself defined another unit, *horsepower (hp)*, which is about 746 J/s or 746 W. Table 1.1 provides some useful conversion factors usable in work, energy, and power (da Rosa, 2013).

1.2.2 Impact of the open circuit voltage and internal resistance of an energy source

Now let us look at a simple energy source, with a constant terminal voltage of V volts, with an energy storage capacity of E watt-hours (Joules $\equiv$ watt-seconds). In any such ESD, there is a finite internal resistance, and if we quantify that by a lumped resistance, r_{int}, we can draw up an equivalent circuit as shown in Figure 1.1(a). Now we can have a load in the form of a resistor depicted by R_L connected to this device which forms the closed circuit as shown in Figure 1.1(b).

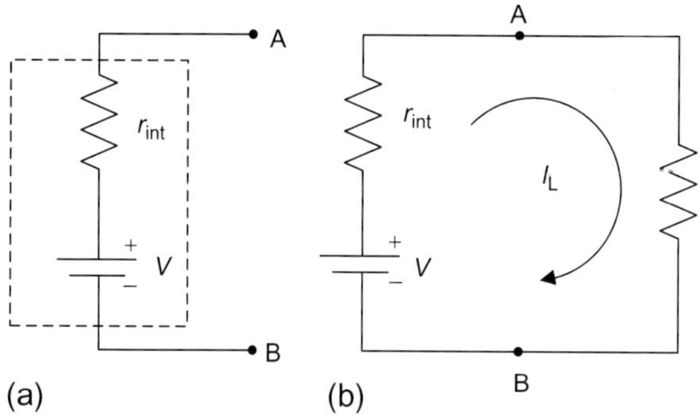

Figure 1.1 Energy storage device (ESD) and a resistive load connected to its external terminals (a) storage device represented by a constant voltage source and a fixed lumped resistance of value r_{int} (b) closed circuit formed by the external load.

This situation can be simply analyzed by applying the ohms law to the closed circuit and this will give us a current of I_L given by

$$I_L = \frac{V}{R_L + r_{int}} \tag{1.6}$$

The actual voltage across the load is given by

$$V_L = \frac{VR_L}{R_L + r_{int}} \tag{1.7}$$

This can be rewritten as

$$V_L = \frac{V}{1 + \frac{r_{int}}{R_L}} \tag{1.8}$$

Equation (1.8) indicates us that useful voltage appearing at the load end keeps dropping as the internal resistance of the energy source keeps increasing. Therefore, a good energy source should not only have a higher open circuit voltage, but also a very low internal resistance. Now we can calculate the power dissipated at the load by

$$P_L = V_L I_L = \frac{V}{1 + \frac{r_{int}}{R_L}} \frac{V}{R_L + r_{int}} = \frac{V^2}{R_L}\left[\frac{1}{\left(1 + \frac{r_{int}}{R_L}\right)^2}\right] \tag{1.9}$$

This indicates us that total power deliverable to a load by an ideal power source with zero internal resistance is V^2/R_L, and in practice the maximum power deliverable to a load is restricted by the internal resistance of the energy source. For instance, in many rechargeable batteries internal resistance keeps increasing with the discharge. Then even if the electrochemistry in the cell allows us to have a constant open circuit voltage, as the cell keeps discharging the available power keeps dropping due to the increase of internal resistance. This is one of the reasons for SCs' ability to supply large amounts of power for short durations, compared to batteries. Many modern SC families have very low internal resistance, for example in the order of fractional milliohms to tens of milliohms, and this allows these devices to drive large amounts of power into a load, when fully charged up to the rated voltage.

1.2.2.1 Maximum power transfer

Given the relationship in Eqn (1.9), we see that due to internal resistance of the ESD, part of the energy is converted and wasted as heat generated in the internal resistance, and only a particular percentage is usable at the external load. In many cases, we need a particular level of power to be delivered to the external load and this is limited by the combination of internal resistance and the external load resistance. Using differentiation of the function given by Eqn (1.9), with respect to R_L we can show that maximum possible load power occurs when $R_L = r_{int}$. Under this condition, where $R_L = r_{int}$ maximum possible power is given by $V^2/4R_L$.

1.2.3 Energy wasted inside a source and its heating effect

From Eqn (1.6), we can derive the amount of power wasted as heat inside an energy source as

$$P_{loss,int} = I_L^2 r_{int} = \left(\frac{V}{R_L + r_{int}}\right)^2 r_{int} \tag{1.10}$$

Using Eqn (1.4), this can be rewritten as

$$P_{loss,int} = P_L \left(\frac{r_{int}}{R_L}\right) \tag{1.11}$$

This indicates us that while we extract energy from an energy source, a quantifiable share is wasted within the source. This energy wastage is in the form of heat, and it could usually have an adverse effect on the life time of the device. In general, most device specifications are temperature sensitive, and higher operational temperatures tend to have adverse effects on the delivery of energy and the lifetime of the device.

1.2.4 Time delays in delivering or transferring energy

In many cases, the amount of useful energy stored in a storage device needs to be transferred into a load within a very short duration. One automotive example is the quick energy required in accelerating a sports car to a particular speed within a very short period of time. Another possible domestic example is to heat up the stored water in a pipe line between the tap and the central heating system, so that when we open the tap we get warm water instantaneously. In such cases, a designer should be able to estimate and quantify the speeds and delays related to a combination of the ESD and the associated load. In this process, time constants associated with the ESD and the load will come into the picture.

To appreciate this aspect, let us consider a simple example of a capacitor of value C with an equivalent series resistance (ESR) of r_C connected to a load of R_L via a switch as in Figure 1.2. If the capacitor is initially charged to a DC voltage of V_{C0}, total energy stored in the capacitor is given by

$$E_{C0} = \frac{1}{2} C V_{C0}^2 \tag{1.12}$$

At time $t=0$ when the switch is activated, instantaneous current through the load resistor is given by

$$I_{R_L, t=0} = \frac{V_{C0}}{R_L + r_C} \tag{1.13}$$

This indicates us that the ESR of the capacitor could play a significant role on the maximum possible current, particularly when the load resistance is pretty small. Considering the combination of the capacitance, external load resistance, and the ESR in series, circuit time constant, τ, is given by

$$\tau = (R_L + r_C)C \tag{1.14}$$

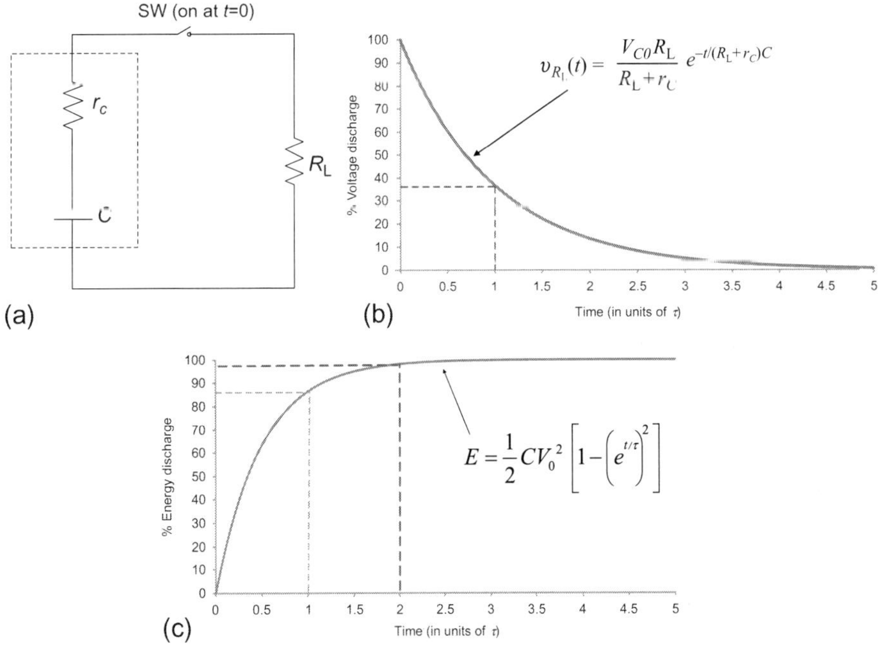

Figure 1.2 Capacitor as an energy storage device: (a) Simplified circuit showing capacitor ESR in the overall circuit; (b) capacitor discharge curve; and (c) percentage energy released to a load as a function of time measured in time constant as a unit.

Using this time constant and the initial voltage across the capacitor, we can derive the time-varying exponential voltage waveform across the capacitor as

$$v_C(t) = V_{C0}\mathrm{e}^{-t/\tau} = V_{C0}\mathrm{e}^{-t/(R_\mathrm{L}+r_C)C} \tag{1.15}$$

Based on the Eqn (1.15), in practical situations within about 5 time constants, capacitor voltage discharges down to almost 98% of its original voltage. Given these relationships, the capacitor releases its stored energy based on the following relationship:

$$E_{C0} - E_{C,t} = \frac{1}{2}C\left[V_{C0}^2 - \left(V_{C0}\mathrm{e}^{-t/(R_\mathrm{L}+r_C)C}\right)^2\right] = \frac{1}{2}CV_{C0}^2\left(1 - \mathrm{e}^{-[2t/(R_\mathrm{L}+r_C)C]}\right) \tag{1.16}$$

$E_{C,t}$ is the remaining energy in the capacitor after a time period of t.

The above indicates that the energy discharged within a period of one time constant, E_D, τ is given by

$$E_{\mathrm{D},\tau} = \frac{1}{2}CV_{C0}^2\left(1 - \mathrm{e}^{-2}\right) \approx 0.86E_{C0} \tag{1.17}$$

Similarly, we can calculate the discharge of useful energy into an external load and the associated loss in the ESR of the capacitor. Given the above situation, if the ESR is significantly large compared to the external load, transfer efficiency will be quite low. Figure 1.2(b) depicts the voltage across the capacitive element of the capacitor versus time and Figure 1.2(c) indicates the total energy discharged into the external load and the ESR. From Figure 1.2(c), we see that within less than 2 time constants, nearly all energy stored could be discharged.

The two examples we have considered, namely, an electrochemical cell (with a constant open circuit voltage over time and temperature) and a capacitor, indicate us that energy delivery capabilities of these two devices are different. In the second case of the capacitor, the amount of energy stored is dependent only on the voltage capability of the capacitor and the value of capacitance. Energy delivery is not constant over time and it tends to deliver over 85% of its energy within the first time constant. Also in an electrochemical device like a battery, to predict the overall stored energy, you need much more information about the device, compared to the electrical parameters applicable to the device operation.

1.2.5 Complex models of ESDs

In the previous section, we considered two extremely simplified models for an electrochemical cell or a capacitor with two lumped parameter elements. However, in the real world of ESDs, the device models are much more complicated than these two sample cases. Figure 1.3 indicates a more realistic model for a lead-acid battery (Grillo et al., 2011) and Figure 1.4 indicates a model for an SC (Musolino et al., 2013).

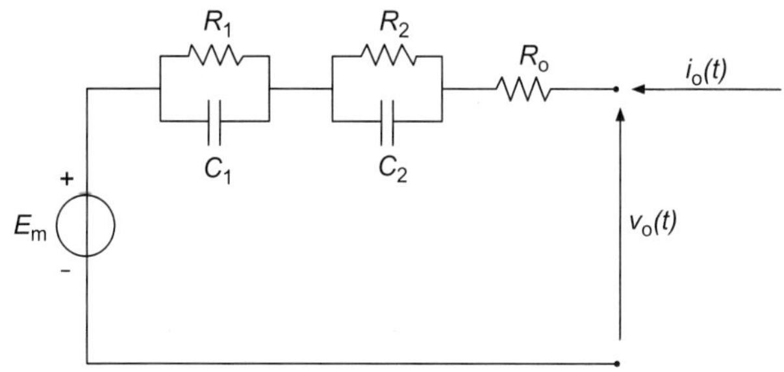

Figure 1.3 Equivalent circuit for a lead acid battery
Adapted from Grillo et al., 2011.

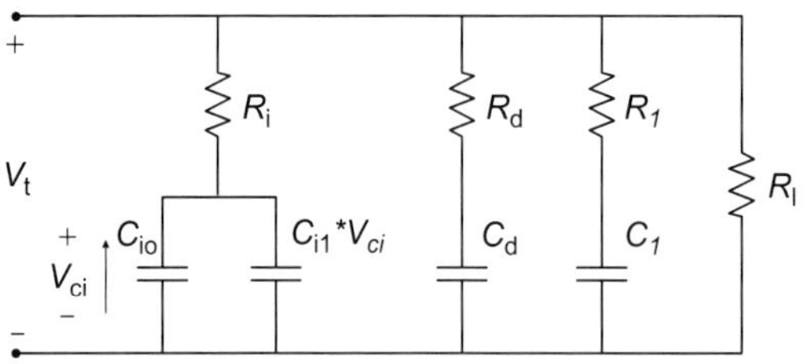

Figure 1.4 Supercapacitor (SC) equivalent circuit
Adapted from Musolino et al., 2013.

Compared to the simplified model indicated in Figure 1.1, the equivalent circuit shown in Figure 1.3 has an electromotive force which varies with the temperature and the state of charge of the battery (Moubayed et al., 2008). Single lumped resistance used to represent the internal resistance of the battery model in Figure 1.1 really comprises three different components based on terminal resistance, and two other components that depend on the state of charge and the discharge current. More details are available in (Moubayed et al., 2008). Two capacitors shown across the two resistances R_1 and R_2 add time delays to make the voltage and current estimations in external circuits more complex, for both charging and discharging. Later chapters in the book will discuss more details on modeling batteries and SCs.

When we look at the SC equivalent circuit in Figure 1.4, to estimate time delays and limitations, such a situation calls for much more analytical- or simulation-type approaches, since the mathematics involved with such a problem is much more complex. In developing these complex models, researchers have combined experimental and simulation work, in order to match the physics in electric double-layer effect. Present day SCs or ultra-capacitors, technically mean the same family of electrical double-layer capacitors (Zubieta and Bonert, 2000). We will discuss more details on these capacitors in Chapter 5.

1.3 Energy storage in electrical systems

In electrical systems and electronic products, energy storage requirements come in two basic forms. One is the short-term storage where components in powered electrical circuits store energy in electrostatic or electromagnetic form. Capacitors and inductors are typical examples. The other case is where we use ESDs for longer term energy supply as a case of a redundant energy source.

When a system is powered by an unreliable energy supply, such as frequently failing AC power line, or, the system is a portable one, we require some form of energy storage to keep the system operational with the expected reliability levels. As an example if the AC mains supply to a critical industrial load such as a manufacturing plant suddenly fails, there should be a suitable standby power system such as a generator set buffered by a UPS. In this scenario, a UPS is expected to have sufficient energy storage in the form of chemical batteries, or by other means to supply energy until the generator sets installed are started. When the generator is coming into operation, energy storage is in the form of the fuel storage available for the generator set. In critical situations such as semiconductor fabrication facilities, where the AC mains supply is primarily fail safe, still there are short-term situations where the quality of the AC supply is not up to the expected standard, and a short autonomy UPS system comes into operation to back up the AC mains. In such a situation, the UPS system may have only limited energy storage in the form of a SC bank, ready to back up the needs of a few AC cycle duration. In such cases, other options could be a fly-wheel or a compressed air supply.

1.3.1 Basic electrical components as in-circuit energy storage

When electrons flow thorough a resistor, energy is dissipated in the resistance and converts that as heat and no ideal resistor could store energy. However, in an inductor within a closed circuit carrying a current of I, stored energy E_{L} is given by

$$E_{\mathrm{L}} = \frac{1}{2}LI^2 \tag{1.18}$$

This energy storage occurs due to the magnetic flux generated by the current within the core of the inductor. However, since the energy is stored in the inductor due to the flux lines created by the current, this energy is not physically transportable with the device unless the rest of the circuit responsible for generating this current is kept undisturbed.

In contrast if we connect a voltage source to a capacitor, its voltage exponentially rises to the value of the voltage source. If the DC source used to charge the device has an open circuit voltage of V, storage energy in the capacitor of value C is given by

$$E_C = \frac{1}{2}CV^2 \tag{1.19}$$

Compared to the case of an inductor, this energy storage occurs in the form of energy associated within an electrostatic field based on electric-charge storage. For this reason, if the capacitor has no significant leakage (which represents an infinite leakage resistance), the device with the energy stored can be transported after disconnecting the voltage source. However, in practical capacitors, the leakage resistance is finite and this creates a leakage current through this equivalent resistance and energy is wasted in the form of heat dissipated in this large resistance. Practical devices such as electrolytic capacitors with capacitance values ranging from nanofarads to few thousand microfarads do retain their charge from microsecond order durations to few minutes, but they cannot be treated as useful ESDs similar to electrochemical devices such as batteries. In modern SC families, capacitance values vary from fractional Farads to few thousands of Farads, and their leakage currents are in the order of few microamperes to milliamperes only. This allows them to be used as short-term energy storage and delivery devices in power electronic systems.

1.3.2 Energy storage options for longer term and infrequent utilization

In the previous paragraphs, we quantified how common energy storage components such as inductors and capacitors store energy based on current through the device or the voltage across the device. These scenarios are mostly used in circuit level design and in very short-term basis. However, in situations such as critical electrical facilities at airports, semiconductor manufacturing and computer server farms there are few longer term options for energy storage. They are

1. Electrochemical devices such as batteries
2. SC banks
3. Fly-wheels
4. Fuel cells and the associated energy source
5. Compressed air supply
6. Superconductive magnetic energy storage (SMES)

Most of these systems are usually used for back up purposes when a power failure or a power quality issue occurs on the incoming AC mains supply. On electrochemical batteries, which allow you to store energy for very short term to few hours of backup, we will have a detailed discussion in Chapters 2 & 3.

SC banks are usually combined with battery packs as hybrid devices to get the best of performance of batteries and the SCs. Given the simple basics in the previous sections, a capacitor to achieve a high stored energy, capacitance should be large or the voltage across the terminals should be large. In general, common forms of capacitors provide the designers with capacitance values in the range of picofarads to several thousand microfarads. Voltage capabilities usually come in the range of 10 V to over 10 kV and the technology capabilities to improve these capacitors grow slowly.

In contrast to commonly used capacitors, SCs come with very low voltage capability, usually in the range of 2 to less than 16 V. However, their capacitance values range from about 0.1 to over 5000 F. Photograph in Figure 1.5(a) depicts a set of electrolytic capacitors in the range of 10-2200 μF. Photograph in Figure 1.5(b) depicts a set of SCs with values varying from 0.2 to 650 F. Photograph in Figure 1.5(c) depicts a very large electrolytic capacitor used in Toyota Prius electrical system. Photograph in

Figure 1.5 Different capacitors: (a) commonly used electrolytic caps; (b) Supercapacitors from three different manufacturers; (c) a very large electrolytic capacitor; (d) an SC module with balancing circuits.
(c) Courtesy of AECS Ltd., Hastings, NZ. (d) Courtesy of ABB Research, Napier, New Zealand.

Figure 1.5(d) indicates an SC module with multiple SCs (of 4000 F value) in series combined with a voltage balancing module. Table 1.2 provides a comparison of different capacitor options for the energy storage capabilities varying from 1 to over 50 J (Kularatna et al., 2011).

SCs usually have very low internal resistance (in the form of ESR) compared to rechargeable batteries, in general. Also their ESR is relatively constant over its

Table 1.2 **Comparison of typical electrolytic capacitors and supercapacitors for their ESR values and other useful specifications**

			Parameters			
Energy storage limit	**Capacitor type**	**Manufacturer**	**Capacitance**	**Terminal voltage (V)**	**Short circuit current (A)**	**ESR (mΩ)**
<1 J	Electrolytic	RS3	2200 μF	16	104	153
1–5 J	Supercap	Maxwell	1 F	2.7	3.85	700
		Cap-xx	2.4 F	2.3	115	20
	Electrolytic	Cornell Dubilier	2200 μF	50	704	71
5-50 J	Supercap	Maxwell	10 F	2.5	14	180
		Cap-xx	1.2 F	4.5	112.5	40
		Nesscap	10 F	2.3	33	70
	Electrolytic	Cornell Dubilier	82,000 μF	16	1441	11.1
		VICOR	270 μF	200	325	614
>50 J	Supercap	Maxwell	350 F	2.7	840	3.2
		Nesscap	120 F	2.3	144	16

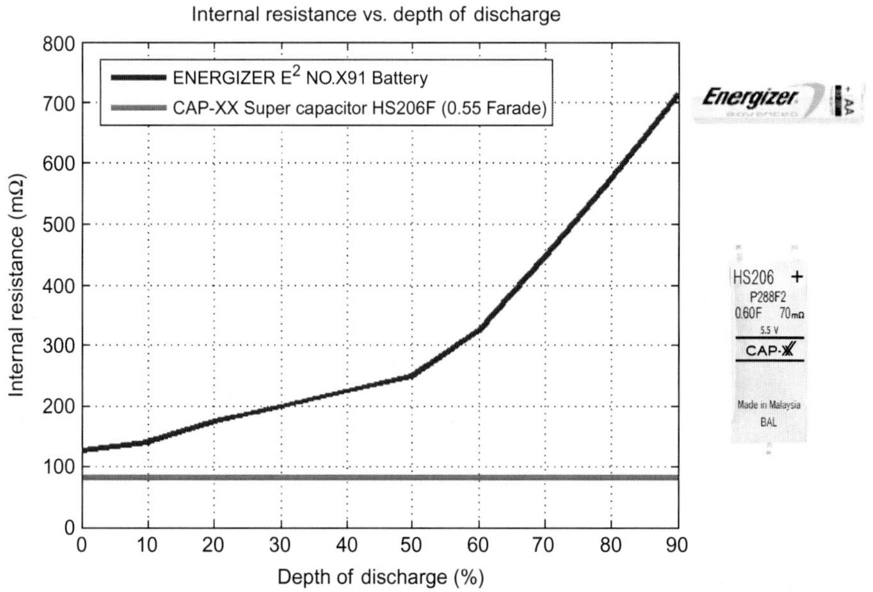

Figure 1.6 Comparison of internal resistance of a supercapacitor and an energizer type AA cell.

discharge period or the calendar life. This situation is illustrated from the example in Figure 1.6 where it indicates the case of discharging these two devices at an approximate current of 1 A and plotting the measured internal resistance against the depth of discharge. Even a small SC such as a 0.6F value (HS206F from Cap-XX) indicates a constant internal resistance of less than 100 mΩ, while the AA energizer cell indicates a case where its internal resistance keeps increasing with the depth of discharge. Figure 1.7 depicts a case of discharging different types of SCs at a constant current of 1 A and comparing it with the behavior of an AA-type cell. While the SCs simply obey the behavior of a capacitor voltage (based on the simple relationship of $\Delta V = \Delta Q/\Delta V$) given by

$$\Delta V = \frac{I_L}{C} \Delta t \tag{1.21}$$

However, due to the complex nature of the internal resistance variation of the AA cell as we saw from Figure 1.6, the terminal voltage variation is not easily predictable by a simple model as we indicated in Figure 1.1.

In general, SCs do not have very high energy storage capability compared to electrochemical batteries. However, for short-term high power requirements, very low and relatively fixed internal ESR of SCs allows them to be used as short-term high-power delivery devices. This complementary-nature of the two families allows designers to develop hybrid-architectures for cost-effective back up energy sources.

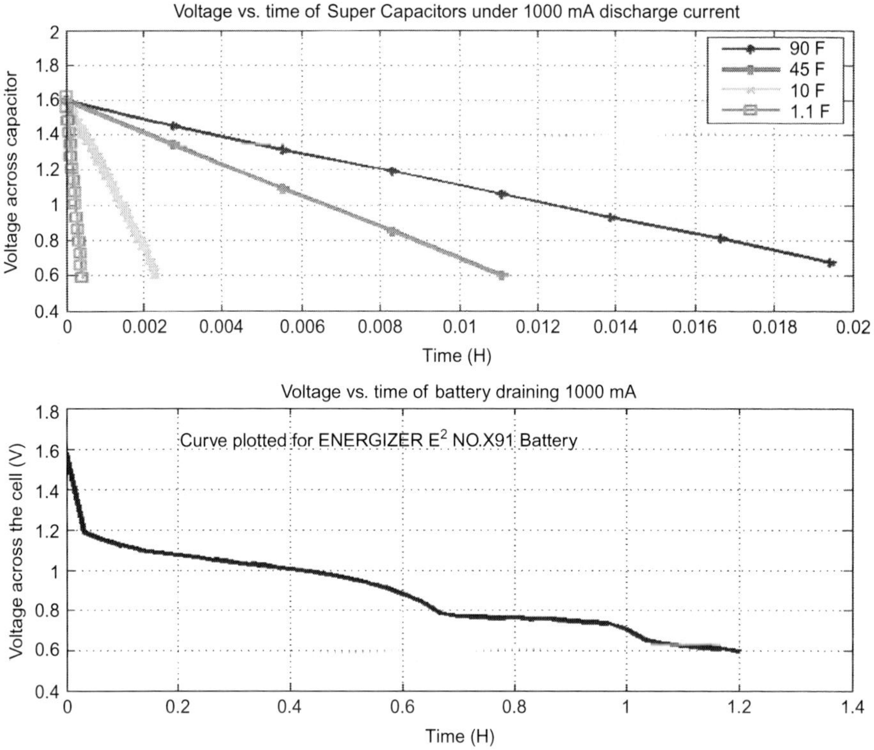

Figure 1.7 Terminal voltage comparison of supercapacitors and batteries under constant current discharge: (a) supercapacitors and (b) a disposable AA-type cell.

1.3.3 Flywheel as an ESD in electrical systems

Flywheels are kinetic ESDs where a rotating mass is used to store energy based on the inertia of the rotating mass and the rotational speed of the rotor. Situation is equivalent to the case of a mass of m moving at a speed of v where the kinetic energy is given by $\frac{1}{2}mv^2$. In the case of a flywheel, usually rotating at a very high speed, the kinetic energy stored in the rotor, E_k is given by

$$E_k = \frac{1}{2} J\omega^2 \tag{1.22}$$

where J is the moment of inertia and ω is the angular velocity.

In practice, the rotor is built up in a way that mass is kept low, but the value of J is maximized. Early generations of flywheels used large steel rotors, but the newer systems use carbon-fiber and other composite material in order to keep the weight low. Magnetic bearings of different complexities are used to minimize friction and high-temperature superconductors (HTSC) are used in very high-speed latest designs.

A relatively new application of flywheels is in the storage of electrical energy, which is achieved by the addition of an electrical machine and a power converter. Electrical machine may be integrated with the flywheel and operates at variable speed, and the power converter is usually provided by a power electronic variable speed drive. The main feature of flywheel energy storage (FES) systems is that they can be charged and discharged at relatively high speed for many AC cycles in an application such as a UPS. Typical state of the art composite rotors have the capability to store 100 Wh/kg (Ruddell, 2003). There are two basic categories of flywheels, low speed and high speed.

Low speed ones are made of steel rotors and speeds reach up to about 600 rev/min. They usually carry conventional bearings. High-speed flywheels coming into commercial use with operating speeds up to 50,000 rev/min use advanced composite material, and magnetic bearings including HTSC techniques (Ruddell, 2003). Main electrical applications of FES are UPS, power quality systems, traction applications, etc. A detailed discussion on these is beyond the scope of this book.

1.3.4 Fuel cells

A fuel cell is galvanic cell in which the chemical energy of a fuel is converted directly into electrical energy by means of electrochemical processes. The fuel and oxidizing agent are continuously and separately supplied to the two electrodes of the cell, where they undergo a chemical reaction. Figure 1.8 depicts the basic arrangement of a fuel cell. An electrolyte is placed between two electrodes in order to conduct the ions from one electrode to the other. Fuel is supplied to the anode, where electrons are released from the fuel under catalyst. The electrons, under the potential difference between

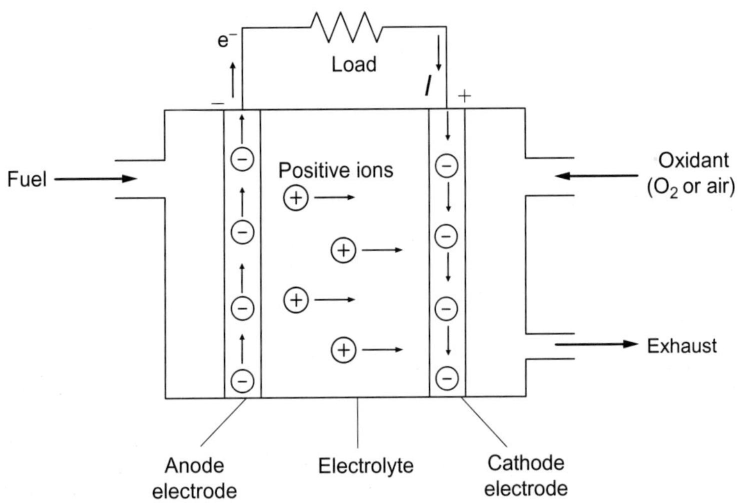

Figure 1.8 Basic operation of a fuel cell.

these two electrodes, flow through the external circuit to the cathode, where combining positive ions and oxygen, reaction products, or exhaust are produced. This process is different to a battery since we have to supply the fuel and oxidant into the fuel cell to produce electrons which flows into the external circuit. More details on theory are available in Chapter 14 of Eshani ct al. (2010).

Fuel cells do not come under the category of electrochemical batteries, since they require an external supply of fuel. Fuel cells can be divided into five subsets: (i) stationery plug-in power (hundreds of megawatts capability), (ii) back-up power (tens to hundreds of kilowatts), (iii) traction power (portable power supplies at 10-100 kW), and (iv) small portable power (1-100 W) and mini- or micropower (10 μW to 1 W). Fuel cells have the potential to provide clean, efficient sustainable power in a range of market segments (Zhao, 2009).

There are many types of fuel cells, with different power capabilities. Six different types are: (i) proton exchange membrane or polymer exchange membrane fuel cells (PEMFC), (ii) Alkaline fuel cells (AFC), (iii) phosphoric acid fuel cells (PAFC), (iv) molten carbonate fuel cells (MCFC), (v) solid oxide fuel cells (SOFC), and (vi) direct methanol fuel cells (DFMC). Table 1.3 summarizes some operating temperature ranges and the state of electrolyte. PEMFCs were developed in the 1960s to help the US manned space program and they are now developed for automotive applications. With a power density of 0.35-0.6 W/cm^2, these are favorable for EV and hybrid electric vehicles (HEV). Low-temperature operation and fast start capability are most desirable characteristics of these for EV and HEVs.

Methanol- and hydrogen-based fuel cells seem to be attractive to automotive applications. Figure 1.9 indicates the characteristics of hydrogen-oxygen fuel cell. Figure 1.9(a) indicates that these fuel cells have terminal voltage ranges varying from 0.5 to 1.0 V. Figure 1.9(b) indicates their efficiency and power capability versus current density. More details are available in Baldauf and Preidel (1999), Berlowitz and Darnell (2000), and Wang (2002). For fuel cells using oxygen as the oxidant, air is usually used as the oxidant. Figure 1.10 depicts the cell voltage, system efficiency and net power density versus net current density applicable to a hydrogen-air fuel cell.

Table 1.3 Operating data for different fuel cell systems

Cell system	Operating temp range (°C)	Electrolyte
PEMFC	60-100	Solid
AFC	100	Liquid
PAFC	60-200	Liquid
MCFC	500-800	Liquid
SOFC	1000-1200	Solid
DMFC	100	Solid

Source: Eshani et al. (2010).

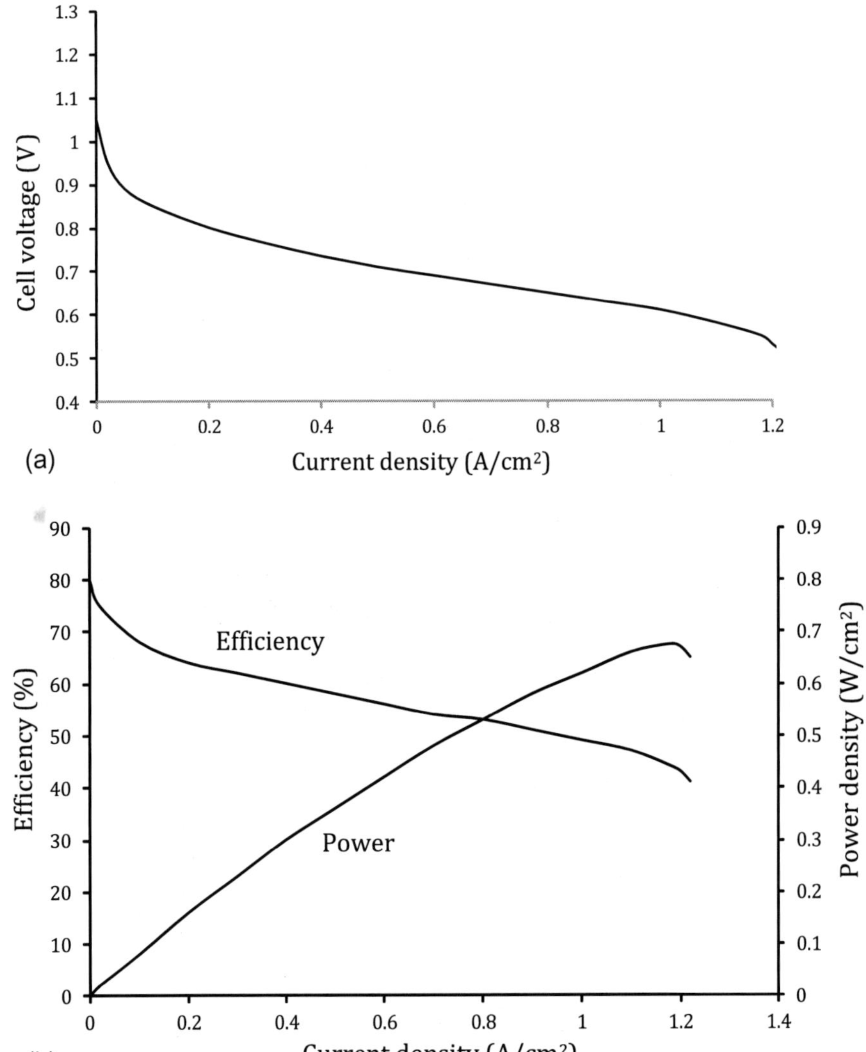

Figure 1.9 Characteristics of hydrogen-oxygen fuel cell: (a) voltage versus current density (b) efficiency and power versus current density.

Usually, a hydrogen-air fuel cell works together with power electronic DC-DC converter and associated auxiliaries. Figure 1.11 indicates this case where the fuel cell exchanges air, water, and hydrogen. More details are available in Eshani et al. (2010).

Compared to the usage of hydrogen-based fuel cells in larger systems such as EVs and HEVs, another very interesting category of fuel cells is the microfuel cells which are mostly based on DMFC. The applications of these are in microelectronic

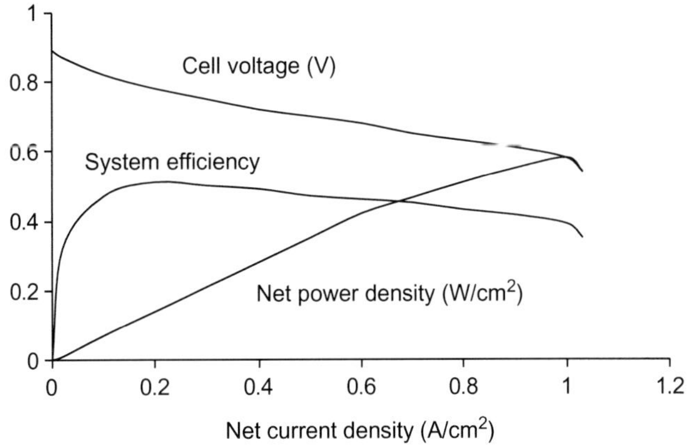

Figure 1.10 Characteristics of hydrogen-air fuel cell at different current densities.

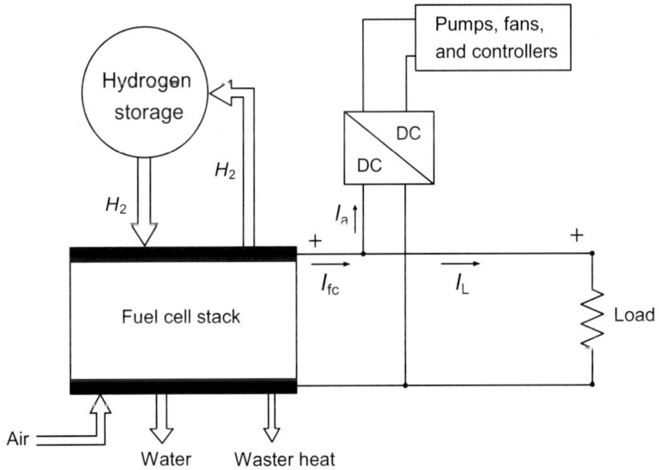

Figure 1.11 A hydrogen-air fuel cell system.
Adapted from Eshani et al., 2010.

environments and wireless sensor nodes. Figure 1.12 depicts the arrangement of a DFMC and its device life versus duty cycle of operation. The graphs in Figure 1.12(b) indicate the ability of these devices to last long in micropower applications such as sensor nodes. For details, Zhao (2009) is suggested.

It is important to recognize that fuel cells are strictly not ESDs, but they are energy converters based on a fuel feed. Therefore, a detailed discussion on this subject is beyond the scope of this book.

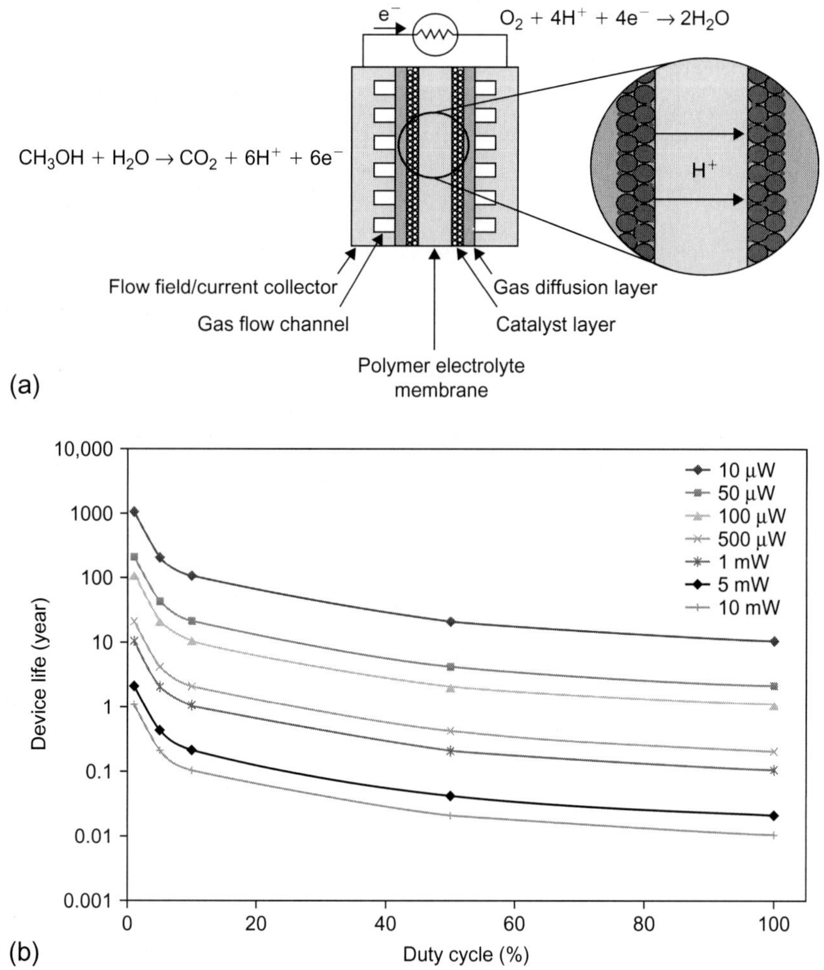

Figure 1.12 Direct methanol fuel cell (DFMC): (a) simplified arrangement and (b) device life versus duty cycle for 12 M methanol fuel cell carrying 1 cubic centimeter of methanol and discharged at 0.5 V.
Source: Zhao, 2009.

1.4 Compressed air energy storage

Compressed air energy storage (CAES) is a technology that stores energy as compressed air for later use. Energy extraction may finally occur at a gas turbine. CAESs have been considered for numerous applications, most notably for electric grid support for load leveling applications. In such systems, energy is stored during periods of low demand and then converted back to electricity when the demand is high. Natural caverns are often used as air reservoirs in order to store large amounts of energy, as these systems are generally used for power levels varying from 35 to over 300 MW (Vazquez et al., 2010). During the last decade, attempts were made to combine CAES and SCs with the possibility to achieve maximum efficiency point tracking (Lemofouet and Rufer, 2006). This subject is not within the scope of this book.

1.5 Superconductive magnetic energy storage

Superconductive magnetic energy storage (SMES) systems make use of the energy stored in a DC magnetic field created by a direct current flowing through a superconducting coil. The coil is cryogenically cooled to achieve extreme conductivity (where it is defined as a superconductor) and no resistive losses. This combination of a DC fed inductor with near zero resistance can be used as a very fast acting ESD. Industrial applications include adjustable speed drives, power quality improvement products, and back-up power supplies. Due to the capability of very fast response, these systems are used in many special circumstances, with power levels ranging from few 100 kW to few MW. More details can be found in Buckles and Hassenzabl (2000) and Ali et al. (2010).

1.6 Rapid energy transfer requirements and fundamental circuit issues

In general, an ESD's electrical capabilities can be summarized by few simple specifications such as:

(i) Energy storage capability
(ii) Internal resistance of the device
(iii) Fundamental delays associated with device properties (time constants and related issues)

Energy storage capability is obviously the prime specification, but when it comes to delivering that energy into an external load, the maximum power that can be transferred depends on the internal resistance of the device. This was discussed as a summary in Section 1.2.2. What we can deliver as maximum power is limited by the minimum possible value of the internal resistance of the device.

This situation is well illustrated in the case of modern SCs with very low ESR in the order of 0.3 milliohms to few 10s of milliohms. In a practical case of an SC with a rated voltage around 2.5 V and an ESR of 1 mΩ, device could deliver an instantaneous maximum power of $V^2/4r_C$, (when $r_C = R_L$, which is the condition for max power delivery) which translates to a value of 1562 W at 2.5 V. Compared to this case, if we are to use an electrolytic capacitor with a voltage capability of 16 V, and with an ESR value of 100 mΩ, the maximum power capability is only 640 W. Table 1.4 summarizes the energy storage capability and maximum power delivery capability of an SC and an electrolytic capacitor.

What is indicated in Table 1.4 provides us an interesting scenario. A 3000 F SC can store over 10 kJ of energy and it can initially supply over 6 kW of power. However, a 2200 μF, 50 V capacitor just cannot store even 3 J of energy. However, its

Table 1.4 Comparing the capability of supercapacitors and electrolytic capacitors

Capacitor type	Manufacturer	Capacitance	Voltage rating (V)	ESR (mΩ)	Total energy storage capability	Maximum possible output power (load resistance = ESR) (kW)
Supercapacitor	Maxwell	3000 F	2.7	0.3	10.9 kJ	6.07
Electrolytic capacitor	Cornell-Dubilier	2200 μF	50	71	2.75 J	8.8

instantaneous power capability is around 8 kW. In a situation like where the two capacitors feed a heating element, SC will do significantly better in heating, despite its low DC voltage rating. Based on the discussion in Section 1.2.4, a capacitor could release a huge instantaneous current, limited only by the total path resistance is indicated in Figure 1.2(b). Also the graph in Figure 1.2(c) indicates that it could release almost over 85% of its charge within a period of one time constant. This in summary indicates us that an SC could deliver huge percentage of its stored energy within a period of time determined by the time constant of the overall circuit.

Now let us consider another scenario of a low-voltage AC supply based on a transformer feeding a heating coil as in Figure 1.13(a). In a situation such as a mains transformer stepping down to a very low voltage, and supplying a very low resistance heater coil, we can configure the system to deliver a large continuous power into the load on continuous basis. However, since a transformer's nonideality leads to

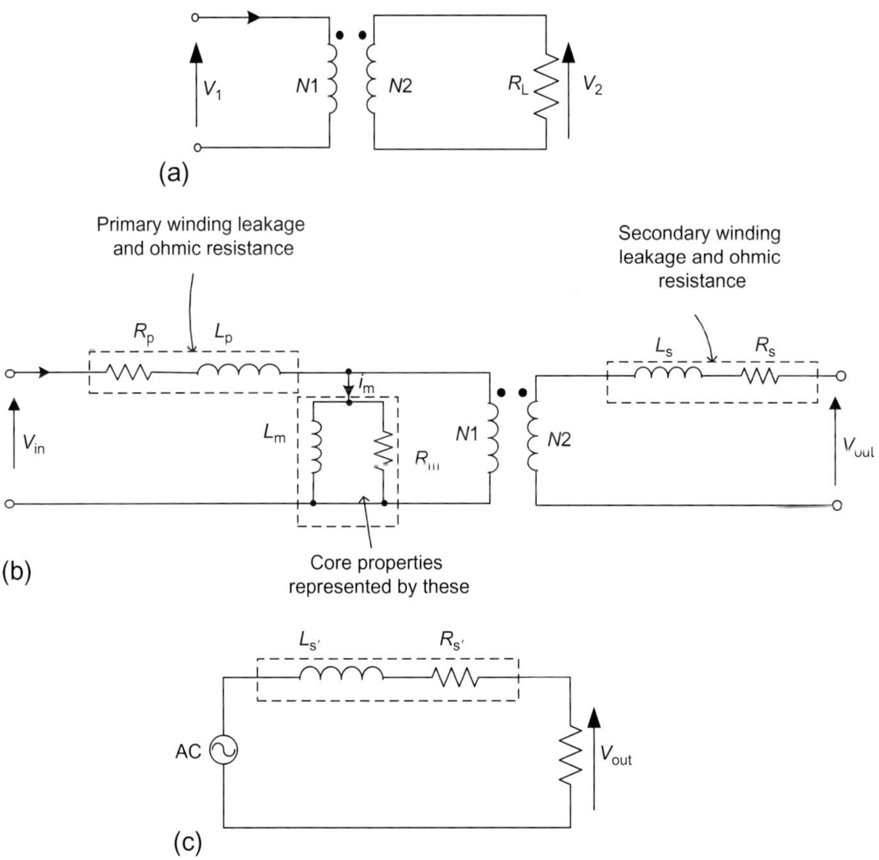

Figure 1.13 Step down transformer-based supply delivering power into a resistive load: (a) basic circuit; (b) general equivalent circuit of transformer indicating lumped components representing nonideal behavior; and (c) Equivalent circuit seen by the resistive load.

an equivalent circuit such as in Figure 1.13(b), load sees an equivalent circuit with an inductive impedance together with an AC source of value nV_p, as depicted in Figure 1.13(c). Now with the transformer (turns ratio of $n=N_2/N_1$) secondary providing a voltage source of approximately nV_p, together with an equivalent series internal impedance based on an inductance and a resistance representing its nonideal behavior, overall equivalent circuit seen by the load will be a combination of a voltage source, an inductance, and a resistance feeding the resistive load.

This creates the opposite of the SC's case where inductive circuit does not allow a sudden change in current at the initial switch on time. Circuit will behave in such a way that, instantaneous AC voltage developed at the secondary side (approximately nV_p) will be equal to the sum of the drops across the inductance and the resistive components. With an *L-R* combination, circuit transients will behave in an opposite manner compared to the case of a capacitor and will take a finite time to reach the steady state of voltage developed across the resistive load. However, once this stage is reached power can be delivered to the load continuously.

This case can be easily illustrated in a water heating example as in Figure 1.14. We supplied cold water at a flow rate of 0.1 L/s at the right-hand side of the glass tube,

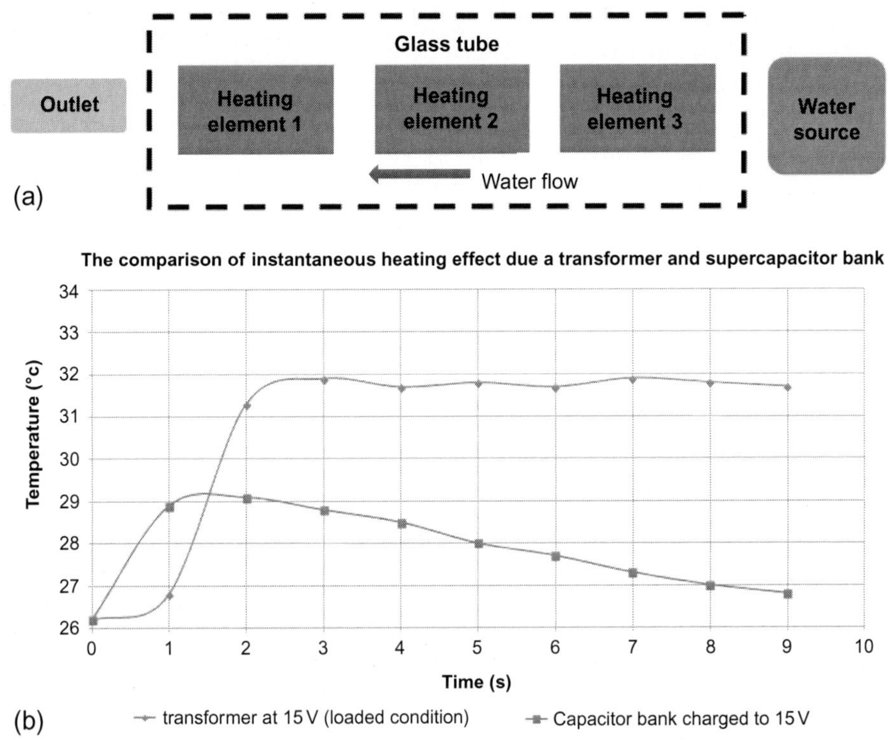

Figure 1.14 Attempt to heat water from a cold water supply using prestored energy in a capacitor bank versus a transformer secondary supply with a similar RMS voltage-capability: (a) arrangement of heating elements and (b) temperature rise versus time for two cases.

and used a series SC bank of 65 F with a maximum voltage capability of over 15 V, and a transformer secondary supplying a similar RMS voltage at the heating coil. Temperature rise of the water at the left-hand side of the tube is shown in Figure 1.14(b). While the SC bank could dump almost all the energy into water within about 1.5 time constants (in this example case, it was 9 s) and increase the temperature of the flowing water, transformer was unable to heat the water as fast as the SC bank. However, the transformer supply could maintain the temperature continuously after its transient delay associated with heating. The case simply illustrates the differences of capacitive and inductive AC energy sources, in an application where instant heating is needed. This is a simple and practical illustration on the capability of SCs for instant energy supply requirements.

1.7 Technical specifications of ESDs

ESDs are enabling technologies for many established and new applications of power electronic systems. Well-established areas are power quality improvement products, UPS systems, and portable devices. New application areas include EV, renewable energy systems and power peak shaving and many more new applications. ESS could be characterized by a set of general specifications and these allow us to quantitatively compare various device family options and also to look at the performance of a particular energy source continuously as it is effectively utilized in given power electronic environment.

1.7.1 Energy and power density

1.7.1.1 Energy density

Energy density of an ESS is its energy divided by its weight or volume. When weight is used it is called the *gravimetric energy density* and *volumetric energy density* when the volume is used. Terms "*energy density*" and "*specific energy*" are sometimes used for volumetric and gravimetric measures, respectively. Units are Watt-hour per liter or Watt-hour per kilogram.

1.7.1.2 Power density

It is the maximum amount of power an ESS can deliver per unit volume or unit weight. When the volume is used, it is also called the *volumetric power density* and usually measured in Watts per liter. When weight is used it is called *gravimetric power density* or *specific power* and measured in Watts per kilogram.

1.7.1.3 Cycle life

Cycle life is a measure of an ESS's ability to withstand repetitive deep discharging and recharging using the manufacturer's cyclic charging recommendations and still provide minimum required capacity for the application. Cyclic discharge testing can be done at any of various rates and depths of discharge to simulate conditions in the application.

1.7.1.4 Cyclic energy density

For purposes of comparison, a better measure of ESS characteristics is a composite characteristic which considers energy density over the service life of the system. A composite characteristic, *cyclic energy density*, is defined as the product of energy density and cycle life at that energy density and has the dimensional units, Watt-hour-cycles per kilogram (gravimetric) or Watt-hour-cycles per liter (volumetric).

1.7.1.5 Self-discharge rate

Self-discharge rate is a measure of how long an ESS can be stored or left unused and still provide minimum required capacity and be recharged to rated capacity. This commonly is measured by placing an ESS (such as batteries) on shelf stand at room (or elevated) temperature and monitoring open circuit voltage over time. In cases such as batteries self-discharge is measured in terms of percentage capacity loss per month or per year in terms of energy (in Watt-hour) lost.

In cases like capacitors, it is usual to indicate a value of a resistor which is placed in parallel to the terminals of the capacitor to denote this issue.

1.7.1.6 Charge acceptance or coulombic efficiency

In ESS systems such as batteries where the open-circuit voltage is relatively constant, charge accumulated or discharged in terms of $\int_0^t i dt$ is used to discuss the capability of the device to accept and deliver current into a given load. The charge delivered to the load, C_{load} will be usually less than the charge fed into the device, C_{charge}. The ratio of these two quantities, measured in coulombs or Ampere-hours, is called the charge-acceptance or the coulomb-efficiency. This measure is usually dependent on the charge (or discharge) rate, temperature, age, and the overall condition of the ESS.

The above specifications are generally applicable to many types of ESS systems, and in batteries there are many other specifications which are used in practical circumstances. These will be discussed in Chapter 2.

1.8 Ragone plot

ESS systems come in different sizes, such as pumped ESSs (used in hydroelectric plants), large and small fuel cells, different types of batteries, SCs, and SMES. In all these systems, two very important specifications used to compare the overall performance are energy density and power density. To compare different energy delivery capability options, a Ragone plot (developed by David V Ragone) is very frequently used.

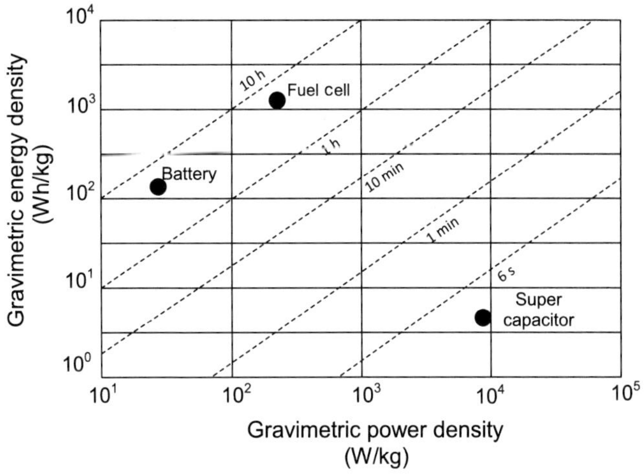

Figure 1.15 Ragone plot with three specific systems placed at relative locations.

Figure 1.15 depicts the concept of the Ragone plot where vertical axis is volumetric or gravimetric energy density, and the horizontal axis is power density. A Ragone plot reveals at a glance the characteristics of different energy deliveries or storage systems. More capable device means a large energy density as well as a large power density. One should try to achieve systems placed at the north-east corner of the chart. da Rosa (2013) provides a summary of how to use the data sheet information and calculate results to place the device on the Rgaone plot. In Figure 1.15, relative positioning of a lead acid battery (of capacity 80 Ah; internal resistance of 37.5 mΩ, weight 31 kg, volume 0.01 m^3), SC (3000 F, internal resistance of 3.7 mΩ, weight 0.55 kg, volume 0.000475 m^3), and a fuel cell (continuous power capability 110 kW, maximum current of 440 A, DC voltage of 250 V, weight 120 kg, volume of 82 L). More details on the calculations related to these are in da Rosa (2013).

References

Ali, M.H., Wu, B., Dougal, R.A., 2010. An overview of SMES applications in power and energy systems. IEEE Trans. Sustain. Energy 1 (1), 38–47.

Baldauf, M., Preidel, W., 1999. Status of development of direct methanol fuel cell. J. Power Sources 84, 161–166.

Berlowitz, P.J., Darnell, C.P., 2000. Fuel cell choices for fuel cell powered vehicles. In: Proceedings of Fuel Cell Transportation 2000 (SAE SP-1505), Warrendale, PA, 2000, pp. 15–25.

Buckles, W., Hassenzabl, W.V., 2000. Superconducting magnetic energy storage. IEEE Power Eng. Rev. 20 (5), 16–20.

da Rosa, A.V., 2013. Fundamentals of Renewable Energy Processes, third ed. Academic Press, Oxford, 884p.

Eshani, M., Gao, Y., Emadi, A., 2010. Modern Electric, Hybrid Electric and Fuel Cell Vehicles. CRC Press, Boca Raton, 534p.

Grillo, S., Martini, L., Musolino, V., Piegari, L., Tironi, E., Tornelli, C., 2011. Management of different energy storage devices using a losses minimization algorithm. In: Proceedings of IEEE-ICCEP Conference, pp. 420–425.

Kularatna, N., Kankanamge, K., Fernando, J., 2011. Supercapacitors enhance LDO efficiency—part 2: implementation. Power Electron. Technol. Mag. 37 (5), 30–33.

Lemofouet, S., Rufer, A., 2006. A hybrid energy storage system based on compressed air and supercapacitors with maximum efficiency point tracking (MEPT). IEEE Trans. Ind. Electron. 53 (4), 1105–1115.

Moubayed, N., Kouta, J., El-Ali, A., Dernayka, H., Outbib, R., 2008. Parameter identification of the lead-acid battery model. In: Proceedings of IEEE-PVSC Conference, May 2008, pp. 1–6.

Musolino, V., Piegari, L., Tironi, E., 2013. New supercapacitor model with easy identification procedure. IEEE Trans. Ind. Electron. 60 (1), 112–120.

Ruddell, A., 2003. Investigation on storage technologies for intermittent renewable energies: evaluation and recommended R & D strategy, CCLRC-Rutherford Appelton Laboratory, Report WP-ST6 Flywheel, 30p.

Vazquez, S., Lukic, S.M., Galvan, E., Franquelo, L.G., Carrasco, J.M., 2010. Energy storage systems for transport and grid applications. IEEE Trans. Ind. Electron. 57 (12), 3881–3895.

Wang, M., 2002. Fuel cell choices for fuel cell vehicles, well-to-wheel energy and emission impacts. J. Power Sources 112, 307–321.

Zhao, T.S., 2009. Micro Fuel Cells—Principles and Applications. Academic Press, London, 300p.

Zubieta, L., Bonert, R., 2000. Characterization of double-layer capacitors for power electronic applications. IEEE Trans. Ind. Appl. 36 (1), 199–205.

Bibiliography

Christen, T., Carlen, M.W., 2000. Theory of Ragone Plots. J. Power Sources, 91, 210–216.

Christen, T., Ohler, C., 2002. Optimizing energy storage devices using Ragone plots. J. Power Sources, 110, 107–116.

Pell, W.G., Conway, B.E., 1996. Quantitative modeling of factors determining Ragone plots for batteries and electrochemical capacities. J. Power Sources, 63, 255–266.

Ter-Garzarian, A.G., 2011. Energy Storage for Power Systems, second ed. IET, London, 276p.

Rechargeable battery technologies: an electronic engineer's view point

2

2.1 Introduction

From the invention of the lead-acid battery by Gaston Planté in 1859, electrochemistry has progressed steadily over the last one and half centuries. With the miniaturization of electronic components proliferating steadily over the last half century, the insatiable demand for smaller and lightweight portable electronic equipment has dramatically increased the need for research on battery chemistries as well as semiconductor components for the optimal management of batteries. Battery chemistries come in two different forms, namely, disposable or primary batteries and secondary or rechargeable batteries. Market for batteries can be found in IDTechEx Ltd. (2012a,b), Advanced Rechargeable Battery Market (2009), and US demand (2009).

Mature rechargeable battery chemistries are: (i) lead acid, (ii) nickel cadmium, (iii) nickel metal hydride, (iv) lithium-ion, (v) lithium-polymer/lithium metal, and (vi) lithium-iron phosphate. With the demand growing from electric vehicles (EV) and portable consumer products, many organizations spend huge amounts of research money on new battery chemistries. Some significant areas include zinc-based chemistries and silicon as a material for improving some properties of batteries (Silicon-air battery touts unlimited shelf life, 2009). Higher energy density, superior cycle life, environmental friendliness, and safe operation are among the general design targets of secondary battery manufacturers. Primary or disposable batteries are a reasonably mature market and a product chemistry range, but still there are attempts to increase the energy density, reduce self-discharge rate (in order to increase the shelf life), and also to improve the usable temperature range. To complement these developments, many semiconductor manufacturers continue to introduce new integrated circuit families for battery management.

This chapter describes the characteristics of batteries such as sealed lead acid, NiCd, NiMH, Li-based chemistries, and zinc-air in an engineer's viewpoint together with modern techniques used in battery management integrated circuits (ICs). This is done without detailing the chemical reactions and associated battery chemistries. Concepts and applications related to smart battery systems and related standards, and IEEE 1625/1725 standards for safety of battery-powered systems are also briefly introduced.

2.2 Battery terminology and fundamentals

2.2.1 Capacity

Battery or cell capacity means an integral of current over a defined period of time.

$$\text{Capacity} = \int_0^t i\,\mathrm{d}t \tag{2.1}$$

This relationship applies to either charge or discharge; that is, capacity added or capacity removed from a battery or cell. The capacity of a battery or cell is measured in milliampere-hours (mAh) or ampere-hours (Ah).

Although the basic definition is simple, many different forms of capacity are used in the battery industry. The distinctions between them reflect differences in the conditions under which the capacity is measured.

2.2.1.1 Standard capacity

Standard capacity measures the total capacity that a relatively new, but stabilized production cell or battery can store and discharge under a defined standard set of application conditions. It assumes that the cell or battery is fully formed, that it is charged at standard temperature at the specification rate, and that it is discharged at the same standard temperature at a specified standard discharge rate to a standard end-of-discharge voltage (EODV). The standard EODV is itself subject to variation depending on discharge rate as discussed.

2.2.1.2 Actual capacity

When the application conditions differ from standard, the capacity of the cell or battery changes, the term *actual capacity* includes all nonstandard conditions that alter the amount of capacity the fully charged new cell or battery is capable of delivering when fully discharged to a standard EODV. Examples of such situations might include subjecting the cell or battery to a cold discharge or a high-rate discharge.

2.2.1.3 Available capacity

That portion of actual capacity, which can be delivered by the fully charged new cell or battery to some nonstandard EODV, is called available capacity. Thus, if the standard EODV is 1.6 V/cell, the available capacity to an EODV of 1.8 V/cell would be less than the actual capacity.

2.2.1.4 Rated capacity

Rated capacity is defined as the minimum expected capacity when a new, but fully formed, cell is measured under standard conditions. This is the basis for C rate (defined later) and depends on the standard conditions used which may vary depending on the manufacturers and the battery types.

2.2.1.5 *Retained capacity*

If a battery is stored for a period of time following a full charge, some of its charge will dissipate. The capacity which remains that can be discharged is called retained capacity.

2.2.2 *Peukert's law and the battery capacity*

The capacity of a battery, sometimes referred to as C_{load} or simply C, is somewhat an inaccurate measure of how much charge a battery can deliver to a load. It is an imprecise number because it depends on temperature, age of the cells, state of the charge, and on the rate of discharge. It has been observed that two identical, fully charged batteries, under the same circumstances, will deliver different charges to a load depending on the current drawn by the load. In other words C is not constant and the value of C is for a fully charged battery is not an adequate description of the characteristic of the battery unless it is accompanied by an additional information, *rated time of discharge* with the assumption that the discharge occurs under a constant current regime. This discussion is based on the Peukerts's law, where in 1897 Peukert showed that for a lead-acid battery,

$$I^n t = \Lambda = \text{constant} \tag{2.2}$$

where t is the time in hours and I is the current in amperes. Number n, the Peukert Number is always somewhat larger than 1. For lead-acid batteries this is approximately 1.2. In most practical situations in engineering we define the capacity of a battery, C, for a constant discharge rate of I amperes as

$$C = It \tag{2.3}$$

Combining the two relationships in Eqns (2.2) and (2.3),

$$C = \Lambda I^{1-n} \tag{2.4}$$

This provides an explanation on the battery discharge curves we discuss later, where the capacity of a battery is dependent on the discharge current as shown in Figure 2.1.

It is important to note that in this form, Peukert's equation (often referred to as a law) presents a unit imbalance. The intent of this equation is to account for the intrinsic losses associated with discharging batteries at elevated currents. In particular when discharging a battery at increasingly higher currents, the internal cell resistance increases and the recovery rate of the cell decreases (Hausmann and Depick, 2013). Later research demonstrated that the decrease in recovery rate is due to the degradation in the number of active sites in the positive active material as well as an increase between in the resistance between the positive material and the electrolyte (Baert and Vervaet, 1999). The exponent, n, known as Peukert's constant is to account for these losses (Doerffel and Sharkh, 2006). This equation is valid only if the discharge current and the cell temperature are constant. More details are available in Hausmann and Depick (2013), Baert and Vervaet (1999), Doerffel and Sharkh

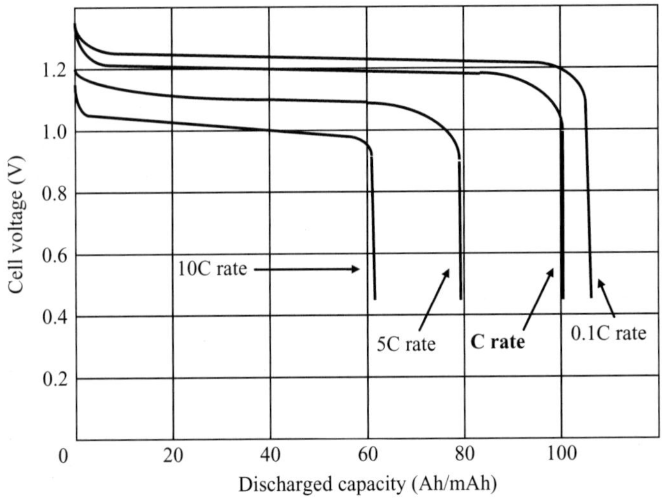

Figure 2.1 Capacity vs. discharge rate of a typical cell.

(2006), and Guoliang et al. (2010) on how this relationship needs to be modified in practical batteries such as lithium-ion and nickel metal hydride batteries.

2.2.3 C rate

The C rate is defined as the rate in amperes or milliamperes numerically equal to the capacity rating of the cell given in Ah or mAh. For example, a cell with a 1.2 Ah capacity has a C rate of 1.2 A. The *C* concept simplifies the discussion of charging for a broad range of cell sizes since the cells' responses to charging are similar if the C rate is the same. Normally a 10 Ah cell will respond to a 1.0 A (0.1C) charge rate in the same manner that a 2 Ah cell will respond to a 0.2 amp (also 0.1C) charge rate. The rate at which current is drawn from a battery affects the amount of energy which can be obtained. At low discharge rates the actual capacity of a battery is greater than at high discharge rates. This relationship is shown in Figure 2.1.

2.2.4 Energy density

Energy density of a cell is its energy divided by its weight or volume. When weight is used it is called the gravimetric energy density and volumetric energy density when the volume is used. Terms "energy density" and "specific energy" are sometimes used for volumetric and gravimetric measures, respectively.

2.2.5 Power density of a battery

It is the amount of power a battery can deliver per unit volume at specified state of charge (SOC), usually 20%. It is also called the volumetric power density and usually measured in watts per liter.

2.2.6 Cycle life

Cycle life is a measure of a battery's ability to withstand repetitive deep discharging and recharging using the manufacturer's cyclic charging recommendations and still provide minimum required capacity for the application. Cyclic discharge testing can be done at any of various rates and depths of discharge (DODs) to simulate conditions in the application. It must be recognized, however, that cycle life has an inverse logarithmic relationship to DOD.

2.2.7 Cyclic energy density

For purposes of comparison, a better measure of rechargeable battery characteristics is a composite characteristic which considers energy density over the service life of the battery. A composite characteristic, *cyclic energy density*, is defined as the product of energy density and cycle life at that energy density and has the dimensional units, watt-hour-cycles/kilogram (gravimetric) or watt-hour-cycles/liter (volumetric).

2.2.8 Self-discharge rate

Self-discharge rate is a measure of how long a battery can be stored and still provide minimum required capacity and be recharged to rated capacity. This commonly is measured by placing batteries on shelf stand at room (or elevated) temperature and monitoring open-circuit voltage over time. Samples are discharged at periodic intervals to determine remaining capacity (RC) and recharged to determine rechargeability.

2.2.9 Charge acceptance

Charge acceptance is the willingness of a battery or cell to accept charge. Charge acceptance can be defined as the ratio $\frac{C_{\text{load}}}{C_{\text{charge}}}$ where C_{charge} is the total coulombs of charge pumped into the battery. Sometimes this is called the coulombic efficiency. This is affected by cell temperature, charge rate, and the SOC. More on this will be discussed later with graphs.

2.2.10 Depth of discharge

DOD is the capacity removed from a battery divided by its actual capacity, expressed as a percentage.

2.2.11 Battery discharge curves and related terminology

Figure 2.2 is typical discharge curve of a secondary battery, applicable to almost all chemistries in general. A fully charged battery usually displays a higher than nominal terminal voltage, which is relatively flat over the discharge curve. However when a battery is discharged up to about 80% of its rated capacity, voltage drop becomes significantly larger, due to its internal resistance increase, etc., and usually display a sharp drop characterized by a knee.

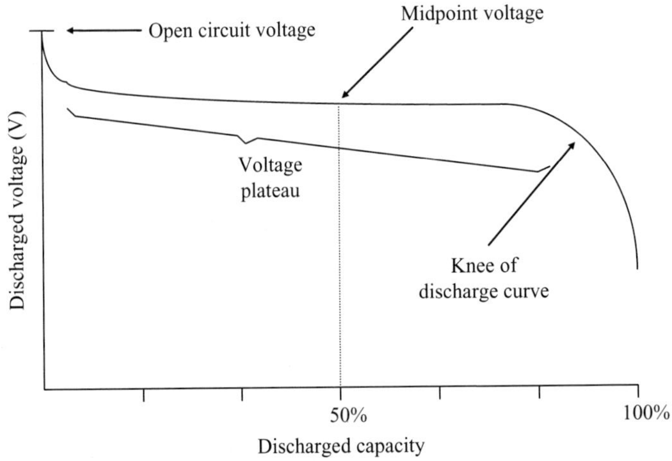

Figure 2.2 Midpoint voltage and the voltage plateau.

2.2.11.1 Voltage plateau

Voltage plateau is the protracted period of very slowly declining voltage that extends from the initial voltage drop at the start of a discharge to the knee of the discharge curve. An example is shown for a typical lead-acid cell in Figure 2.2.

2.2.11.2 Midpoint voltage

Midpoint voltage is the battery voltage when 50% of the actual capacity has been delivered (see Figure 2.2). This is considered as the nominal cell voltage of the battery for design purposes.

2.2.12 Overcharge

Overcharge is defined as continued charging of a cell after it has become fully charged. When a cell is not yet fully charged, the electrical energy of the charge current is converted to chemical energy in the cell by the charging reactions. But, when all of the available active material has been converted into the charged state, the energy available in the charging current goes to produce gases from the cell or to activate other nonuseful chemical reactions. Usually this results in a temperature rise in the cell.

2.2.13 State of charge (SoC)

SOC of a battery pack is defined as the percentage of the RC (or RC divided by its rated capacity). In other words, SOC is the ratio between the saved energy in the battery and the total energy that can be saved in the battery. If the rated capacity is measured in coulombs or Ah, RC should also be estimated in the same units. In cases where, remaining or the rated energy content is estimated in watt-hours, SOC can be estimated based on the energy storage capability. The SOC provides the current state of the battery, and enables the battery be safely be charged and discharged at a level suitable for battery life enhancement.

It is common to estimate the SOC percentage using charge transfer in and out of the battery, and in this case typical coulomb-counting methods are used. In this case, we can use the rated capacity of the battery, Q_{rated}, which is not a function of temperature or aging (since it is usually estimated at a standard laboratory conditions at a controlled temperature). However to accommodate the SOC variation with aging, it is necessary to use a different term, Q_{dicahrge}. This is defined as the maximum total electrical charge, expressed in Ah that a battery can deliver from the fully charged state (SOC of 100%) to a fully discharged state (SOC of 0%) as indicated in Eqn (2.5),

$$Q_{\text{discharge}} = \int_0^{\text{total}} I_b(t)\,\mathrm{d}t \tag{2.5}$$

Then the SOC can be expressed using the ratio or percentage of the capacity that has been discharged relative to Q_{dicahrge}

$$\text{SOC}(t) = \text{SOC}(t_0) - \frac{\int_0^{t} I_b(t)\,\mathrm{d}t}{Q_{\text{discharge}}} \tag{2.6}$$

where I_b is the battery current.

In applications where energy transferred to and from the battery is considered, in Wh or kWh, one needs to use the integration of terminal voltage multiplied by the battery current, and this may be useful in applications where the true energy transferred between system elements are important. In applications such as EV, hybrid electric vehicles (HEV), and plug-in HEV (PHEV), since the engine capacities are measured in kW, battery pack SOC estimates are estimated based on the initial energy capacity $E_0(t)$ based on the relationship given in Eqn (2.7).

$$\text{SOC}(t) = \text{SOC}(t_0) - \frac{1}{E_0(0)} \int_0^{t} P_i(t)\,\mathrm{d}t \tag{2.7}$$

where $E_0(0)$ and $P_i(t)$ are the initial energy capacity and the instantaneous power fed from the battery into the load, respectively. When $P_i(t)$ is negative, battery is charged from the external energy source. In this case, the system should have monitoring capability for the battery voltage and the current both. Kim and Cho (2011) and Ebbesen et al. (2012) provide some details on estimation of this useful parameter, and also some insight into dealing with practical inaccuracies in estimating the SOC and other useful parameters such as state of health (SOH).

2.2.14 State of health

In practical circumstances, designers are challenged to predict the life of a battery pack, and for this purpose SOH is a useful parameter to estimate the remaining useful life of a battery pack.

SOH is the ability of a cell to store energy, source and sink high currents, and retain charge over extended periods, relative to its initial or nominal capacity or the capability. The available charge stored within a fully charged cell is expected to fall with cell usage, as active material on cell plates degrades due to multiple reasons within the battery chemistry. Early detection of SOH degradation would allow a "smart" battery pack to take remedial action, thereby restoring cell capacity (Bhangu et al., 2005).

In estimating the fading of the nominal (energy) capacity of a battery over time due to several complex chemical processes taking place within the battery cells, several types of capacity fade models are usable, namely, (i) electrochemical models, (ii) event-based models, and (iii) energy throughput-based models. Electrochemical models are derived from first principles and they are far superior than the other two models. However to use them in engineering battery packs with predictable performance is very difficult in real time due to limitations in computational resources.

In order to estimate the SOH based on a throughput-based capacity model, an assumption is made that, under constant operating conditions, a battery can withstand a certain amount of energy throughput, equivalent to a number of charge/discharge cycles, before it reaches the end of life. *By definition, the end of life of a battery is reached when the battery capacity has dropped by 20% from its initial capacity.* Based on these conditions, similar to the SOC, we can define SOH of a battery as,

$$\mathrm{SOH}(t) = \mathrm{SOH}(0) - \frac{1}{2N \cdot E_0(0)} \int_0^t P_{\mathrm{i}}(t)\mathrm{d}t \tag{2.8}$$

$\mathrm{SOH}(0) = 1$ and N are the initial SOH and the total number of cycles before end of life, respectively. The factor 2 in the denominator accounts for both charging and discharging. When $\mathrm{SOH} = 0$ end of life for the pack is reached. In general, N is not constant but depends on operating conditions of the battery such as DOD and C rate, etc. More details are available in Ebbesen et al. (2012).

2.3 Battery technologies: an overview

Most commonly available primary batteries (the disposable types) offer significantly greater energy density together with very low self-discharge rates. They are usually cheaper than the rechargeable batteries. Most common primary chemistries include: (i) carbon-zinc (Leclanche cells), (ii) alkaline-MnO_2, (iii) lithium-MnO_2, (iv) lithium-sulfur dioxide, (v) lithium-iron disulphide, (vi) lithium-thionyl chloride ($LiSOCl_2$), (vii) silver-oxide, and (viii) zinc-air. Mercury primary batteries have gradually gone out of the market place due to toxicity and environmental reasons. More details on these chemistries can be found in Quinnell (1991) and Schimpf (1996).

Many types of rechargeable chemistry are used in electronic systems. Common rechargeable chemistries are based on variations of lead acid, nickel-based and lithium-based systems mainly, while limited zinc-based systems and rechargeable alkaline batteries are also available. The choice of a particular battery technology is limited by size, weight, cycle life, operating temperature range, and cost. Comparison of basic characteristics of major chemistries is depicted in Table 2.1.

Table 2.1 Secondary battery chemistry characteristics

Parameter	Units/conditions	Sealed lead acid	NiCd	NiMH	Li-Ion	Li-polymer	Li-iron phosphate	Rechargeable alkaline
Average cell voltage	V	2.0	1.2	1.2	3.6	1.8-3.0	3.2-3.3	1.5
Relative cost	NiCd = 1	0.6	1	1.5-2.0				0.5
Internal resistance		Low	Very low	Moderate	High	High	High	
Self-discharge	%/month	2-4%	15-25%	20-25%	6-10%	18-20%		0.3%
Cycle life	Cycles to reach 80% of rated capacity	500-2000	500-1000	500-800	1000-1200		1500-2000	<25
Overcharge tolerance		High	Med	Low	Very low			Med
Internal resistance		Low	Very low	Moderate	High (coke electrode) Highest (graphite electrode)			
Energy by volume (volumetric energy density)	watt-hour/liter	70-110	100-150	200-350	200-330	230-410	200	220
Energy by weight (gravimetric energy density)	watt-hour/kg	30-45	40-60	60-80	120-160	120-210	100	80

NiCd batteries were used to power most rechargeable consumer appliances before 1990. It is a mature, well-understood technology. However, cadmium was coming under increasing regulatory scrutiny (including mandatory recycling in some jurisdictions), and the maturity of NiCd technology also means most of the capacity and life cycle improvements have already been made. NiMH chemistry which entered the market in 1989 offers incremental improvements in energy density both by weight and volume over NiCd. Li-ion is better still, offering over twice the watts per liter and per kilogram of NiCd batteries. As always, this higher performance comes at a higher price.

Using liquid lithium, Li-ion cells exhibit a much higher energy density than Ni-based chemistries, with a relatively higher terminal voltage than Ni-based cells, and are the commonly used energy source for most portable systems after 2000. The advantages of Li-ion chemistry, however, also come at the cost of greater electrical fragility. Li-ion particularly is more easily and extensively damaged by less than optimal battery management so much so that fail-safe circuits are required to disconnect the cells from the load under over-current or over-temperature conditions. Therefore, these electronic circuits are usually built into the battery pack.

In general, with an increase of the battery energy densities, and any possibilities of quantum jumps in new chemistries, one important concern is the safety and the transportability. Table 2.2 indicates the energy densities of selected fuels and batteries compared with high explosives with transportation capability. A value of 1000 Wh/kg energy density is essentially the practical maximum with its own oxidant for safe transportation (Hamlen et al., 1995). A conventional battery in general is having its own oxidant, similar to this situation and therefore the maximums shown in Table 2.2 applies in development options. However this condition does not apply to metal-air batteries, such as zinc-air, etc., since the oxidant is supplied by oxygen from the air.

With Li-based chemistries with high-energy densities entering the market, and with some fires and explosions occurring in the past, Li-battery manufacturers are not

Table 2.2 Energy densities for selected fuels and batteries

Fuel or battery	Volumetric energy density (Wh/kg)
Diesel fuel	10,000
Methanol	5000
High explosive	1000
Primary battery (maximum estimated around 1995)	500
Rechargeable battery (maximum estimated around 1995)	200
$LiSO_2$ battery	175
Alkaline battery	80
NiCd battery	40
Li-ion battery	120

Source: Hamlen et al. (1995).

supposed to sell them to uncertified battery pack assemblers (BPA) and strict charge and discharge control practices are in place.

During the last decade, newer lithium-based chemistries entered the market. Important ones are the Li-polymer (sometimes known as Li-ion polymer) and Li-iron phosphate chemistries which have better performance than the Li-ion (Chu, 2009; Kim, 2001). Note that Table 2.1 is only a guide and for more precise details individual manufacturer data sheets need be consulted.

Lead-acid batteries are most familiar in automobiles because they are the most economical chemistry for delivering large currents. Lead acid also has a long trickle life and therefore serves well for classic "floating" applications. While flooded lead-acid technology being popular for automobile and similar applications, sealed lead-acid batteries serve the electronic engineering environments. On the downside, lead acid has the least capacity by volume and weight.

2.4 Lead-acid batteries

Lead-acid batteries are the oldest and most widely used rechargeable battery systems today, mostly due to the automotive industry's attachment to the lead-acid chemistry. There are basically two types of lead-acid batteries, namely, the (i) flooded lead-acid type and (ii) sealed lead-acid type.

2.4.1 *Flooded lead-acid batteries*

The flooded lead-acid battery today basically uses the design developed by Faure in 1881. It consists of a container with multiple plates immersed in a pool of dilute sulfuric acid. In most industrial batteries, the cell stack is flooded with high purity, aqueous sulfuric acid at 30-40% concentration. The traction batteries use more concentrated solutions to increase the energy density. The separators are basically nonconductive, microporous structures with good wettability, small to medium pore size, good strength and flexibility, and low electrical resistance. These can be made of rubber, cellulose, PVC, PE, glass fiber, or microglass. In recent years, leaf separators have often being replaced by envelope separators, which can prevent dendritic growth and improve the retention of active material (Broussely and Pistoia, 2007).

Recombination is minimal, so water is consumed through the battery life and the batteries can emit corrosive and explosive gases when experiencing overcharge. So-called maintenance-free forms of flooded batteries provide excess electrolyte to accommodate water loss through a normal life cycle. Most industrial applications for flooded batteries are found in motive power, engine starting, and large system power backup. Today, other forms of battery have largely supplanted flooded batteries in small- and medium-capacity applications, but in larger sizes flooded lead-acid batteries continue to dominate.

The applications of lead-acid batteries can be grouped into three main categories: (i) starting, lighting, and ignition (SLI) or automotive; (ii) traction (deep cycle); and (iii) stationary. In Table 2.3, a summary of comparison is given.

Table 2.3 **Comparison of lead-acid battery applications**

Application	Duty type	Design features
SLI: Cars, motor cycles, trucks, aircrafts, and ships	High-power discharge performance over a wide temperature range. Shallow cycling duty with service life >3 years	Thin plate, mostly 12 V, maintenance free
Traction: Electric trucks, road vehicles, and golf carts	Deep-cycle discharges (1-5 h rates). Service life of 500-2000 cycles	Tubular or flat plate designs, 24-96 V assemblies
Stationary: Telecom, UPS, emergency lighting, and energy storage	Reliability over extended periods. Discharge duty: 1-6 h rates. Service life of 5-25 years on continuous charge. Limited cycling capability	Variety of designs, including tubular, flat plate, Planté, and VRLA cell types

Source: Adapted from Broussely and Pistoia (2007).

By far, the biggest application for flooded batteries is SLI service on automobiles and trucks. Large flooded lead-acid batteries also provide motive power for equipment ranging from forklifts to submarines and provide emergency power backup for many electrical applications, most notably the telecommunications network. More details on different types of lead-acid batteries are available in Broussely and Pistoia (2007).

2.4.2 Sealed lead-acid batteries

Sealed lead-acid batteries first appeared in commercial use in the early 1970s. Although the governing reactions of the sealed cell are the same as other forms of lead-acid batteries, the key difference is the recombination process that occurs in the sealed cell as it reaches full charge. In conventional flooded lead-acid systems, the excess energy from overcharge goes into electrolysis of water in the electrolyte with the resulting gases being vented. This occurs because the excess electrolyte prevents the gases from diffusing to the opposite plate and possibly recombining. Thus, electrolyte is lost on overcharge with the resulting need for replenishment. The sealed lead cell, like the sealed nickel-cadmium, uses recombination to reduce or eliminate this electrolyte loss.

Sealed lead-acid batteries for electronics applications are somewhat different from the flooded type commonly found in the automobile. There are two types of sealed lead-acid batteries: the original gelled electrolyte and retained (or absorbed) system. The gelled electrolyte system is obtained by blending silica gel with an electrolyte, causing it to setup in gelatin form. The retained system employs a fine glass fiber separator to absorb and retain liquid electrolyte. Sometimes the retained system is named absorbed glass mat (AGM). AGM is also known in the industry as "starved design." Starved refers to the absorption limits of the glass separator creating a limitation to the

AGM design relating to diffusion properties of the separator. In certain cases, the AGM battery must be racked and trayed in a specific position for optimum performance. Both these types, gelled and AGM, are called valve-regulated lead acid (VRLA) systems. Today, sealed lead cells are operating effectively in many markets previously closed to lead-acid batteries. For a detailed account of lead-acid cells, Gates Energy Products Inc. (1992), Hirai (1990), Moore (1993), Moneypenny and Wehmeyer (1994), and Nelson (1997) are suggested.

2.4.2.1 Discharge performance of sealed lead-acid cells

The general shape of the discharge curve, voltage as a function of capacity (if the current is uniform), is shown in Figure 2.2. The discharge voltage of the sealed lead-acid battery typically remains relatively constant until most of its capacity is discharged. It then drops off sharply. The flatness and the length of the voltage plateau relative to the length of the discharge are major features of sealed lead cells and batteries. The point at which the voltage leaves the plateau and begins to decline rapidly is often identified as the knee of the curve.

Sealed lead-acid batteries may be discharged over a wide range of temperatures. They maintain adequate performance in cold environments and may produce actual capacities higher than their standard capacity when used in hot environments. Figure 2.3 indicates the relationships between capacity and cell temperature. Actual capacity is expressed as a percentage of rated capacity as measured at 23 °C. Accounting for the temperature effect is rather difficult, because the capacity also depends on a number of factors: design, method of manufacture, storage, service history, and most importantly current density. However in a limited temperature around 30 °C, the International Electrochemical Commission has recognized the validity of Eqn (2.9),

$$C_T = C_{30}(1 + 0.008[T - 30]) \tag{2.9}$$

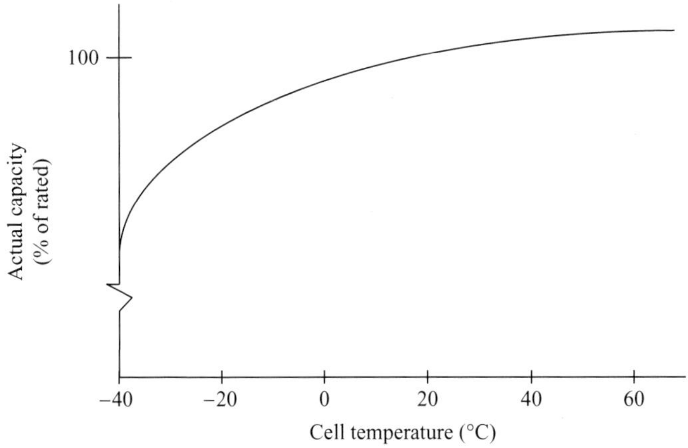

Figure 2.3 Typical discharge capacity as a function of cell temperature.

where C_T and C_{30} are respective capacities in Ah at each temperature in Celsius. Also for the effects of low and high discharge rates n (as in Peurkert's equation) varies between 1 and 2.

2.4.2.2 Capacity during battery life

The initial actual capacity of sealed lead-acid batteries is almost always lower than the battery's rated or standard capacity. However, during the battery's early life, the actual capacity increases until it reaches a stabilized value which is usually above the rated capacity. The number of charge-discharge cycles or length of time on float charge required to develop a battery's capacity depends on the specific regime employed. Alternatively, if the battery is on charge at 0.1C, it is usually stabilized after receiving 300% (of rated capacity) overcharge. The process may be accelerated by charging and discharging at low rates.

Under normal operating conditions, the battery's capacity will remain at or near its stabilized value for most of its useful life. Batteries will then begin to suffer some capacity degradation due to their age and the duty to which they have been subjected. This permanent loss usually increases slowly with age until the capacity drops below 80% of its rated capacity, which is often defined as the end of useful battery life. Figure 2.4 shows a representation of the capacity variation with cycle life that can be expected from sealed lead-acid batteries.

2.4.2.3 Effect of pulse discharge on capacity

In some applications, the battery is not called upon to deliver a current continuously. Rather, energy is drawn from the battery in pulses. By allowing the battery to "rest" between these pulses, the total capacity available from the battery is increased. Figure 2.5 represents typical curves representing the voltage delivered as a function of discharged capacity for pulsed and constant discharge at the same rate.

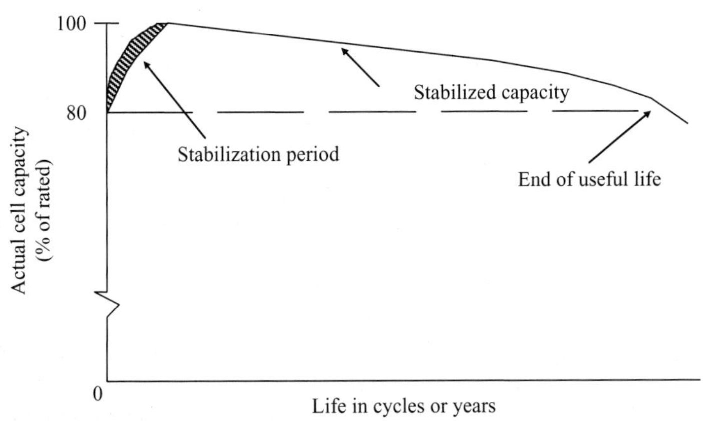

Figure 2.4 Typical cell capacity during its life.

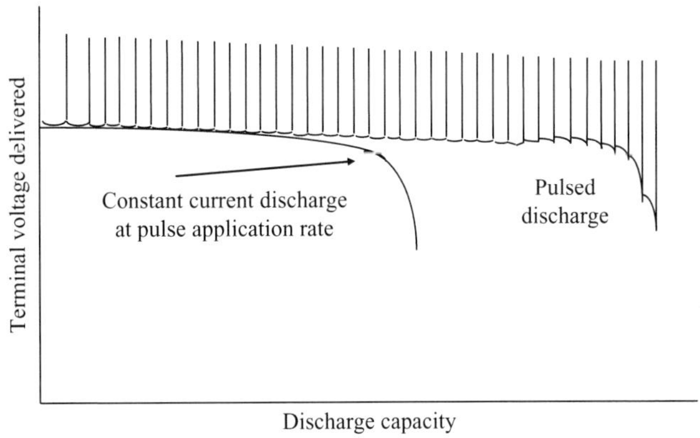

Figure 2.5 Typical pulsed discharge curve.

For the pulsed curve, the upper row of dots represents the open-circuit voltage and the lower sawtooth represents the voltages during the periods when the load is connected. The use of discharged capacity as the abscissa eliminates the rest periods and shows only the periods of useful discharge.

2.4.3 Charging

In general, experience with sealed lead-acid chemistry indicates that application problems are more likely to be caused by undercharging than by overcharging. Since the starved-electrolyte cell is relatively resistant to damage from overcharge, designers may want to ensure that the batteries are fully charged, even at the expense of some degree of overcharge. Obviously, excessive overcharge, either in magnitude or duration, should still be avoided.

The charge acceptance of sealed lead-acid batteries in most situations is quite high, typically greater than 90%. A 90% charge acceptance means that for every ampere-hour of charge introduced into the cell, the cell will be able to deliver 0.9 Ah to a load. Charge acceptance is affected by a number of factors including cell temperature, charge rate, cell SOC, age of the cell, and the method of charging.

The SOC of the cell will dictate to some extent the efficiency with which the cell will accept charge. When the cell is fully discharged, the charge acceptance initially is quite low. As the cell becomes only slightly charged it accepts current more readily and the charge acceptance jumps quickly, approaching 98% in some situations. The charge acceptance stays at a high level until the cell approaches full charge.

As mentioned earlier when the cell becomes fully charged, some of the electrical energy goes into generating gas which represents a loss in charge acceptance. When the cell is fully charged, essentially, all the charging energy goes to generate gas except for the very small current that makes up for the internal losses which otherwise would be manifested as self-discharge. A generalized curve representing these phenomena is shown in Figure 2.6(a). As with most chemical reactions, temperature does

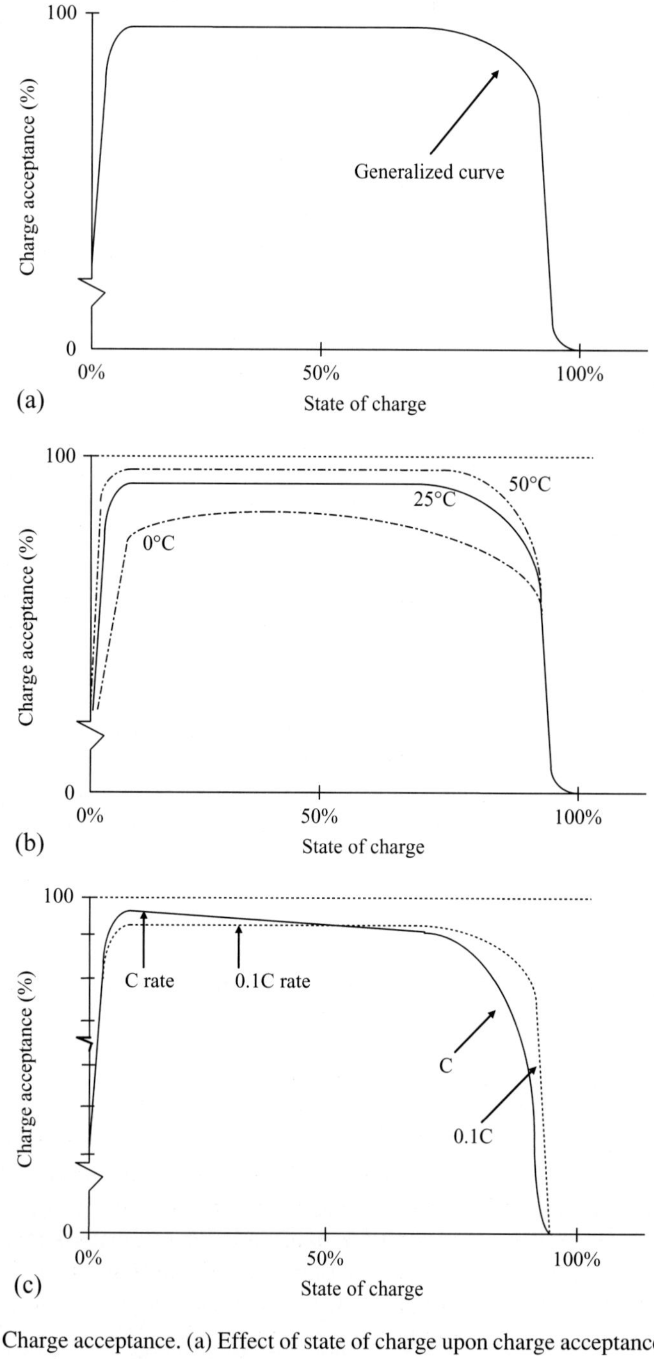

Figure 2.6 Charge acceptance. (a) Effect of state of charge upon charge acceptance, (b) charge acceptance at various temperatures, and (c) charge acceptance at various charge rates.

have a positive effect upon the charging reactions in the sealed lead-acid cell. Charging at higher temperatures is more efficient than it is at lower temperatures, all other parameters being equal, as shown in Figure 2.6(b).

The starved-electrolyte sealed lead cell charges very efficiently at most charging rates. The cell can accept charge at accelerated rates (up to the C rate) as long as the SOC is not so high that excessive gassing occurs. And the cell can be charged at low rates with excellent charge acceptance.

Figure 2.6(c) shows the generalized curve of charge acceptance now further defined by charging rates. When examining these curves, one can see that at high states of charge, low charge rates provide better charge acceptance.

2.5 Nickel-cadmium batteries

Nickel-cadmium (NiCd) batteries belong to a group of five rechargeable batteries NiCd, NiMH, Ni-H_2, Ni-Zn, and Ni-Fe, having in common a Ni-based positive electrode and an alkaline solution. Even though some known drawbacks can be associated with the NiCd battery, especially low-energy density and environmental impact (due to Cd content), still it can be advantageously be used in a number of industrial applications. Cd is used as the negative electrode and β-NiOOH as a positive. The electrolyte is a solution of potassium hydroxide (KOH) of 22% concentration with some LiOH for improving life cycle and temperature performance. Two broad categories of NiCd cells available are vented and sealed types. Vented batteries can have four different electrode constructions, namely, (i) pocket plate, (ii) sintered plate, (iii) fiber nickel cadmium (FNC), and (iv) plastic-bonded plate. Pocket plate type is the oldest with capacities up to 1450 Ah; sintered plate type was developed to increase the energy density by about 50% for high-power requirements; FNC types allow a wide power output capability; and the more recent plastic-bonded type allows reduced weight and volume. Sintered plate types can go up to 100 Ah, while fiber- and plastic-bonded types can reach up to about 500 Ah. More details are found in Broussely and Pistoia (2007).

The operation of the sealed NiCd batteries is similar to VRLA type lead acids. They can come in three different forms: (i) sintered, (ii) fiber, and (iii) foam. First two types are similar to the cases of vented versions, and in the foam version the Ni electrode is obtained by Ni plating porous fibers followed by pyrolysis. More details are found in Broussely and Pistoia (2007).

Sealed NiCd batteries are well suited to applications where a self-contained power source increases the versatility or reliability of the end product. Among the significant advantages of NiCd families are: higher energy density and discharge rates, fast recharge capability, long operating and storage life, etc. These properties place used to place NiCd families at the top of usage in the portable products. In addition, the NiCds are capable of operating over a wide temperature range and in any orientation with reasonable continuous overcharge capability.

In NiCd batteries, a reaction at the negative electrode consumes the oxygen gas that generates at the positive electrode during overcharge. The design prevents the negative electrode from generating hydrogen gas, permitting a sealed structure. NiCd batteries mainly adopt cylindrical or prismatic type configurations. Because NiCd batteries contain cadmium, an environmentally hazardous substance, their disposal has become controversial. This has spurred research into other alternative chemistries.

2.5.1 Discharge characteristics

The discharge voltage of a sealed NiCd cell typically remains relatively constant until most of its capacity is discharged. It then drops off rather sharply. The flatness and length of the voltage plateau relative to the length of discharge are major features of sealed NiCd cells and batteries. The discharge curve, when scaled by considering the effects of all the application variables, provides a complete description of the output of a battery. Differences in design, internal construction, and conditions of actual use of cell affect the performance characteristics. For example, Figure 2.7(a) illustrates the typical effect of discharge rate.

2.5.2 Charge characteristics

Nickel-based batteries are easily charged by applying a controlled current. The charging current can be pure direct current or it may contain a significant ripple component such as half-wave or full-wave rectified current.

This section on charging sealed nickel-cadmium batteries refers to charging rates as multiples (or fractions) of the C rate. These C rate charging currents can also be categorized into descriptive terms, such as standard-charge, quick-charge, fast-charge, or trickle-charge as shown in Table 2.4.When a nickel-cadmium battery is charged, not all of the energy input is converting the active material to a usable (chargeable) form. Charge energy also goes to converting active material into an unusable form, generating gas, or is lost in parasitic side reactions.

Figure 2.8 shows the charge acceptance of NiCd cells. The ideal cell, with no charge acceptance losses, would be 100% efficient. All the charge delivered to the cell could be retrieved on discharge. But nickel-cadmium cells typically accept charge at different levels of efficiency depending upon the SOC of the cell, as shown by the bottom curve of Figure 2.8.

Figure 2.8 describes this performance for successive types of charging behavior (zones 1, 2, 3, and 4). Each zone reflects a distinct set of chemical mechanisms responsible for loss of charge input energy.

In zone 1, a significant portion of the charge input converts some of the active material mass in to a nonusable form; that is, charged material not readily accessible during medium- or high-rate discharges, particularly in the first few cycles. In zone 2, the charging efficiency is only slightly less than 100%; small amounts of internal gassing and parasitic side reactions are all that prevent the charge from being totally efficient. Zone 3 is the transition region.

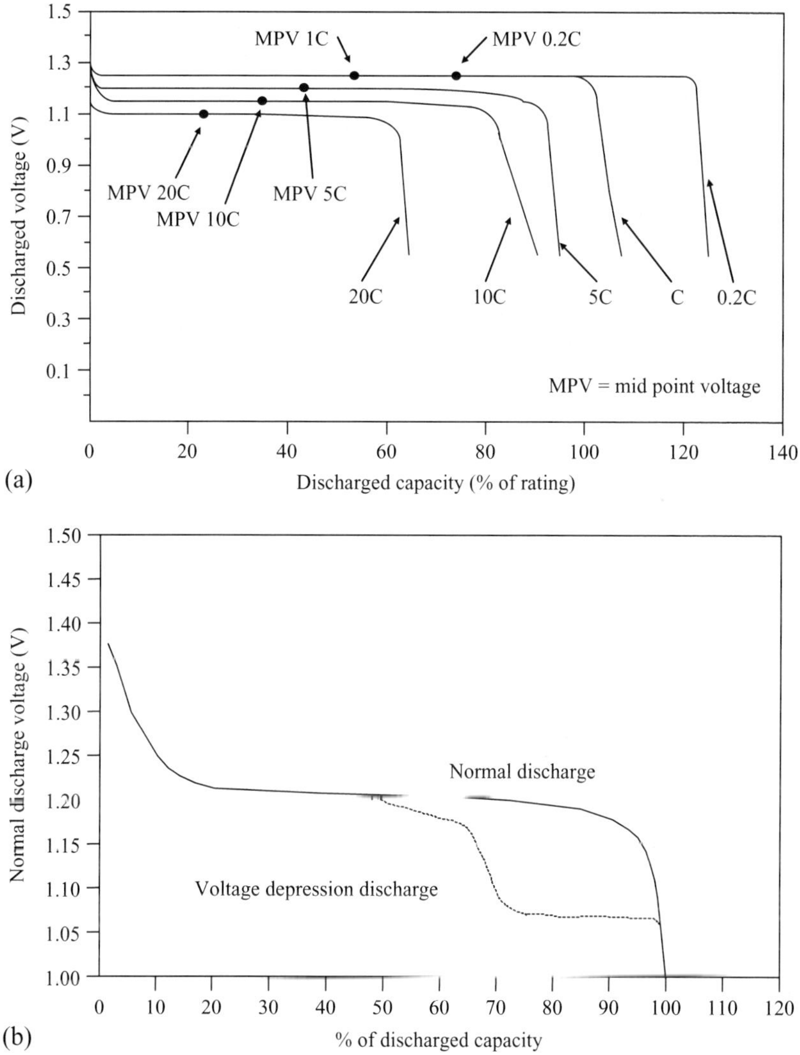

Figure 2.7 Discharge curves for NiCd cells. (a) Typical curves at 23 °C and (b) voltage depression effect.

As the cell approaches full charge, the current input shifts from charging positive active material to generating oxygen gas. In the overcharge region, zone 4, all of the current coming into the cell generates gas. In this zone the charging efficiency is practically none.

The boundaries between zones 1, 2, 3, and 4 are indistinct and quite variable depending upon cell temperature, cell construction, and charge rate. The level of charge acceptance in zones 1, 2, and 3 is also influenced by cell temperature and charge rate. For details Gates Energy Products Inc. (1992) is suggested.

Table 2.4 **Definition of rates for charging NiCd cell**

Method of charging	Charge rate Multiples of C rate	Recharge time (h)	Charge control
Standard	0.05	36-48	Not required
	0.1	16-20	
Quick	0.2	7-9	Not required
	0.25	5-7	
	0.33	4-5	
Fast	1	1.2	Required
	2	0.6	
	4	0.3	
Trickle	0.02-0.1	Used for maintaining charge of a fully charged battery	

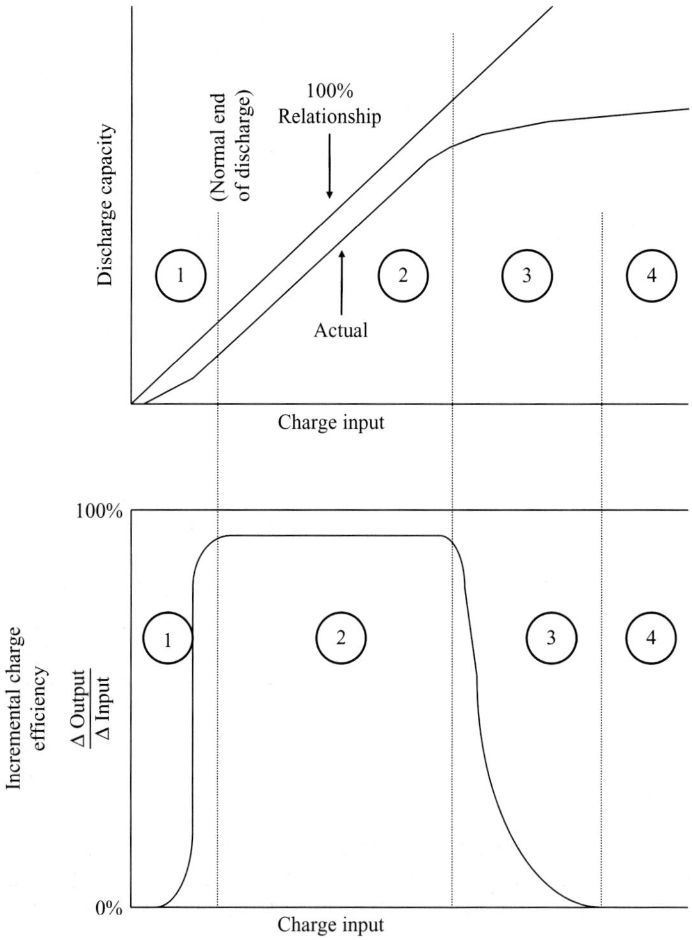

Figure 2.8 Charge acceptance of a sealed NiCd cell at 0.1C and 23 °C.

2.5.3 Voltage depression effect

When some NiCd batteries are subjected to numerous partial discharge cycles and overcharging, cell voltage decreases below 1.05 V/cell before 80% of the capacity is consumed. This is called the *voltage depression effect* and the resultant lower voltage may be below the minimum voltage required for proper system operation, giving the impression that the battery has worn out (see Figure 2.7(b)). Cells exposed to overcharge, particularly at higher temperatures, are quite common and the voltage may be about 150 mV lower than the normal cell voltage. Voltage depression is an electrically reversible condition and disappears when the cell is completely discharged and charged. This process is sometimes called conditioning. This effect is sometimes erroneously called the "memory effect."

Broussely and Pistoia (2007) provides a significant discussion on NiCd battery chemistry with useful details of construction and applications with chemical reactions.

2.6 Nickel metal hydride batteries

While NiCd battery performance was rapidly advancing after 1980, nickel metal hydride (NiMH) cells introduced in early 1990s have shown nearly 170% increase in energy density. By 2000, volumetric energy density of NiMH cells has increased over 300 Wh/liter (Powers, 2000). These extensions of the Ni-based chemistries have become popular with product applications such as note book computers and cellular phones, etc. The first practical NiMH batteries entered the market in early 1990s. In these cells, environmentally unfriendly cadmium negative electrode was replaced by an alloy that could reversible absorb and desorb hydrogen. This chemistry has been the choice for EV till about 2005 (Powers, 2000; Stempel et al., 1998).

2.6.1 Construction

In many ways, nickel metal hydride (NiMH) batteries are the same as NiCd types, but they use nickel for the positive electrode, and a recently developed material known as a hydrogen-absorbing alloy, for the negative electrode. When a NiMH cell is charged, hydrogen generated by reaction with cell electrolyte is stored in the metal alloy (M) in the negative electrode. Meanwhile at positive electrode, which consists of nickel oxyhydroxide loaded into a nickel foam substrate, a hydrogen ion is ejected and nickel is oxidized (Stempel et al., 1998). With an operating voltage of 1.2 V, they provide high capacity, large energy density characteristics comparable to those of NiCd models.

2.6.2 A comparison between NiCd and NiMH batteries

The NiCd cell is more tolerant of fast recharging and overcharging than NiMH cells. NiCd cells hold their charge longer than do NiMH cells. NiCd cells will withstand between 500 and 2000 charge/discharge cycles compared to about 500-800 cycles for NiMH cells. Further NiCd cells will withstand a wider temperature range than NiMH cells. On the other hand, NiMH cells do not exhibit the notorious “memory effect” that NiCd cells sometimes do. As with any new technology at the time of entering the market, NiMH’s prices were higher than those of NiCds (Briggs, 1994; Small, 1992).

Voltage profile of NiMH cells during discharging is very similar to that of the NiCd cells. NiMH cells’ open-circuit voltage is 1.3-1.4 V. At moderate discharge rates, NiMH cells’ output voltage is 1.2 V. Both NiCd and NiMH cells have relatively constant output voltage during their useful service. Figure 2.9 is a typical graph from a battery company comparing the output voltage of 700-mAh NiCd and 1100-mAh NiMH AA cells while under load. Note that the NiMH cell’s greater capacity results in approximately 50% longer service life.

NiMH chemistry advances helped introducing batteries suitable for EV and aerospace applications. These bipolar NiMH batteries were meeting the volumetric and gravimetric energy density needs of these applications during mid 1990s to mid 2005 (Cole et al., 2000; Reisner and Klein, 1994).

Figure 2.10 is another typical battery-company graph showing that NiCd and NiMH batteries and cells charge in similar fashions as well. However, the little bumps at the end of the two cells’ charge curves bare closer examination. You will always see these negative excursions even though absolute cell voltages vary significantly with temperature.

The negative excursions signal a fully charged cell more or less independently of temperature, a useful quirk that sophisticated battery chargers exploit. Note that the NiCd cell’s negative-going voltage excursion after reaching full charge is more pronounced than the NiMH cells.

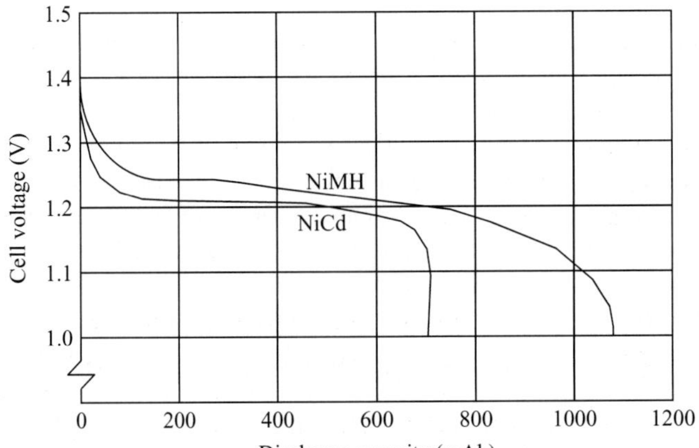

Figure 2.9 Comparison of discharge characteristics of NiCd and NiMH batteries.

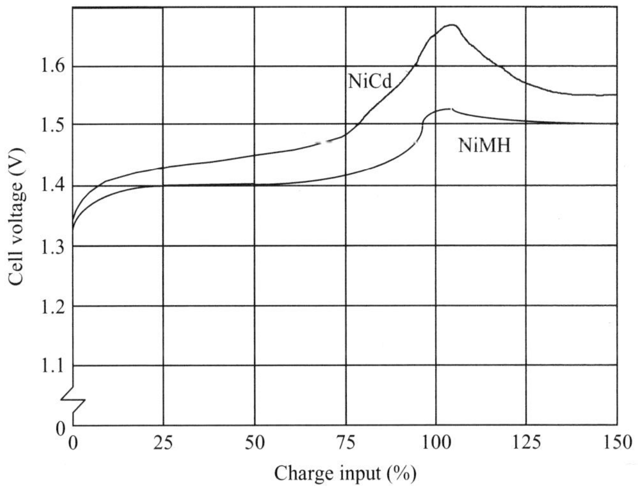

Figure 2.10 Battery voltage at the achievement of 100% charge.

2.7 Lithium-based rechargeable batteries

Toward the latter part of 1990s, the demand for portable systems was increasing at a dramatic rate. Meeting these goals required improvements in battery technologies beyond traditional NiCd and NiMH systems. Newer lithium-based rechargeable battery systems have overcome the safety and environmental obstacles posted by early efforts and are, in general, the most efficient rechargeable battery packs available. With an energy density by weight about twice that of nickel-based chemistries (Table 2.1), Li-ion batteries can deliver lighter weight packs of acceptable capacity. Li-ion also has about three times the cell voltage of NiCd and NiMH batteries; therefore fewer cells are needed for a given voltage requirement. Li-ion batteries have become the choice in notebook PCs and many other portable systems because of their high-energy density, declining costs, and readily available management circuits.

Rechargeable lithium cells come in several different chemistries, namely, (i) lithium-ion, (ii) lithium-polymer, (iii) lithium metal, and (iv) lithium-iron phosphate. Most of these chemistries have come into commercial use after 1992 onwards (Dan, 1997; Morrison, 2006a,b, 2007). Figure 2.11(a) indicates the progress of Li-ion chemistry based on the commonly used 18650 type cells, which is the most matured out of all the lithium cells. Figure 2.11(b) indicates the relative capabilities of the four common rechargeable chemistries.

The first noticeable difference between Li-based chemistries and nickel-based chemistries is the higher internal impedance of the lithium-based batteries. Figure 2.12 shows this by graphing the actual discharge capacity of a Li-ion cell at different discharge currents compared to a NiCd cell. At a 2 A discharge rate (2C), less than 80% of the rated capacity is available for Li-ion compared to nearly 95% of the rated capacity for NiCd (Freeman and Heacock, 1995). For systems with

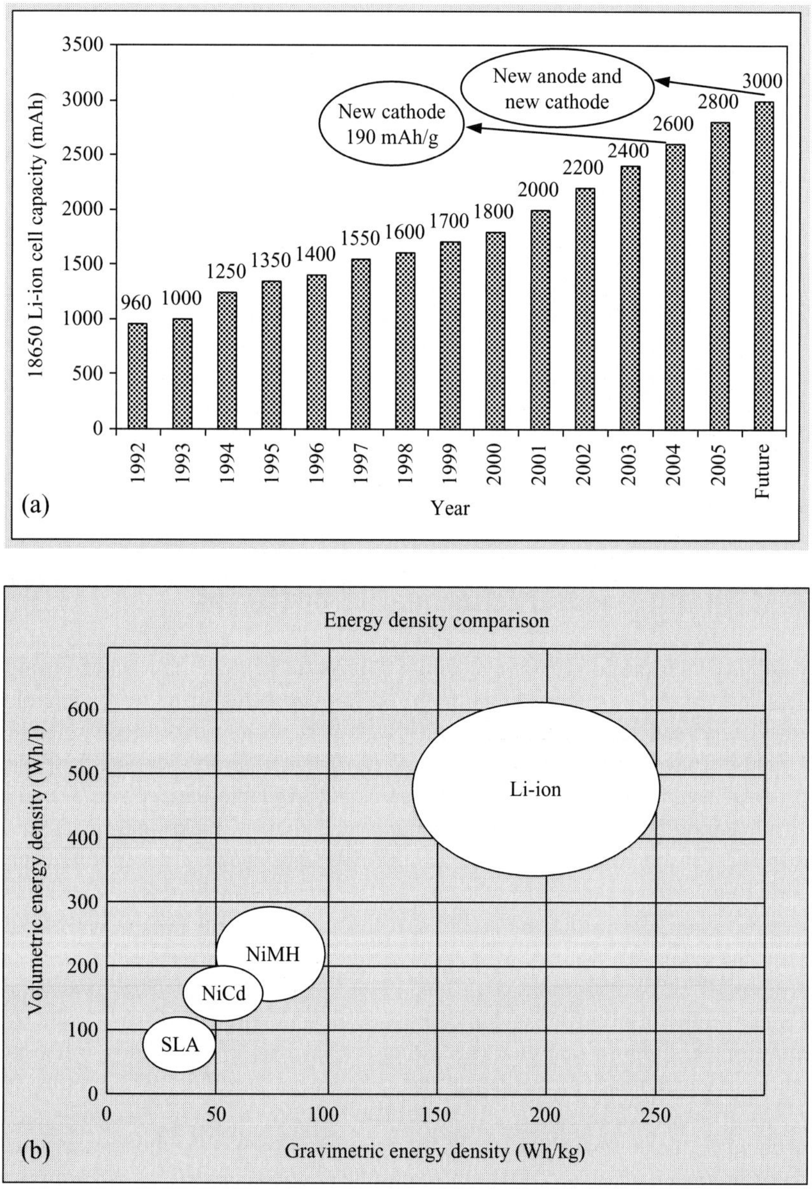

Figure 2.11 Progress of the Li-ion cells and comparison with other chemistries. (a) mAh capacity increases over the period from 1992 for 18650 cells and (b) comparison of Li-ion with other chemistries.
Source: Power Electronics Technology (Morrison, 2006a).

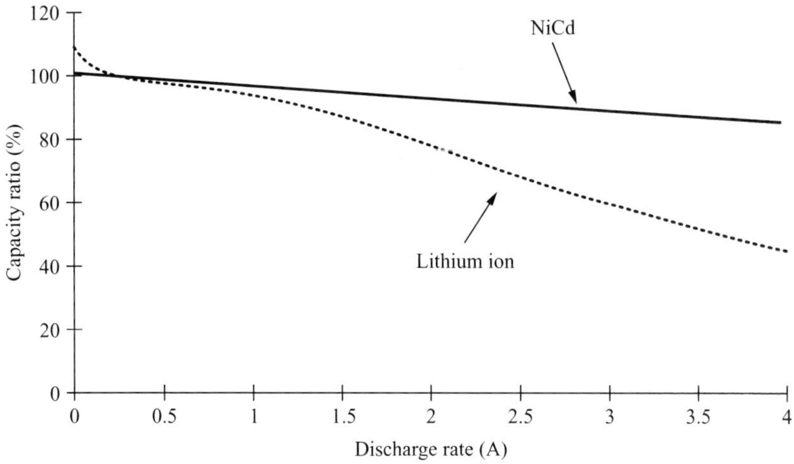

Figure 2.12 Li-ion and NiCd capacity vs. discharge current.

discharge currents greater than 1 A, the capacity realized from the Li-ion battery may be less than expected. Parallel battery stack configurations are often used in Li-ion battery packs to help reduce the severity of this problem. Due to the nature of the lithium chemistries, lithium batteries cannot tolerate overcharge and overdischarge.

Given the comparison of energy densities of fuel, batteries and explosives (Figure 2.2), the most important factor in using lithium chemistries in a portable consumer product is safety. Fortunately, battery safety is comprehensively addressed by cell and BPA in concert with the semiconductor manufacturers. For details Bennett and Brawn (1997) is suggested. Commercially available Li-ion packs have an internal protection circuit that limits the cell voltage during charge to between 4.1 and 4.3 V/cell, depending on the manufacturer. Voltages higher than this rating could permanently damage the cell. A discharge limit of between 2.0 and 3.0 V (depending on the manufacturer) is necessary to avoid reducing the cycle life of the battery and damaging the battery.

2.7.1 Construction

The anode, or negative electrode, in a Li-ion cell is comprised of a material capable of acting as a reversible Li-ion reservoir. This material is usually a form of carbon, such as coke, or graphite, or pyrolytic carbon. The cathode, or positive electrode, is also a material which can act as a reversible lithium-ion reservoir. Due to lithium ions shuttling back and forth between these two reservoirs, these batteries are sometimes called the rocking chair cell (Fuller et al., 1994)

Preferred cathode materials are currently $LiCoO_2$, $LiNiO_2$, or $LiMn_2O_4$ because of their high oxidation potentials of about 4 V versus lithium metal. Commercially available Li-ion cells use a liquid electrolyte made up of mixtures which are predominately organic carbonates containing one or more dissolved lithium salts (Levy, 1995). While $LiCoO_2$ was the preferred cathode material at the early stages of development, in newer systems $LiMn_2O_4$ cathodes are used. Chemical process details are summarized in Levy (1995). In the latest generation of Li-ion cells, a new cathode material known as Nickel oxide-based New Platform is used by Panasonic batteries (Morrison, 2006a), extending the commonly used 18650 series to 2.9 Ah.

In lithium-polymer batteries, the liquid electrolyte is replaced with a polymer in gel or solid form. The polymer electrolyte provides the required electrode stack pressure, so the metal can is no longer required and it becomes possible to package the cells easily. A laminate of aluminum foil and plastic, the pouch occupies less space and weighs less than the metal can. For these reasons lithium-polymer cells came in compact prismatic cell formats around early 2000 (Morrison, 2000). Also they are considered safer than liquid Li-ion, as they do not leak when they are punctured. As a result, simplifications are possible for in-pack protection circuits (Morrison, 2000).

$LiFePO_4$ batteries, also known as LFP battery systems were introduced around late 1990s, and recently they have gone into mass production, due to demand from EVs, hybrid EVs, electric bikes, and power tools. In these batteries, cathode material, $LiFePO_4$, is environmentally benign, inexpensive, and relatively abundant (Jiayuan et al., 2009).

2.7.2 Charge and discharge characteristics

Today, the predominant Li-ion technologies use coke or graphite for an anode material. Figure 2.13 illustrates the differences in the two types of cells during discharge. The graphite anode discharge voltage is relatively flat during a majority of the discharge cycle, while the coke anode discharge voltage is more sloped (Juzkow and St. Louis, 1996).

The energy available from the graphite anode cell is higher for a given capacity due to the higher average discharge voltage. This may be useful in systems that need the maximum watt-hour capacity for a given battery size. Also, the charge and discharge cutoff voltages between the two Li-ion systems vary among manufacturers.

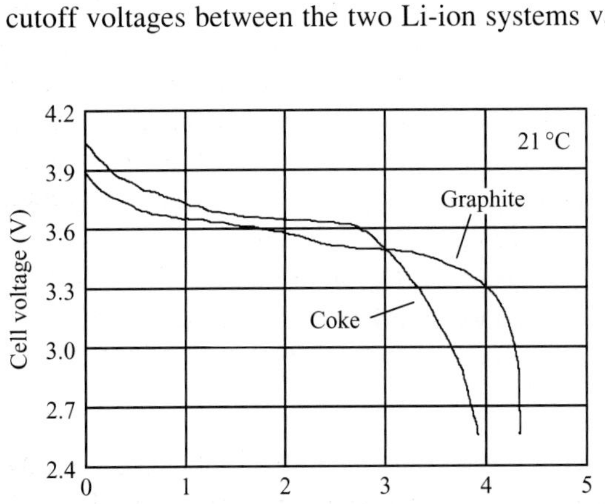

Figure 2.13 Li-ion discharge profile for different electrodes. *Source:* Adapted from (Juzkow and St. Louis, 1996), Moli Energy Limited, USA.

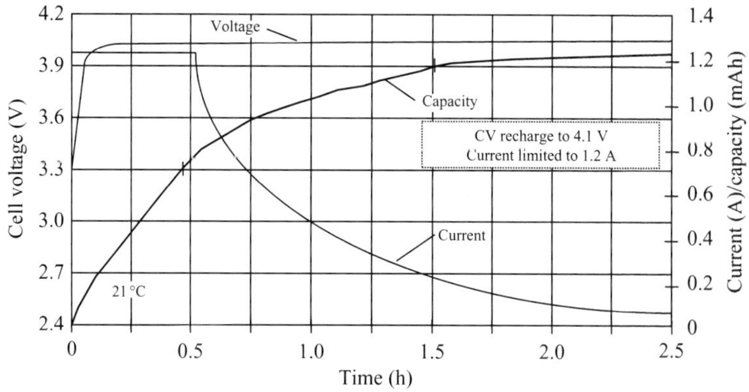

Figure 2.14 Li-ion charge profile at constant potential charging at 4.1 V and current limited to 1.2 A.
Source: Moli Energy Ltd., USA.

Figure 2.14 shows the typical charge profile for Li-ion batteries. The charge cycle begins with a constant current limit, transitioning to a constant voltage limit, typically specified between 4.1 and 4.3 V $\pm$ 1%, depending on manufacturer's recommendation. This allows maximum charge capacity without cell damage. Charging to a lower voltage limit does not damage the cell, but the discharge capacity will be reduced. A 100 mV difference could change the discharge capacity by more than 7%.

2.7.3 Li-ion micro batteries

Li-ion chemistry has matured vastly during the recent years and coin cells of volume less than 1 cm^3 with volumetric energy densities up to about 100 Wh/l are available. Better miniaturization can be achieved by solid-state thin-film batteries (Bates et al., 2000). These devices with solid-state electrodes and a thin-film layer of electrolyte have achieved high cycle stability and specific capacity. However, the overall capacity of these cells are quite small, and to overcome this 3D silicon wafer processing combined with state of the art lithium battery electrodes are under development (Hahn et al., 2012). This research ongoing at Fraunhofer IZM, Germany, is expected to yield 2000-10,000 micro batteries on a 300 mm silicon wafer, based on cavities developed on silicon for electrodes, current collectors, and contacts. More details are available in Hahn et al. (2012). These micro batteries are aimed at powering miniaturized sensor nodes, RFID devices, and medical devices, etc.

2.8 Reusable alkaline batteries

Alkaline technology has been used in primary batteries for several years. With the development of the reusable alkaline manganese technology, secondary alkaline cells have quickly made their way into many consumer and industrial applications. In many applications, reusable alkaline cells can be recharged from 75 to over 500 times, and initially have three times the capacity of a fully charged NiCd battery. These cells do not compete with NiCds in high-power applications, however.

Intensive research and development activities carried out at Battery Technologies Inc. (BTI), Canada and at the Technical University in Graz, Austria in the late 1980s and early 1990s resulted in the successful commercialization of the rechargeable alkaline manganese dioxide zinc (RAM™) system. BTI has chosen to sell licenses and production equipment, where necessary, for the manufacturing and worldwide marketing rights of its proprietary RAM technology. For example, Rayovac Corporation, one of the Licensees, launched its line of reusable alkaline products under the name RENEWAL™ in the USA, Pure Energy Battery Corporation in Canada (PURE ENERGY™), and Young Poong Corporation in South Korea (ALCAVA™). For details Nossaman and Parvereshi (1995), Sengupta (1995), and Ivad and Kordesch (1997) are suggested.

The chemistry behind the reusable approach depends on limiting the zinc anode to prevent overdischarge of the MnO_2 cathode. Additives are also incorporated to control hydrogen generation and other adverse effects on charge. Rated cycle life is around 25 cycles to 50% of initial capacity. Longer cycle life is possible depending on drain rate and DOD. To take advantage of the reusable alkaline cell and increase its life, a special "smart charger" is required.

2.8.1 Cumulative capacity

Using reusable alkaline cells can drive down the total battery cost to the consumer. This cost saving can be determined by looking at the cumulative capacity of a reusable cell versus the onetime use of a primary alkaline cell. Figure 2.15 illustrates the capacity of AA cells being discharged down to 0.9 V at 100 mA. It shows that although the initial use of the reusable alkaline is almost that of primary alkaline, the reusable one can be recharged for continued use. Table 2.5 shows the increase in cumulative capacity by limiting the DOD and achieving more cycles.

Overcharging also affects the cycle life of reusable alkaline. Reusable alkaline is not tolerant of overcharge and high continuous charge currents, and may be damaged if high current is forced into them after they have reached a partially recharged state. Proper charging schemes should be used to prevent an overcharged condition.

2.9 Zn-air batteries

Primary Zn-air batteries have been in existence for over 50 years with applications such as hearing aids and harbor buoys. The light weight and high-energy content

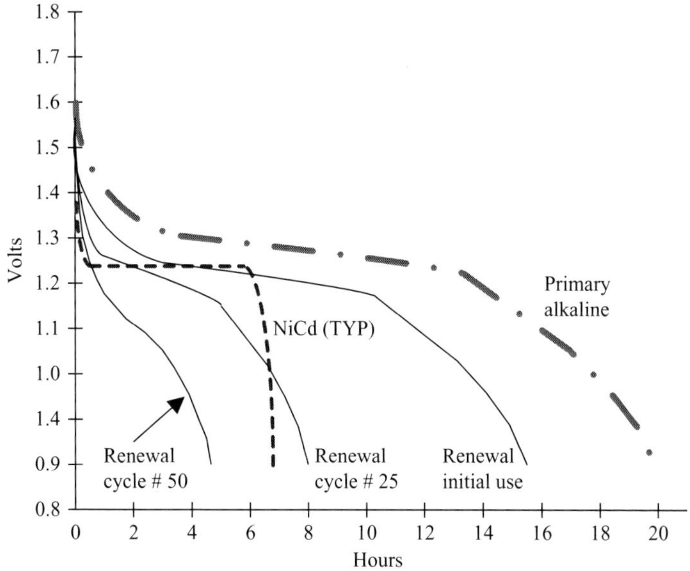

Figure 2.15 100 mA discharge curve comparison for NiCd, primary alkaline, and reusable alkaline.
Source: Benchmarq Microelectronics/BTI Technologies.

Table 2.5 Capacity of "AA" cells at various depths of discharge (DODs) (values in mAh)

Condition	125 mA to 0.9 V		
	100% DOD	**30% DOD**	**10% DOD**
Cycle 1	1500	450	150
Cycle 50	400		
Cumulative 50	33,000	22,000	7000
Cumulative 100		44,000	15,000
Cumulative 500			73,000

in Zn-air technology have promoted research on Zn-air rechargeable chemistry by companies such as AER Energy Resources, USA in the late 1990s, and, around 2009 by companies such as RWE Innogy (Germany) and ReVolt Technology AS (RWE Innogy, 2009). Focus is on EV and portable appliances. Another application is solar powered rural telecom systems.

Rechargeable Zn-air technology is an air-breathing technology where the oxygen in ambient air is used to convert zinc into zinc oxide in a reversible process. Cells use air-breathing carbon cathode to introduce oxygen from air into potassium hydroxide electrolyte. Cathode is multilayered with a hydrophilic layer and anode is comprised of metallic zinc.

The characteristic voltage of zinc-air systems is nominally around 1 V. For example, during discharge, they will operate within a voltage between 1.2 and 0.75 V. The current and power capability of the system is proportional to the surface area of the air-breathing cathode. For more current and power, a larger surface area cell is required. For less current and power, a smaller cell may be used. Compared to other rechargeable chemistries, Zn-air needs an air manager for an intake and exhaust of air to allow the chemical process.

Figure 2.16 compares the performance of recent developments on zinc-air chemistry (RWE Innogy, 2009) with other common rechargeable chemistries. This clearly indicates that Zn-air batteries require less weight and volume. Discharge and charge characteristics of Zn-air batteries by AER Energy Resources are shown in Figure 2.17.

The cells exhibit a flat voltage profile over the discharge cycle. Typical charge voltage is 2 V/cell using a constant voltage/current taper approach. Life cycle varies between 50 and 400, depending on the DOD. Cost per Wh is apparently the lowest compared to Ni- and Li-based chemistries. For details Cutler (1997) is suggested.

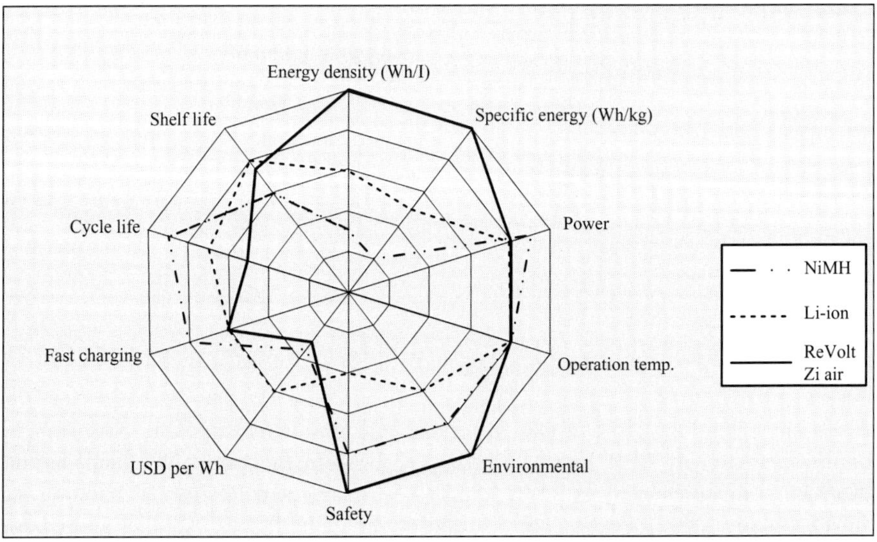

Figure 2.16 Performance comparison of rechargeable Zn-air chemistry with other chemistries. *Source:* RWE Innogy (2009)/Green Car Congress.

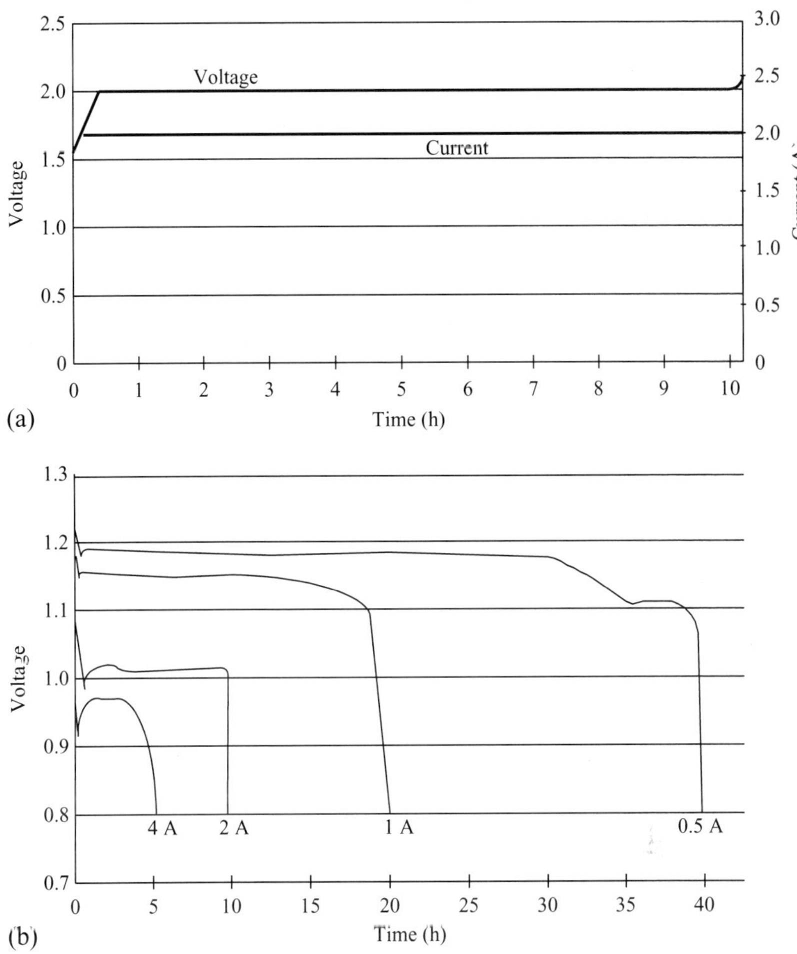

Figure 2.17 Charge/discharge characteristics of a Zn-air chemistry. (a) Charge and (b) discharge.
Courtesy: AER Energy Technologies.

References

Advanced Rechargeable Battery Market: Emerging Technologies and Trends Worldwide, March 1, 2009, 245 p., marketresearch.com (Pub. ID: SB1933124).

Baert, D., Vervaet, A., 1999. Lead-acid battery model for the derivation of Peukert's law. J. Power Sources 44, 3491–3504.

Bates, J., et al., 2000. Thin-film lithium and lithium-ion batteries. Solid State Ionics 135 (1-4), 33–45.

Bennett, P.D., Brawn, G.W., 1997. Introduction to applying Li-ion batteries. In: Proceedings of Portable by Design Conference, USA, pp. 125–134.

Bhangu, B.S., Bentley, P., Stone, D.A., Bingham, C.M., 2005. Nonlinear observers for predicting state-of-charge and state-of-health of lead-acid batteries for hybrid electric vehicles. IEEE Trans. Veh. Technol. 54 (3), 783–794.

Briggs, A., 1994. NiMH technology overview. In: Portable by Design Conference Proceedings, pp. BT-42–BT-45.

Broussely, M., Pistoia, G., 2007. Industrial Applications of Batteries—From Cars to Aerospace and Energy Storage. Elsevier, Amsterdam, pp. 691-736.

Chu, B., 2009. $LiFePO_4$ batteries help consumer devices come to life. Power Electron. Technol. 35, 10–15.

Cole, J.H., Eskra, M., Klein, M., 2000. Bipolar nickel-metal hydride batteries for aerospace applications. IEEE AES Mag. 15 (1), 39–45.

Cutler, T., 1997. Rechargeable zinc-air design options for portable devices. In: Proceedings of Portable by Design Conference, USA, p. 112.

Dan, P., 1997. Recent advances in rechargeable batteries. Electron. Des. 45 (3), 112–116.

Doerffel, D., Sharkh, S.A., 2006. A critical review of using the Peukert equation for determining the remaining capacity of lead-acid and lithium-ion batteries. J. Power Sources 155 (2), 395–400.

Ebbesen, S., Elbert, P., Guzzella, L., 2012. Battery state-of-health perspective energy management for hybrid electric vehicles. IEEE Trans. Veh. Technol. 61 (7), 2893–2900.

Freeman, D., Heacock, D., 1995. Lithium-ion battery capacity monitoring within portable systems. In: HFPC Conference Proceedings, pp. 1–8.

Fuller, T.F., Doyle, M., Newman, J., 1994. Simulation and optimization of the dual lithium ion insertion cell. J. Electrochem. Soc. 141 (1), 1–10.

Gates Energy Products Inc., 1992. Rechargeable Batteries Applications Handbook. Butterworth- Heinemann, Boston.

Guoliang, W., Rengui, L., Chunbo, Z., Chan, C.C., 2010. Apply a piece-wise Peukert's equation with temperature correction factor to NiMH battery state of charge estimation. J. Asian Electr. Veh. 8 (2), 1419–1423.

Hahn, R., Höppner, K., Eisenreich, M., Ferch, M., Wilke, M., Lang, K.D., 2012. Development of rechargeable micro batteries based on micro channel structures. In: IEEE International Conference on Green Computing and Communications, pp. 619–623.

Hamlen, R.P., Christopher, H.A., Gilman, S., 1995. US Army battery needs—present and the future. IEEE AES Mag. 10, 30–33.

Hausmann, A., Depick, C., 2013. Expanding the Peukert equation for battery modeling through inclusion of a temperature dependency. J. Power Sources 235, 148–158.

Hirai, T., January 1990. Sealed lead-acid batteries find electronic applications. PCIM, pp. 47-51.

IDTechEx Ltd., October 1, 2012a. Batteries & Supercapacitors in Consumer Electronics 2013-2023: Forecasts, Opportunities, Innovation, 345 p. (SKU: CGAQ4888235).

IDTechEx Ltd., December 1, 2012b. Traction Batteries for Electric Vehicles Land, Water & Air 2013-2023, 333 p. (SKU: CGAQ4862363).

Ivad, J.D., Kordesch, K., 1997. In-application use of rechargeable alkaline manganese dioxide/zinc (RAM™) batteries. In: Proceedings of Portable by Design Conference, USA, pp. 119–124.

Jiayuan, W., Zechang, S., Xuezhe, W., September 2009. Performance and characteristic research in $LiFePO_4$ battery for electric vehicle applications. In: IEEE Vehicle Power and Propulsion Conference, 2009 (VPPC '09), pp. 1657–1661.

Juzkow, M.W., St. Louis, C., 1996. Designing lithium-ion batteries into today's portable products. In: Portable by Design Conference, pp. 13–22.

Kim, S., January 2001. Lithium-ion polymer batteries promise improved size, safety, energy density. PCIM, pp. 30-39.

Kim, J., Cho, B.H., 2011. State-of-charge estimation and state-of-health prediction of Li-ion degraded battery based on an EKF combined with a per-unit system. IEEE Trans. Veh. Technol. 60 (9), 4249–4260.

Levy, S.C., 1995. Recent advances in lithium ion technology. In: Portable by Design Conference Proceedings, pp. 316–323.

Moneypenny, G.A., Wehmeyer, F., April 1994. Thinline battery technology for portable electronics. In: HFPC Conference Proceedings, pp. 263–269.

Moore, M.R., October 1993. Valve regulated lead acid vs flooded cell. In: Power Quality Proceedings, pp. 825–827.

Morrison, D., 2000. Thinner Li-ion batteries power next generation portable devices. Electron. Des. 95–106.

Morrison, D., 2006a. New materials extend Li-ion performance. Power Electron. Technol. 50–52.

Morrison, D., 2006b. Li-ion cells build better batteries for power tools. Power Electron. Technol. 32, 52–54.

Morrison, D., 2007. Cathode modeling builds better batteries for power tools. Power Electron. Technol. 52.

Nelson, B., 1997. Pulse discharge and ultrafast recharge capabilities of thin-metal film technology. In: Proceedings of Portable by Design Conference, USA, pp. 13–18.

Nossaman, P., Parvereshi, J., 1995. In systems charging of reusable alkaline batteries. In: Proceedings of HFPC Conference, USA.

Powers, A.R., 2000. Sealed nickel cadmium and nickel metal hydride cell advances. IEEE AES Mag. 15 (12), 15–18.

Quinnell, R.A., December 5, 1991. The business of finding the best battery. EDN, pp. 162-166.

Reisner, D.E., Klein, M., 1994. Bipolar nickel-metal hydride battery for hybrid vehicles. IEEE AES Mag. 9 (5), 24–28.

RWE Innogy Invests €5.5M in ReVolt; Rechargeable Zinc-air Storage Systems, January 16, 2009, http://www.greencarcongress.com/2009/09/revolt-20090901.html.

Schimpf, M., 1996. Choosing lithium primary-cell types. Electron. Des. 44, 141–144.

Sengupta, U., 1995. Reusable Alkaline™ Battery Technology: applications and system design issues for portable electronic equipment. In: Portable by Design Conference, pp. 562–570.

Silicon-air battery touts unlimited shelf life, EE Times Asia, November 25, 2009. http://www.eetasia.com/login.do?fromWhere=/ART_8800590556_765245_NT_17302c65.HTM.

Small, C.H., December 10, 1992. Nickel-hydride cells avert environmental headaches. EDN, pp. 156-161.

Stempel, R.C., Ovshinsky, S.R., Gifford, P.R., Corrigan, D.A., 1998. Nickel metal hydride: ready to serve. IEEE Spectr. 35 (11), 29–34.

2009. US demand for batteries to reach $16.8 billion in 2012. Power Electron. Technol. 35, 9.

Dynamics, models, and management of rechargeable batteries

3

3.1 Introduction

In Chapter 2, we had a simple discussion on battery chemistries and their basic performance specifications in an electronic engineer's viewpoint. To get the best out of any rechargeable battery chemistry, a battery management circuit designer should have an intrinsic understanding of the battery chemistry and the internal behavior in an electrical viewpoint. In a modern application, such as a wireless communication product or an electric vehicle, device power consumption fluctuates widely and the energy source has to deal with highly varying current deliveries to electronic loads, and this demands a fast responding battery pack with an optimally designed battery management system (BMS). In addition to these characteristics determined by the load behavior, in a rechargeable battery pack, state of charge (SOC), state of health (SOH), and end of life (EOL) become important overall quality parameters of an installed pack. In determining optimal management criteria for the longest lifetime and the longest run time of the battery pack, circuit designers have to clearly understand battery dynamics and use optimal battery models to design the correct BMSs.

This chapter provides some fundamentals of battery dynamics, different types of modeling to achieve usable and practical equivalent circuits, and an overview of management systems for different applications. In this chapter, electrochemistry-based modeling approaches are highlighted to better understand the behavior of popular rechargeable chemistries, without going into mathematics. For details of electrochemistry-related modeling, the chapter cites adequate and recent references.

3.2 Simplest concept of a battery

As discussed in Chapter 1, simplest model of a battery (single cell case) is based on a constant voltage source representing the chemical behavior of the electrode-electrolyte combination together with a constant resistance, and this Thevenin equivalent circuit provides us an excellent starting point. However, this equivalent circuit as shown in Figure 1.1 is an oversimplified case, and it does not provide us any idea of practical battery behavior, particularly applied to situations of charging and discharging with time-varying currents occurring between the charging system and/or the connected load, respectively.

3.3 Battery dynamics

To understand the fundamentals of battery dynamics, we should start with a simple concept applied to dynamic systems, where variables within the system are heavily dependent on time. Such a dynamic system can be represented by a simple block diagram shown in Figure 3.1, where the system input receives a time-dependent excitation signal and when it is propagated through the system it gets modified and/or converted to another output parameter that is also time dependent. If the system is excited with a time-dependent signal *u*(*t*), with the system modifying the input signal, based on a transfer function that may be mathematically defined, and this situation creates a time-varying output signal *y*(*t*). Dynamic system can be described by a set of differential equations:

$$\dot{y}(t) = f[y(t), u(t)] \tag{3.1}$$

Input and output signals *u*(*t*) and *y*(*t*) are generally vectors, and if the system is nonlinear, Equation (3.1) will also be nonlinear.

For batteries, it is common to define the current and ambient temperature as input signals. The output signals could be (i) battery terminal voltage, (ii) battery temperature, (iii) SOC, and (iv) SOH. While other parameters such as internal resistance and open-circuit voltage (OCV) are also useful output signals, from a dynamic viewpoint these additional output signals yield only limited additional information. As discussed by Jossen (2006), a simple example of the dynamic characteristic in a battery pack is shown in Figure 3.2, where a NiMH battery undergoes a pulsed discharge after a charging period. The discharge regime is in-line with the GSM standard with a pulse duration of 577 μs and a period of 4.81 ms. Pulse current changes from a 0.2 A during the rest period to a maximum of 2 A during the transmission pulse. The discharge voltage shows a ripple voltage of 40-50 mV and if this is assumed to be due to internal resistance, the battery internal resistance can be estimated as 22-28 mΩ.

In a situation like this, since the discharge current is in pulse mode, compared to a constant DC discharge current operation, we may face additional questions such as:

- How can we quantify the transient voltage fluctuations with short-term ripple superimposed on the longer term terminal voltage profile?

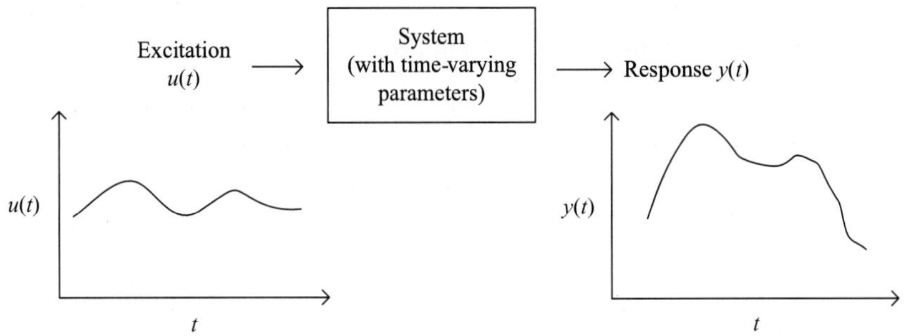

Figure 3.1 A dynamic system.

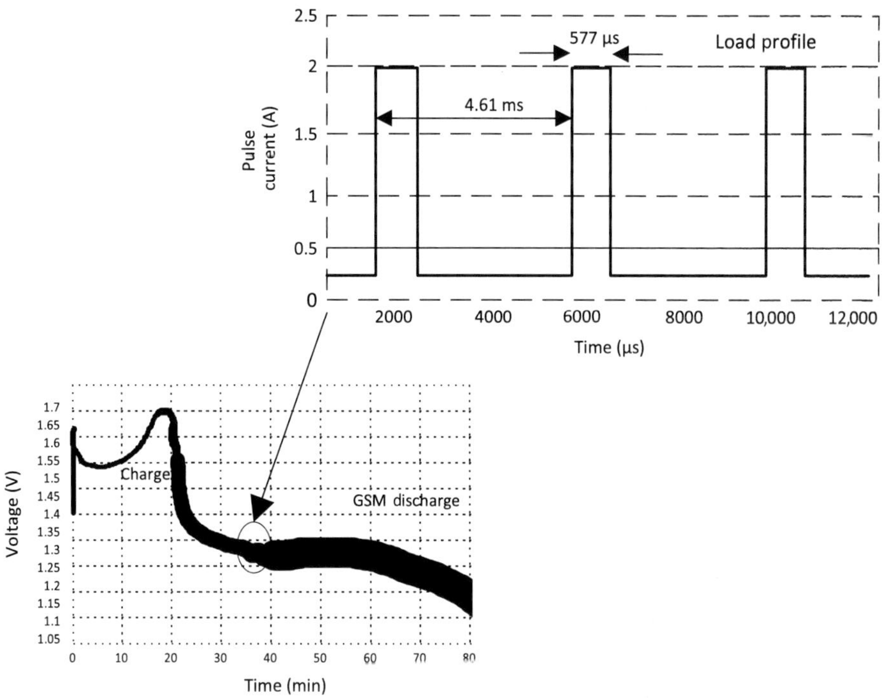

Figure 3.2 Battery discharge with a pulsed current in a typical GSM cellular phone. *Source:* Jossen, 2006.

- Any added heat generated inside the battery?
- What will be the performance of the battery and how does it affect the lifetime estimated?

A battery's dynamic performance can be influenced by both internal parameters (such as SOC, SOH, DC/AC components of internal resistance, and other design parameters) as well as external parameters (such as ambient temperature, DC current, and the short- and long-term history of the battery). Conversely, the dynamic characteristics contain information on these parameters. As indicated by Jossen (2006), there has been significant amount of research work done on battery dynamics applicable in SOC/SOH determination, and the dynamic behavior of a battery is a combination of several effects, which occurs within microsecond to several years.

As depicted in Figure 3.3 different dynamic effects in batteries occur within a wide range of time and is a combination of several effects such as:

(i) Electrical and magnetic
(ii) Electric double layer
(iii) Mass transport
(iv) Cycling and SOC
(v) Reversible
(vi) Aging

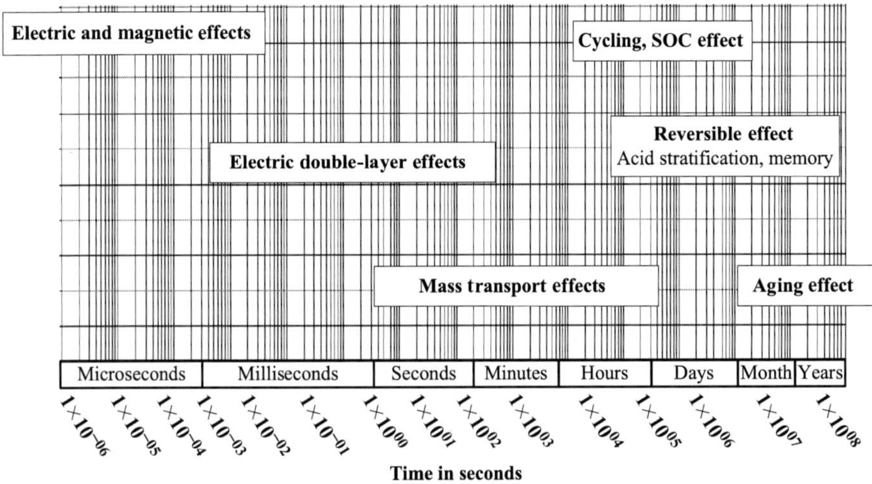

Figure 3.3 Typical time ranges or dynamic effects within batteries.
Source: Jossen, 2006.

Figure 3.3 depicts only different time domains applicable to these effects, which are quite dependent on the chemistry, battery design, temperature, SOC, and SOH. Temperature effects are not shown in the figure, and the dynamics of the battery temperature depends on the heat capacity of the pack, dissipated heat, and the heat generated within the battery. Dynamics of these temperature variations can be in the range seconds to several hours in general.

3.3.1 Long-term effects

Aging, reversible, cycling, and SOC effects are considered longer term phenomena, and as we see from Figure 3.3, their dynamics occur within minutes to years of time durations. These are related to the operation regime of the battery.

3.3.1.1 Aging effects

Aging influences the battery performance and hence the output parameters significantly, and the time domain for battery aging is typically in the range of months to years. For primary cells, aging is considered due to storage, but through the discharge process.

3.3.1.2 Reversible effects

Some electrochemical storage systems show reversible effects. These occur during cyclic operations and can be regenerated by special charge-discharge regimes. For example, vented lead-acid batteries show acid stratification that can be removed by an extended charge. In NiCd cells, memory effect is another example, which can be cleared by applying one or more full charge-discharge cycles. Time domain for such reversible effects lies between the values of aging effects and the cycling effects, as seen in Figure 3.3. Few hours to 1 year time constants are effective in these processes.

3.3.1.3 Cycling and SOC effects

If a battery is charged or discharged, its SOC is changed and this itself is a dynamic characteristic of a battery. Additionally, the open circuit or the terminal voltage itself is a function of SOC. Time domain of SOC change or cycling depends on the operational conditions with time domain from several minutes to days.

During the cycling process, the battery can heat up due to internal heat sources, which can be due to ohmic heating or any other chemical reactions. The generated ohmic heat can be quite influenced by the charge/discharge current profiles. If we assume that there will be only ohmic heating, heating power can be estimated by

$$P_{\text{ohmic}} = R_{\text{int}} I_{\text{eff}}^2 \tag{3.2}$$

where P_{ohmic} is the power dissipated due to ohmic resistances due to an overall internal effective resistance of R_{int} and the I_{eff} is the effective current based on the wave shape. In the example we cited in Figure 3.2, for the GSM waveform, RMS value of the current is estimated as 0.73 A and the average current is 0.42 A. With a form factor of the waveform of 1.73 approximately, the heat generated due to the RMS current of the GSM pulse transmission, it generates three times bigger power, calculated based on the average current of the waveform. More discussion on the subject of waveform parameters discussed here can be found in Kularatna (2008).

3.3.2 Mass-transport effects

Within a chemical system such as a battery, the transport of ions is due to diffusion and migration. Diffusion is caused by the gradient in concentration, while migration is caused by forces of electric field present within the chemical system of the electrodes and the electrolyte. Two cases may have different directions. Migration is often hindered by solvated molecules that cover the ion.

In most cases, diffusion is largely responsible for the mass transport. Diffusion can occur at different inside locations of an electrochemical cell such as the following (Jossen, 2006):

1. In the free electrolyte or the separator, due to ions produced on one electrode are consumed by the opposite polarity electrode.
2. With a porous electrode electrochemical reaction can occur anywhere within the active mass (AM) of the porous electrode.
3. The reaction products can move within the AM to their final location by diffusion.
4. If there is a film on the electrode (e.g., in Li-ion cell there is a film, the solid electrolyte interface (SEI) on the surface of the anode), diffusion can happen through the SEI.

Inside a battery, there are three separate primary parts which are mainly responsible for the chemical reactions. They are the positive electrode, the negative electrode, and the

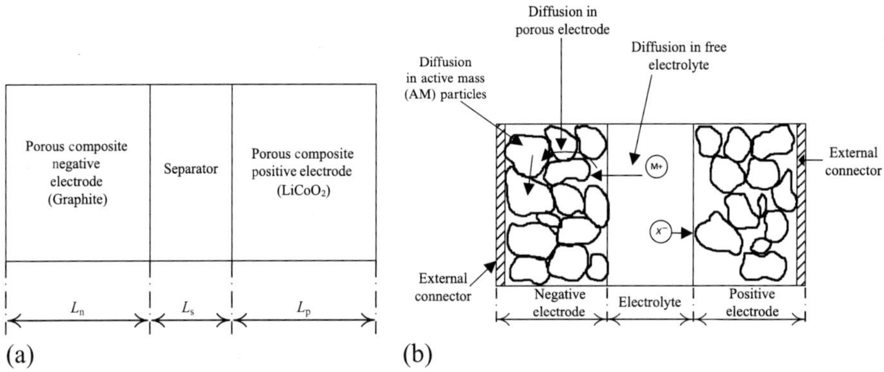

Figure 3.4 Schematic of typical Li-ion cell sandwich: (a) basic structure and (b) diffusion process in such a cell.
Adapted from Ramadass et al., 2004; Jossen, 2006.

electrolyte. In a battery, there is a separator which is in the middle of the battery in between these two electrodes. Figure 3.4(a) depicts a typical case of a Li-ion cell where a porous composite negative electrode made of graphite and a porous composite positive electrode with a poly-propylene separator. The two electrodes and the separator are immersed in the electrolyte. More details of such a simple schematic, used in estimation of capacity fade models for Li-ion batteries is discussed in Ramadass et al. (2004).

Figure 3.4(b) portrays the three locations where diffusion can take place. As the battery has two electrodes, the diffusion within the porous electrode and within the AM is in both electrodes. In such a case, diffusion occurring can be described by Fick's first law (Jossen, 2006), which is dependent on the diffusion coefficient and the rate of change of concentration along the direction of the mass transfer. The diffusion coefficient is material specific; in gases, it is typically 10^{-1} cm^2/s, in liquids 10^{-5} cm^2/s, and in solids 10^{-10} cm^2/s. Temperature is a key influencing factor in the diffusion coefficient.

The limitations of the diffusion of the ions result in locally changed ion concentrations, and from an electrical perspective, it results in an overpotential caused by reduced or increased ion concentration at the location of charge transfer. Time constants representing the diffusion depend on electrode thickness and the structure to a great degree, with typical time constants in the range of seconds to minutes (Jossen, 2006). Impedance spectroscopy, or measuring the response of a battery to an excitation of a small amplitude signal of a voltage or a current to estimate the small signal impedance of the cell, is frequently used to analyze the dynamic behavior of the batteries due to mass-transport effects (Huet, 1998; Jossen, 2006). More discussion on the subject of impedance spectroscopy will be presented later in Section 3.4.

3.3.3 Double-layer effects

Electrical double-layer effect is a phenomenon present both in electrical double-layer capacitors (more commonly known as supercapacitors) and in electrochemical batteries. Figure 3.5 depicts the basic idea of a double-layer capacitor in relation to a

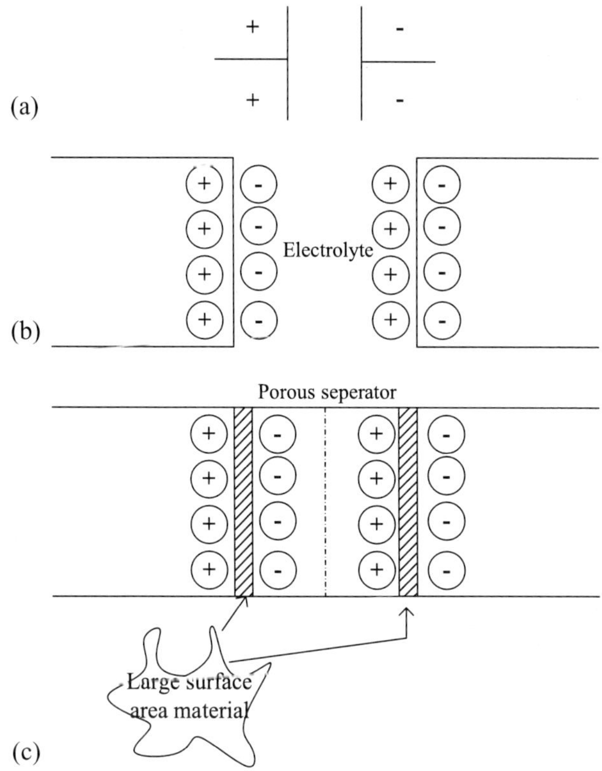

Figure 3.5 Simplified concept of electrical double-layer capacitance: (a) parallel plate capacitor, (b) liquid double-layer capacitor, and (c) double-layer capacitor formed by a porous separator in an electrolyte.
Adapted from Lai and Rose, 1992.

parallel plate capacitor. Figure 3.5(a) shows the case of simple parallel plate capacitor where capacitance is given by kA/d, where k is the dielectric constant, A is the area of the plates, and d is the distance between the two parallel plates. When two electrodes are placed across a liquid electrolyte, due to the action of the electrolyte, two charged layers are present on both ends of the electrolyte, and these are equivalent to two capacitors in series, and this is a liquid double-layer capacitor. In a case like a rechargeable battery, a porous separator is also placed in the middle, and if the active material used in the cell, which are the anode/cathode elements, they can be large area material used similar to supercapacitors (Lai and Rose, 1992). Some materials, known as large surface area materials, typically usable in supercapacitors, have 50-1500 m^2/g surface density. When such a material with extremely low thickness is combined with suitable electrodes, a large capacitor is formed due to large A/d ratio. Inside batteries, such a high capacitance surface layer is formed next to electrodes and creates a double-layer effect.

In a rechargeable battery, due to large surface area in porous electrodes, combined with the short distance between the electrodes and electrolyte, electrochemical double layer is formed similar to the example of Figure 3.5(c). This creates a double-layer capacitance on the electrode surface, with significantly large value, in parallel to the electrochemical charge-transfer reaction.

Recent research publications (Li et al., 2014) and the citations there indicate that electrochemistry-based impedance models for battery chemistries such as Li-ion can be developed to achieve a better understanding of battery dynamics. In this process, the charge-transfer reaction happening on the electrode/electrolyte surface can be simplified by several assumptions such as:

(i) Electrodes consist of multiple spherical particles of identical size and dynamics.
(ii) The material inside the particle is spatially homogeneous.
(iii) Electrodes have perfect conductivity (i.e., no voltage drop inside a particle or among particles).
(iv) The SEI film growth caused by aging only happens on the anode surface.
(v) The electrolyte, separator, and current collectors are treated as lumped pure resistors.

Such a set of simplifications allow us to consider the cell behavior based on identical multiple spherical particles as in Figure 3.6(a) which can lead to a simplified analysis based on single particle pair separated by the separator used in case like Li-ion chemistry.

This kind of analysis allows the developing of a set of mathematical relationships between the internal dynamics and external measurements, to include the electrical double-layer effect with other effects discussed. As the two electrodes of a battery are not equal, dynamic characteristics of both electrodes are also different. In a lead-acid battery typical double layer capacitance values are 7-70 F/Ah and 0.4-1 F/ Ah for the positive and negative electrodes respectively. Lot more analytical details are available in Li et al. (2014) and an overview of different dynamic effects is discussed by Jossen (2006). Figure 3.7 depicts the frequency response of the two electrodes of a lead-acid battery caused by the double-layer effect.

As depicted in Figure 3.7, the cutoff frequency for the positive electrode is approximately 10 Hz and for the negative electrode it is 100 Hz. This means that the

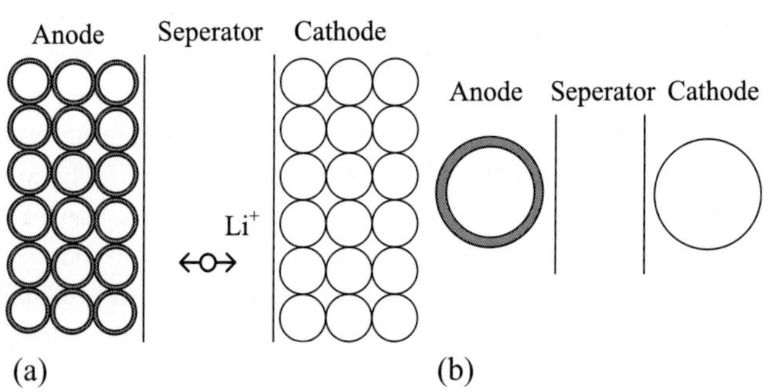

Figure 3.6 Simplification of Li-ion battery electrodes: (a) based on identical multiple particles and (b) single particle at each electrode.
Source: Li et al., 2014.

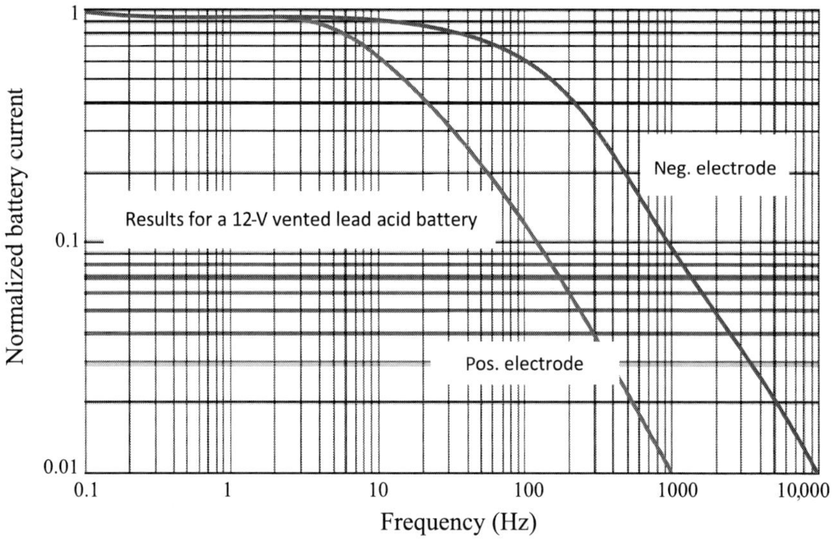

Figure 3.7 Frequency response of the two electrodes of an automotive lead-acid battery caused by the double-layer effect.
Source: Jossen, 2006.

alternative currents with a frequency above 100 Hz do not flow through the charge-transfer reaction, since they are filtered by the double-layer capacitor.

3.3.4 Effects caused by porous electrodes

Porous electrodes as used in batteries are complex components. The behavior is different from planar electrodes and they have a significant influence on the dynamic characteristics. The diffusion of ions through the porous electrodes is one of the main limiting factors in batteries. Additionally, the porous structure changes during discharging, as the volume of the discharged AM in most systems is much higher than the volume of the charged AM. This results in a reduction in the porosity during discharging, which in turn leads to lower diffusion. In many systems, the conductivity of the discharged AM differs from that of the charged AM. This changes the ohmic resistance of the AM during discharging. These effects result in an inhomogeneous discharge of porous electrode, especially at higher currents. The effects of porous distributed electrodes indicate that the dynamic characteristic is more blurred for the porous effect in comparison to other effects.

3.3.5 Electric and magnetic effects

Compared to the slow effects of mass transport and the electrochemical double layer, much faster phenomena are found in a rechargeable battery. Ohmic resistance of the cell is the sum of the electrolyte resistance, resistance of the current collector, the AM, and the transition resistance between the AM and the current collector. In addition to this, each cell has a serial inductance based on the geometry of the cell. For a lead-acid battery values between 10 and 100 nH/cell for 100 Ah, cells are reported (Jossen, 2006). This is in addition to the wiring of the cell. This inductance limits the slew rate of the current, particularly in large batteries such as lead-acid systems. In large lead-acid battery packs, this inductance can be of interest above 1 kHz frequencies. For small batteries, with much smaller inductance, much higher frequencies such as 10-100 kHz show the effect of these inductive components.

With increased frequencies, the penetration depth of the ions in the porous structure decreases and the electrodes start behaving like planar electrodes. At high frequencies, electrode pair becomes a simple plate capacitor equivalent. A typical value of this high-frequency interelectrode capacitance for a lead-acid cell is around 10 nF/cell (Jossen, 2006).

Another important high-frequency effect is the skin effect. Caused by electromagnetic field effects, the penetration depth of AC current in conductive materials is limited. This penetration depth d for cylindrical materials is calculated by $d = 1/\sqrt{k\mu\pi f}$, where k is the conductivity of the material and μ is the permeability of the material. Depending on the depth of materials used for current collectors, this effect can come into play. More details on materials are given in Jossen (2006).

3.3.6 Battery equivalent circuits based on various dynamic effects

For different dynamic effects, a battery pack may be represented by an electrical equivalent circuit, as if only that effect is in charge of the battery dynamics. For example, Figure 3.8(a) shows the equivalent circuit for diffusion effects, Figure 3.8(b) shows the equivalent circuit applicable to double-layer effect, Figure 3.8(c) shows the equivalent circuit applicable for porous electrode effect, and Figure 3.8(d) indicates the high-frequency equivalent circuit of a battery.

3.4 Electrochemical impedance spectroscopy for batteries

Electrochemical impedance spectroscopy (EIS) is a widely used technique in several areas such as nondestructive diagnostics, prognostics, and modeling in electrochemical systems. The EIS involves empirically determining the small signal impedance of an electrochemical material or a device such as battery, supercapacitor, or a fuel cell. In recent years, the technique has been applied to a host of next generation direct-energy conversion technologies such as fuel cells, photovoltaic systems, as well as energy storage devices such as batteries and supercapacitors (Lindahl et al., 2012).

In EIS, an electrochemical device such as battery or a supercapacitor is perturbed from steady-state operation by a potentiostatic (constant voltage) or galvanostatic (constant current) excitation signal over a desired frequency range. The resulting current and voltage response is then analyzed based on the two signals' inter-related

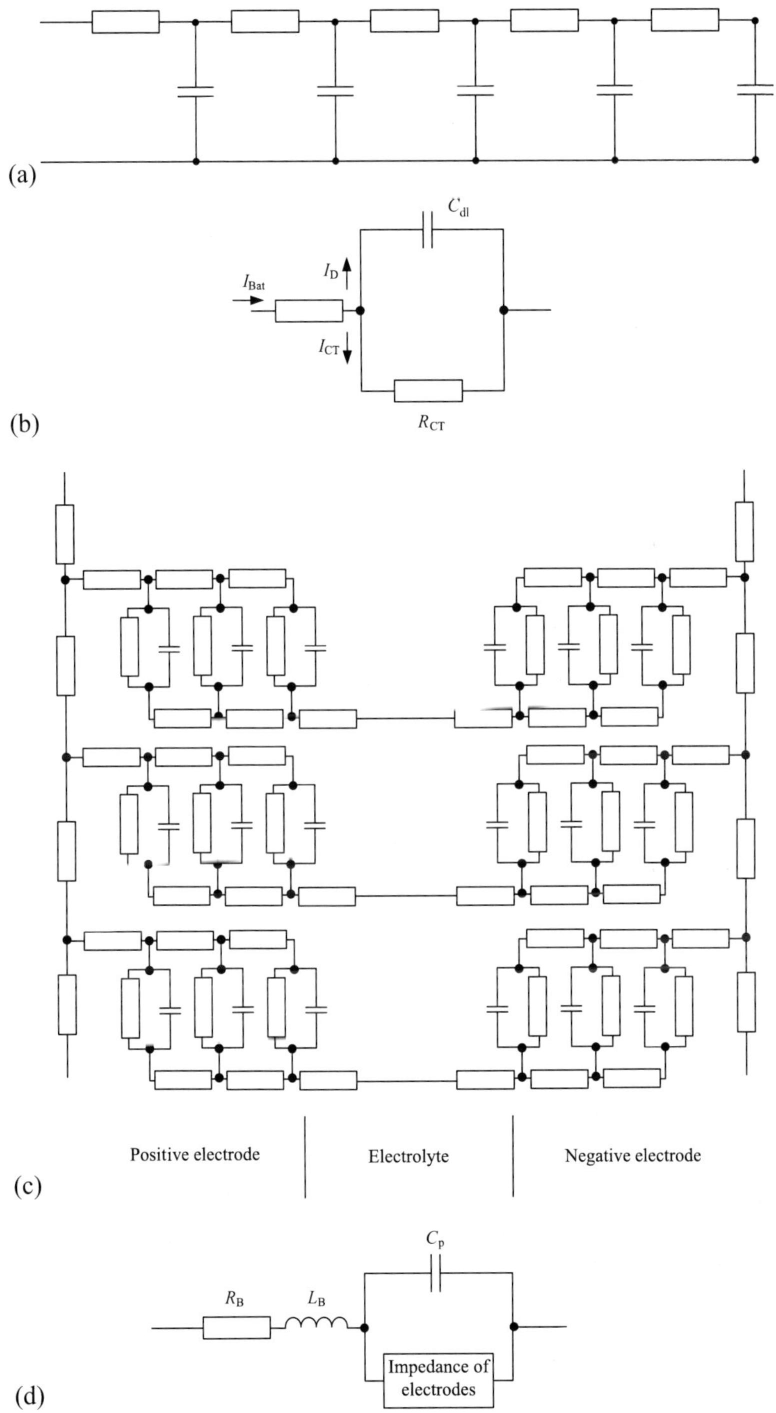

Figure 3.8 Equivalent circuits applicable to different dynamic effects of the batteries: (a) diffusion effects; (b) double-layer capacitor effect; (c) porous electrode case; and (d) high-frequency operation.

magnitude, phase and frequency content determining the localized impedance of the device, on small signal basis, as a function of the frequency. The attraction of EIS lies in the direct correlation between the observable electrical performance (at the terminal level) and its nonobservable electrochemical and physicochemical processes. Thus EIS allows bulk monitoring of the internal processes of an electrochemical device with the results indicative of the device's dynamic performance such as the SOC and the overall health such as SOH (Huet, 1998; Lindahl et al., 2012).

3.4.1 Simple basics used in EIS and sample results for different chemistries

The electrochemical impedance (or simply the AC impedance) of a battery characterizes its dynamic behavior around a selected operating DC voltage across its terminals or at a particular DC current (charging or discharging). In general, any type of excitation signal (e.g., sine wave, step, noise, etc.) can be used. Most commonly a sine-wave is used and in galvanostatic mode where a DC current of I (polarization current), which could be charging or discharging the battery and this sinusoidal excitation current can be written as

$$\Delta I = I_{\text{peak}} \text{Sin}(2\pi f t) \tag{3.3}$$

where I_{peak} is the maximum value of the superimposed excitation current and f is the frequency. This excitation yields a sinusoidal voltage response of ΔV in the form,

$$\Delta V = V_{\text{peak}} \text{Sin}(2\pi f t + \varphi) \tag{3.4}$$

where V_{peak} is the maximum value of the superimposed excitation voltage around a DC terminal voltage of V and φ is the phase angle related to two superimposed voltage and the current. Similar two relationships can be developed also for potentiostatic mode. The two equations (3.3) and (3.4) define an electrochemical impedance for the battery which is given by

$$Z(f) = \frac{V_{\text{max}}}{I_{\text{max}}} e^{j\varphi} \tag{3.5}$$

In general, voltage amplitude must not exceed about 10 mV to ensure that the impedance measurements are performed under linear conditions. In that case, the excitation and response signals are actually sine waves, and the value of $Z(f)$ does not depend on the amplitude of the excitation signal. With large capacity batteries with very low internal impedances, AC current required may be in the order of several amperes (Huet, 1998). Several standard signal processing techniques, including frequency response analysis, fast Fourier transform, and phase-sensitive detection may be used (Lindahl et al., 2012). For all these techniques, fundamentals of the impedance-extraction process are quite similar and more details can be found in Orazem and Tribollet (2008). The value of $Z(f)$ usually depends on the temperature and the SOC.

Also due to mass-transport effects, dynamic behavior of a battery, represented by $Z(f)$, could be quite different during continuous charging and discharging.

In representing the dynamic behavior of an electrochemical system such as a battery or a supercapacitor, based on the value of $Z(f)$ *at* different frequencies, either a Bode plot or a Nyquist plot can be used. It is more common to use the Nyquist plot with the opposite of the imaginary part of the impedance on the ordinate axis, so that the capacitive loops are appear in the upper quadrants. Figure 3.9 shows the typical impedance spectra of a Li-ion battery at different SOC and a constant temperature of 25 °C (Buller et al., 2005).

In Figure 3.9, impedance data have been recorded for different frequencies, and at approximately 286 Hz, impedance becomes a pure resistance of approximately 42 mΩ, with the graphs showing the intersection of the real axis. More discussion on this is in (Buller et al., 2005). Figure 3.10 depicts the typical EIS results for a lead-acid automotive battery (Buller et al., 2003).

A discussion on use of EIS techniques to predict SOC for a NiMH battery is available Bundy et al. (1998).

3.4.2 Specific requirements for EIS and its limitations

Problems that arise when applying the methods of EIS to batteries may be grouped into three categories (Karden et al., 2000).

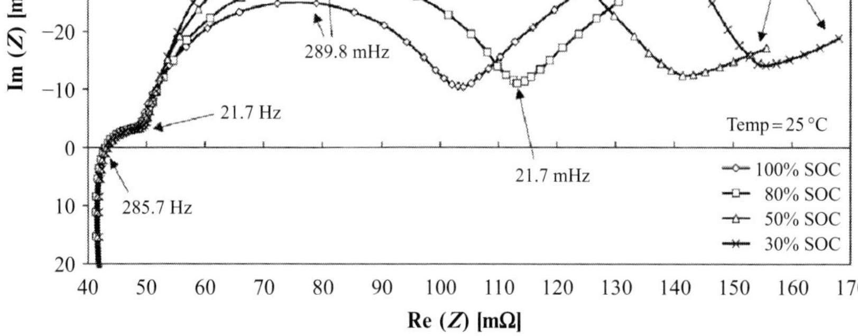

Figure 3.9 EIS results: typical impedance of a Li-ion battery of the type SAFT LM 176065, 3.6 V/5 Ah at $I_{dc} = 0$ A.
Source: Buller et al., 2005.

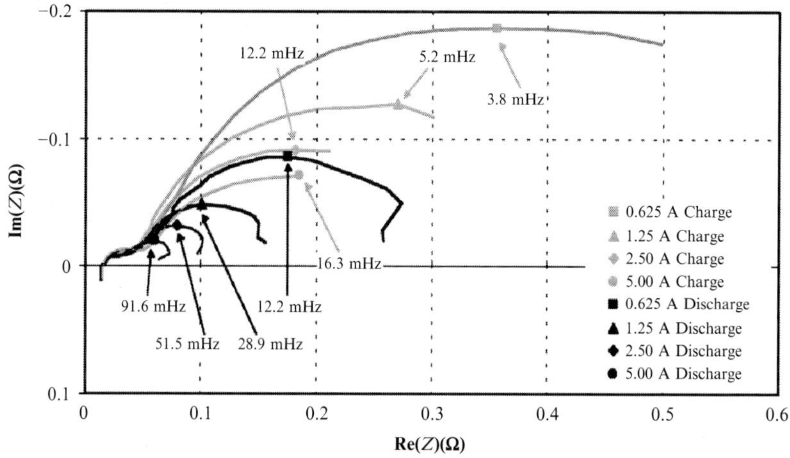

Figure 3.10 EIS Results for a 36-V-AGM-type lead-acid battery at 70% of SOC at 25 °C. *Source:* Buller et al., 2003.

(i) Nonlinearity of the transport and reaction process occurring at large currents.
(ii) Nonstationary nature of the process—since batteries change their structure while charging or discharging, where at least one species of reactants deplete.
(iii) Nonideality of a practical battery (compared to laboratory cell) due to geometry, reaction kinetics and mass transport in porous electrodes.

The above problems make clear that the scope of possible results of an EIS performed is restricted with commercial batteries, compared to experimental cells. However, EIS provides a unique tool for analysis of the dynamical behavior of the batteries. More details are available in Karden et al. (2000).

3.5 Battery equivalent circuit models and modeling techniques

In an electronic engineer's viewpoint, it is most useful to have an equivalent circuit for a battery with any variations of important parameters due to process variables such as temperature and SOC. However, due to various dynamic effects we have discussed in Section 3.3, it is very hard to propose simple equivalent circuits, with linear circuit elements. Following discussion is an attempt to discuss various components within a battery equivalent circuit and how they relate to internal processes occurring within the cells.

3.5.1 Randles' equivalent circuit

First of all, every electrochemical cell is formed by pairing two suitable electrodes and placing them within a suitable electrolyte, and with the necessary separator to get the optimum conversion of chemically stored energy in the cell in the form of electrical energy. For this reason, every battery can be treated as two electrochemical half cells with relatively identical circuit elements, but with different values, which are dependent on several parameters. Randles (1947) proposed a circuit with four components as shown in Figure 3.11(a) to show the combination of electrode-electrolyte reaction, double-layer capacitance, and the electrolyte resistance. In an electrochemical cell, there are two electrodes separated by a single electrolyte, and hence we can apply the Randles circuit to each half cell as shown in Figure 3.11(b). In this case, double-layer capacitance is indicated by C_d, diffusion-related performance is indicated by the charge-transfer resistance R_d combined with the Warburg diffusion element (Jossen, 2006; Warburg, 1899) to model the diffusion process occurring at the electrode.

The *Warburg diffusion element* is a common diffusion circuit element (Jossen, 2006) that can be used to model semi-infinite linear diffusion, that is, unrestricted diffusion to a large planar electrode. A Warburg impedance element can be difficult to recognize because it is nearly always associated with a charge-transfer resistance and a double layer, but is common in all battery chemistries. The Warburg diffusion element (Z_w) is a constant phase element, with a constant phase of 45° (phase independent of frequency) and with a magnitude inversely proportional to the square root of the frequency by

$$Z_W = \frac{\sigma}{\sqrt{\omega}} - j\frac{\sigma}{\sqrt{\omega}} \tag{3.6}$$

where σ is the Warburg coefficient (or Warburg constant) and ω is the angular frequency ($2\pi f$). Amplitude of the Warburg element is given by

$$|Z_W| = \sqrt{2}\frac{\sigma}{\sqrt{\omega}} \tag{3.7}$$

The presence of the Warburg element can be recognized if a linear relationship on the log of a Bode plot ($\log|Z|$ vs. $\log(\omega)$) exists with a slope of value $-1/2$. In Figure 3.11

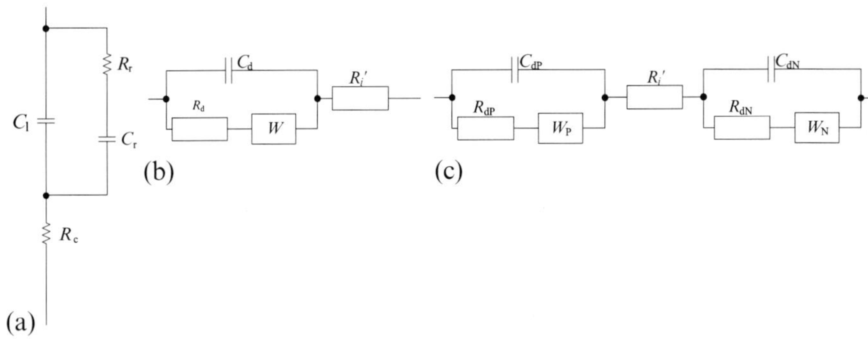

Figure 3.11 Randles' model extended to two electrode half cells: (a) Randles circuit; (b) basic equivalent circuit of the electrochemical half cell; and (c) full cell with two electrodes.

(b), R_i' indicates the electrolyte resistance of the half cell. Figure 3.11(c) shows the combination of the two half cells where the suffixes N and P shows negative and positive electrode effects. By EIS methods (Karden et al., 2000), one can estimate values for these parameters of the cell.

3.5.2 More detailed models based on electrochemistry

It is important to understand that the electrochemistry process within a battery is a complex process where several physical phenomena are combined in a complex way. As shown in Figure 3.6, in a case such as a Li-ion cell with several assumptions as discussed in Section 3.4, current flows and the effects at the anode and cathode could be indicated by Figure 3.12, where discharging of the cell is indicated. In Figure 3.12, $J_{FD,i}$ represents the Faradaic current density due to charge-transfer reactions, $Z_{FD,i}$ is the Faradaic impedance of electrodes caused by charge-transfer reactions (i=n for anode and i=p for cathode), determined from the Faradaic current density and the potential across the interface, $\Phi_{1,i}$ and $\Phi_{2,i}$ represent the potential on the two sides of electrode/electrolyte interface, $J_{dl,i}$ represents the double-layer current density, $C_{dl,i}$ represents the double-layer capacitance, C_{film} represents the capacitance due to the SEI film insulation on anode, J_{film} is the current density through the SEI film, and R_{film} is the SEI film resistance. More detailed discussion and the mathematical derivations based on Butler-Volmer equation, Nernst equation, and Fick's second law related to the process are available in Li et al. (2014). A detailed discussion on these is beyond the scope of this chapter. For a detailed physical explanation on the processes within an electrochemical cell and related mathematical relationships, article by Grimnes and Martinsen (2008) is suggested.

Figure 3.13 shows the circuit structure of the whole battery. The structure has an anode impedance and a cathode impedance connected in series. The impedance of electrolyte, separator, and current collector are represented by a pure ohmic resistor R_0, placed between anode and cathode. At high frequency, battery current collectors and cables introduce inductance, which is represented by a pure inductor L. As per work by Li et al. (2014), Figure 3.13(b) and (c) indicates the Nyquist plots for the cell after 200 and 800 cycles of charge, respectively, based on models related to the circuit

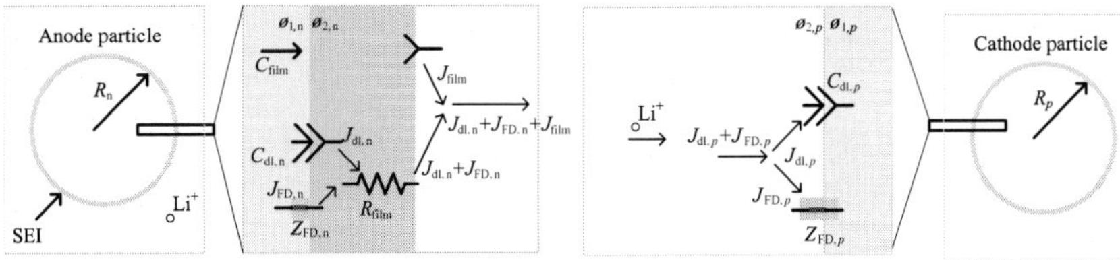

(a) (b)

Figure 3.12 Current flows at the anode and the cathode of an Li-ion cell: (a) anode and (b) cathode.
Source: Li et al., 2014.

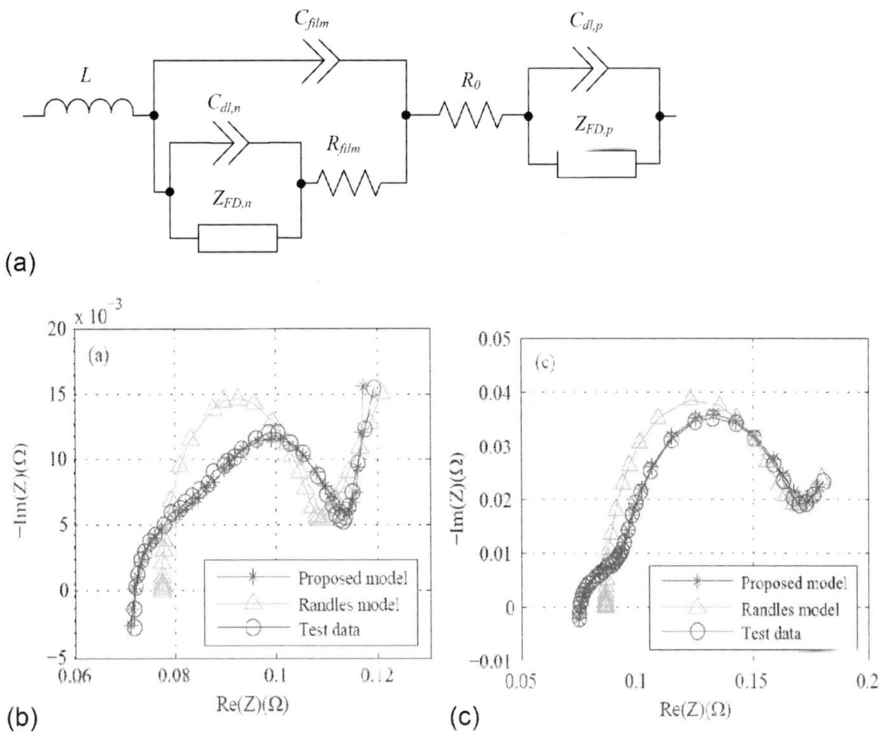

Figure 3.13 Overall circuit structure of a Li-ion cell and its Nyquist plots: (a) overall impedance model of the cell; (b) Nyquist plots after 200 charge cycles, and (c) Nyquist plots after 800 charge cycles.
Source: Li et al., 2014.

structure shown in Figure 3.13(a). The graphs compare results from the proposed model and test data in relation to data generated from a Randles' model.

3.5.3 Frequency domain behavior of battery models and components

In general, product and systems designers do all their measurements in the time domain, but most individual model elements as discussed in Section 3.5.2 provide us more information on the frequency domain. Also EIS techniques are primarily the frequency domain methods where we fix the DC operating current in the cell at a constant value and do estimations of the behavior of components in the frequency domain. For example, in Figures 3.9 and 3.10, we indicate the battery behavior with real and imaginary components together to provide an overview of frequency domain performance. Figure 3.14(a) indicates the equivalent circuit based on the combination of the two half cells with Warburg elements, and Figure 3.14(b) shows the shape of the Nyquist diagram of a lead-acid battery (Huet, 1998). Frequencies indicated are in Hz

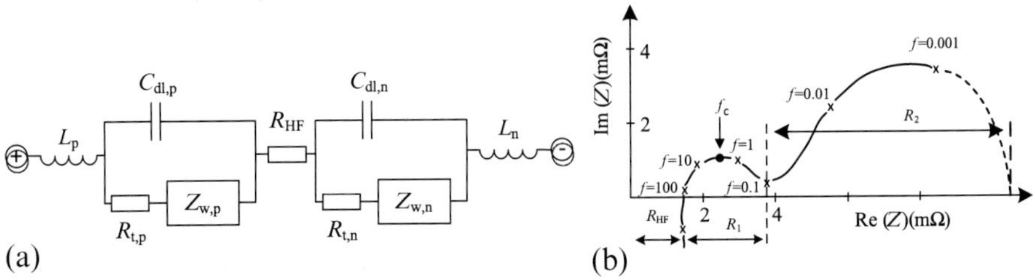

Figure 3.14 Lead-acid battery: (a) comprehensive equivalent circuit and (b) the Nyquist Pot.
Source: Huet, 1998.

and plot easily identifies the values of ohmic resistance of the battery (R_{HF}) and the effect of small capacitive loop of size R_1 of the battery. R_{HF} indicates the effect of connections, the separator, electrolyte resistivity, and the effect of surface coverage by crystallized lead sulfate. Small capacitive loop is related to the porosity of the electrodes. Large loop of size R_2 at lower frequencies depends on sulphatation reactions controlled by mass transport of Pb^{2+} ions (Huet, 1998).

Figure 3.14 depicts the battery impedance of a 940-mAh Li-ion battery at different cycles of charge-discharge and full charged conditions. In this, an electrochemistry-based model as indicated in Figure 3.13 is compared with Randles' model and the test data. In lower frequency regions on the right-hand ends of the figures, effect of Warburg elements is prominent at lower frequencies.

3.5.4 Practical simplifications of battery models for engineering applications

As we have indicated, most of the dynamic effects inside batteries are quantified based on many assumptions and simplifications. In commercial batteries, many of these criteria are only partially valid and most of the components in the models cannot be accurately estimated. Also with nonlinear components such as the Warburg element and

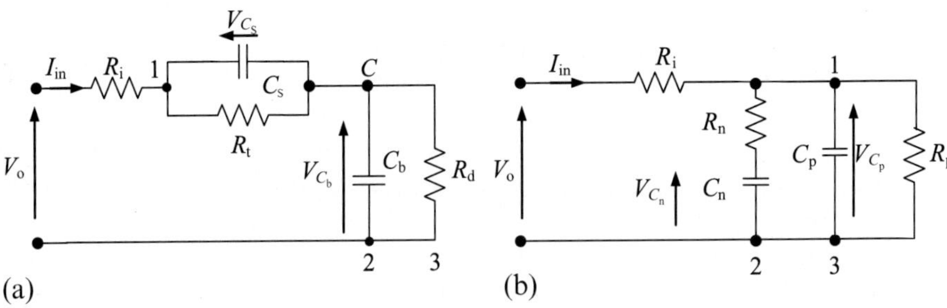

Figure 3.15 Modified Randles' circuits for lead-acid battery: (a) basic model and (b) remapped circuit.
Source: Gould et al., 2009.

the double-layer capacitor, it is difficult to use these models directly in practical systems to estimate SOC, SOH, and remaining useful life (RUL) of the battery pack. In order to achieve much simplified and less-algorithm-intensive models, many simplifications are used with electrochemistry-based models.

3.5.4.1 Simplified models for lead-acid chemistry

In addition to automotive area, lead-acid chemistry is still frequently used in electric vehicles such as airport light transport systems. One example is Urban Light Transport (ULTRA) system used in London Heathrow airport (http://www.ultraglobalprt.com/?page_id=24). This vehicle relies on 4×12 V lead-acid VRLA batteries of capacity nominally rated at 45-50 Ah (Gould et al., 2009). In this kind of a case, battery pack could be discharging at a high rate such as 150 A and maximum charging current may be around 50 A. In predicting the state of function (SOF) of such batteries online, modified and remapped equivalents of Randles' model are used. SOF estimation (combined SOC and SOH) is done on real-time basis in light transport systems such as ULTRA. Figure 3.15 shows the modified version of Randles' circuit shown in Figure 3.11(a). In this five-component-based simplified Randles' circuit description of a 12-V lead-acid battery, C_b is considered as the main charge storage and it is equivalent to a 1×10^5 F capacitor when the battery is fully charged. R_d represents the self-discharge resistance with an approximate value of 5 kΩ. R_i models the battery's terminals and intercell connections (between 5 and 100 mΩ), whereas R_t (in the range of 10-500 mΩ) and C_t (1000-20,000 F) describe mass-transport effects.

The voltage across the capacitor C_b is considered to be a suitable indicator for SOC, whereas SOH is inferred by observing a significant change in C_b over time due to aging effects such as AM degradation (Gould et al., 2009). Chemical aspects of lead-acid battery degradation are discussed by Ruetschi (2004). According to this paper, major aging processes leading to gradual loss of performance, and eventually to the end of service life, are (i) anodic corrosion, (ii) positive plate AM degradation and loss of adherence to grid, (iii) irreversible formation of lead sulfate in the AM, (iv) short circuits, and (v) loss of water.

Randles' model can be remapped into a different equivalent circuit such as in Figure 3.15(b) using star-delta transformation (Gould et al., 2009). It can be shown that

$$C_n = \frac{C_b^2}{C_b + C_s} \tag{3.8}$$

$$C_p = \frac{C_b C_s}{C_b + C_s} \tag{3.9}$$

$$R_n = \frac{R_t (C_b + C_s)^2}{C_b^2} \tag{3.10}$$

$$R_p = R_d + R_t \tag{3.11}$$

By using such a transform to remap the battery parameters into a more useful model and applying Kalman filter (KF), one can get good estimates of SOC and SOH of lead-acid batteries (Gould et al., 2009). More details on these techniques such as KF approach will be presented as an overview later.

3.5.4.2 Hybrid models to consider mass-transport effects

Commonly used impedance-based models assume that the battery is in a quasi-stationary state during measurements. However, the battery packs such as in electric vehicles are subjected to very high continuous, quasi-continuous, or medium-term discharge processes. In situations such as these, to accurately estimate the SOC/SOH, you require improved impedance-based nonlinear simulation models. In these situations, more accurate electrolyte transport models are used, which describe the generation and transport of the sulfuric acid inside the porous electrodes (Thele et al., 2005). In Thele et al. (2005), Matlab/Simulink-based implementation and parameterization in the time domain are given for high Ah throughput automotive and similar battery packs. In this situation, hybrid modeling approach is used which combines the impedance-based model and a basic electrolyte transport model. Equivalent circuit commencing from the Randles' model is extended to show the effects of (i) generalized capacitance elements, (ii) a generalized DC voltage source, and (iii) a parallel current path for gassing as shown in Figure 3.16.

For the parameter determination, several EIS measurements are performed at different working points considering the dependencies not only on the SOC and the temperature but also on the current rate and the nonlinear resistances (Thele et al., 2005). In case like this, measurements in Figure 3.10 can be used. The model extension discussed by Thele et al. (2005) replaces the generalized DC source shown in Figure 3.16 using a voltage source based on the actual acid concentration in both electrodes. This discussion is beyond the scope of this book that aimed at electronic engineers.

In general, at thermodynamic equilibrium, the terminal voltage of a lead-acid battery adjusts to an equilibrium voltage that depends mainly on the electrolyte concentration and thus on the SOC. Use of EIS for dynamic modeling of a lead-acid battery to estimate this equilibrium voltage which depends on SOC is discussed by Mauracher and Karden (1997). Analysis of uncertainties in an EIS measurement for lead-acid battery is discussed by Stevanatto et al. (2014). Also in Stevanatto et al. (2014), a

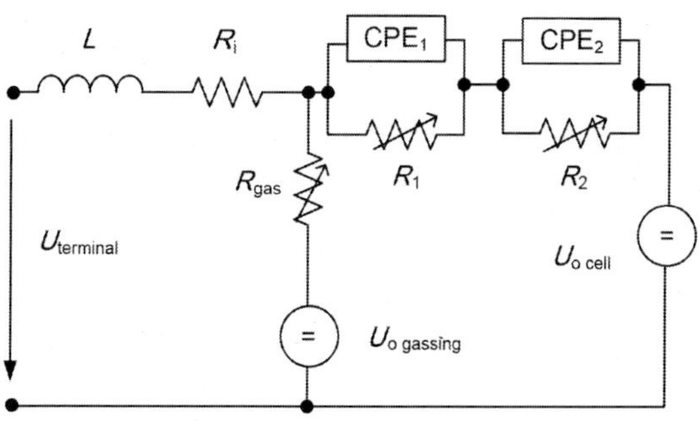

Figure 3.16 Equivalent circuit of a lead-acid battery to include the effect of gassing and the generalized open-circuit voltage combined with impedance elements.
Source: Thele et al., 2005.

discussion on Randles' parameter variations over service life is available. In general, while resistance in a Randles' equivalent circuit keeps increasing, double-layer capacitance keeps decreasing over service life (Stevanatto et al., 2014).

3.5.5 Nickel metal hydride battery models

The NiMH chemistry is an attractive replacement for the NiCd chemistry due to its high energy density, high-current capability, long cycle life, and lack of poisonous materials. NiMH is a popular option for batteries in electric vehicles and this demands relatively accurate models for the chemistry.

A NiMH cell is a dual-intercalation electrochemical system in which proton insertion in the positive solid nickel electrode and the hydrogen de-insertion in the negative metal hydride electrode occur during discharge and vice versa during charge. Figure 3.17 schematically illustrates these processes (Gu et al., 1999). The open-circuit potentials of both electrodes are functions of local SOC, and the cell behavior is strongly dependent on the utilization of the active materials. In addition, discharge or charge of a NiMH cell is controlled by a number of factors, including the finite rate of solid-state diffusion inside active material particles of either one or both electrodes, charge-transfer kinetics at the electrode/electrolyte interface, and ohmic drop through the electrolyte phase. Any of these factors may also result in low cell potential before the active materials loaded in the cell are completely used up. As a result, the inefficient utilization of active materials becomes a critical issue, particularly for electric and hybrid-electric vehicle (HEV) batteries (Gu et al., 1999).

Operation of a sealed NiMH cell is also accompanied by oxygen reactions. Oxygen may be generated at the nickel electrode and cause an excessive build-up of pressure during charge and overcharge. This gassing strongly affects the performance and life of NiMH batteries as well as controls the charge acceptance at high charging rates. It also raises safety concerns in practical applications. A battery must be carefully designed so that the oxygen is effectively transported to and recombined at the metal hydride electrode, preventing the cell pressure from rising drastically. It is thus imperative to develop predictive capabilities for not only the active material utilization but also the cell conditions resulting from the oxygen reactions in advanced NiMH batteries (Gu et al., 1999). Modeling of the NiMH devices uses different approaches such as impedance model (Valoen et al., 1997), combination of numerical and experimental approaches (Mauracher and Karden, 1997), nonisothermal model (Wu et al., 2001), isothermal models (Pan et al., 2002), fuzzy logic (FL) modeling (Singh et al., 2004), and artificial neural networks (ANNs) (Fang et al., 2012).

In general, the number of details to be considered in a NiMH cell will determine the complexity of the model to be achieved. For NiMH cells, the following hierarchy of models could be developed (Wu et al., 2001):

- **(i)** Empirical model—fitted equations from experimental data.
- **(ii)** Lumped model—equations of mass balance, charge balance, electrochemical kinetics, and energy balance.

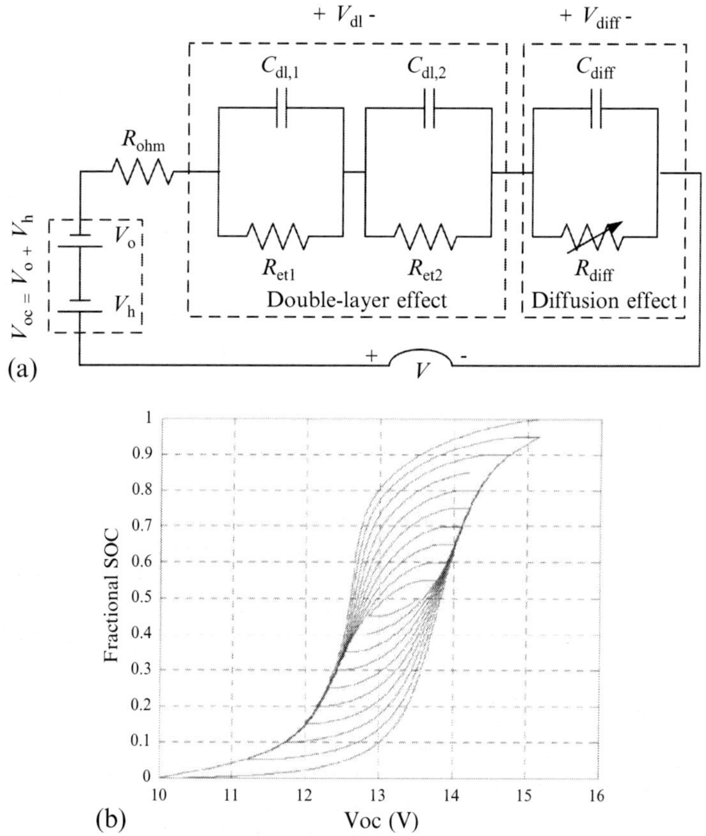

Figure 3.17 NiMH battery equivalent circuit for hysteresis estimation: (a) circuit used and (b) measured hysteresis behavior of a 12-V NiMH battery module (COBASYS Series 1000) at C/20 rat.
Source: Tang et al., 2008.

(iii) Distributed model (with 1D or 2D porous electrode models)—case in item (ii) and equations for the distributed process.
(iv) Coupled electrochemical/thermal model—combination of (i)-(iii) plus heat transfer equations.

In Wu et al. (2001), a lumped model for NiMH cell is developed with attention to many important characteristics, including the hysteresis potential behavior.

While battery models and equivalent circuits similar to the case of lead-acid chemistry are developed with various complexities in these publications, one important characteristic of NiMH cells for electric vehicles is the hysteresis behavior (Tang et al., 2008). As discussed by Tang et al. (2008), using a simplified model such as shown in Figure 3.17, NiMH hysteresis can be developed to include both the double-layer effect and the diffusion effect.

General Motors, USA has developed battery state estimator (BSE) algorithms suitable for HEVs and these are based on two different SOC estimates combined: a current-based SOC and a voltage-based SOC. The current-based SOC is based on conventional coulomb integration, which offers limited accuracy, and is sensitive to the initial value of SOC (Tang et al., 2008). On the other hand, voltage-based SOC inferred from OCV is independent of the initial SOC and more robust to aging effects. The BSE algorithms are based on an equivalent electrical circuit model that represents the dynamic relationship between the battery terminal voltage and current. For NiMH batteries, as illustrated in Figure 3.17, the electrical circuit model involves eight parameters, depicting a static resistance R_{ohm}, a static potential V_{OC} that represents the OCV, and two dynamic processes that model the double-layer effect and diffusion effect, respectively. The V_{OC} is the indicator of SOC. Two RC pairs are used to describe the double-layer effect for cathode and anode, respectively. Double layer is a dynamic process with a time constant ranging from hundreds of milliseconds to several seconds. Additional RC pair with a varying diffusion resistance is used to approximate the diffusion effect, whose time constant can be several seconds to hours. Using this model, Tang et al. (2008) present a phenomenological model based on the Preisiah Operator which is a widely used mathematical tool to model the hysteresis in many areas such as transformers, sensors, relays, and smart materials (Tan and Baras, 2005). Figure 3.17(b) depicts the behavior of the sample battery used in this exercise. For more details, articles Tang et al. (2008) and Tan and Baras (2005) are suggested.

In developing more useful equivalent circuits to relate the dynamic behavior as well as the energetic behavior of a battery pack, Khun et al. (2006) uses Cauer and Foster structures. These structures are well adapted to predict the behavior of batteries in real time. The work begins with a modified Randles' equivalent circuit as in Figure 3.18, where R_{Ω} indicates the electrolyte and connection resistances, parallel pair of R_{tc} and R_{dl} double-layer effect-based charge-transfer phenomenon. A Warburg impedance Z_{w} stands for the diffusion phenomenon. As shown in Figure 3.18(b) in a NiMH battery, charge-transfer and diffusion phenomena occur in distinct frequency areas (Khun et al., 2006). Nyquist plot in Figure 3.18(b) shows the approximate semi-circle for the double-layer effect and the approximate 45° slope for the diffusion phenomenon represented by a Warburg element. In Khun et al. (2006), the Cauer and Foster structures together with Mittag-Leffler theorem are used to develop the equivalent circuit components and to compare the dynamic behavior of the battery at different SOC values based on theory versus experimental results.

3.5.6 Li-ion battery modeling, equivalent circuits and aging issues

Li-ion batteries usually consist of a metal-oxide positive electrode (cathode), a carbon-based negative electrode (anode), and an organic electrolyte based on dissolved lithium salts. Cathode materials used in early versions were based on $LiCoO_2$, $LiMn_2O_4$, or $LiNiO_2$ and with the electric vehicle technology advancing fast, there

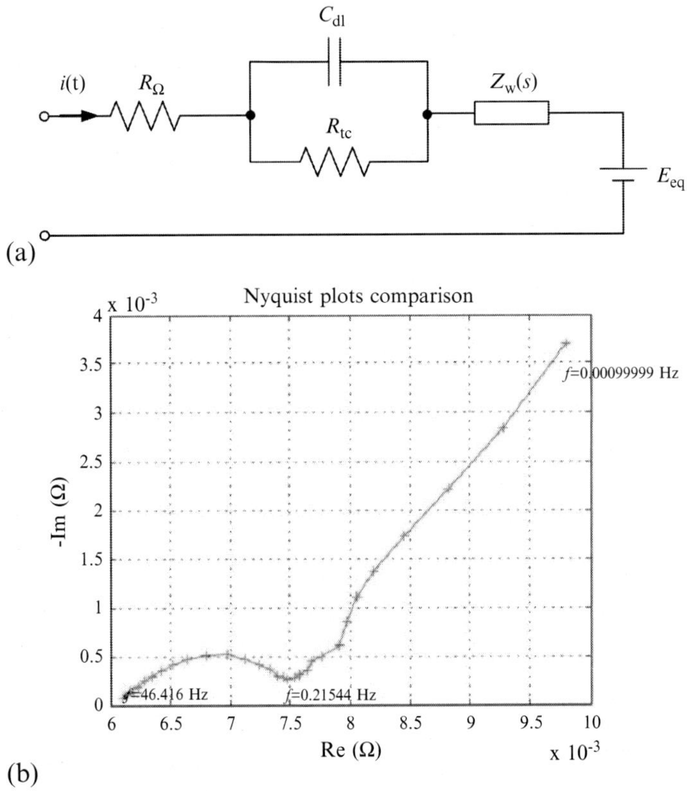

Figure 3.18 Randles' circuit adapted for a NiMH battery and its Nyquist plot: (a) modified Randles' circuit and (b) Nyquist plot.
Source: Khun et al. (2006).

were many variations in these base materials which are used in batteries, aimed at automotive industry (Belt et al., 2011; Bloom et al., 2006; Gao and Manthiram, 2009; Stiaszny et al., 2014). Although there are large variety of possible material combination for electrodes in this chemistry, combining cycle stability (with higher cycle life) and high volumetric or gravimetric energy density reduces the choices to a small number (Stiaszny et al., 2014). Promising candidates for cathodes are layered oxides like $Li(Ni_{1/3}Co_{1/3}Mn_{1/3})O_2$, which exhibits high specific capacity (Belt et al., 2011), but discharge at high rates is limited. To increase safety and rate capability in the cell, $LiMn_2O_4$ is added to the cathode because it exhibits the fast kinetics that is necessary for high power applications (Belt et al., 2011) while at the same reducing the battery's cost. On the other hand, Mn-rich compounds are preferred compared to Ni- or Co-rich materials, although their specific capacity is lower and $LiMn_2O_4$ suffers from manganese dissolution. The best anode material considering both specific capacity and cycle life is graphite (Stiaszny et al., 2014).

Since their introduction in 1991, Li-ion batteries have found many uses in consumer applications, where smaller battery packs are required. With the US government and the US council for automotive research establishing a partnership for new generation electric vehicles (PNGV) in 1993 (Nelson et al., 2002), Li-ion chemistry received a boost for developing higher capability battery packs and several US national laboratories started collaborating in studying the long-term performance of Li-ion chemistry (Bloom et al., 2002, 2005, 2010; US Department of Energy, 2010; Wright et al., 2002).

As discussed and summarized in Section 3.5.2, Li-ion chemistry current flows in anode and cathode are based on several complex processes, and developing a comprehensive model requires expert knowledge in electrochemistry. It is commonly known that the lifetime of lithium-ion batteries is limited by unwanted side reactions (Stiaszny et al., 2014). These side reactions may affect all parts of a battery including the electrolyte, the active material, binder, conducting agents, current collectors, and the separator, which results in capacity decrease and/or increase of the overall cell impedance (Stiaszny et al., 2014). Understanding the substantial aging mechanisms in lithium-ion batteries is of great importance to make correct lifetime predictions. To study the aging phenomena, several analytical methods as well as EIS are used. More details can be found in Stiaszny et al. (2014).

3.5.6.1 Modeling of aging of high power Li-ion batteries

For plug-in hybrid-electric vehicle (PHEV) applications, performance targets for rechargeable batteries indicated in US department of energy (DOE) documents (US Department of Energy, 2010) include calendar lifetimes of 15 years, with survival temperatures ranging from −46 to +66 °C. In these PHEV battery packs, specifications are aimed at defining the conditions for *recharging* (from a continuous power source for sustained replenishment of energy) as well as *regen* (or regeneration) where short duration high-charge current is fed back to a battery pack, typically during braking. During a regen-process during braking, vehicle's kinetic energy is absorbed into the battery pack and this can only persist for a short time due to physical limitations such as heating (US Department of Energy, 2010).

Given this kind of PHEV requirements specified by PNGV, in PHEV battery testing a specific test profile such as in Figure 3.19(a) is recommended. The procedures outlined by the US DOE supported US Advanced battery Consortium test manuals such as in US Department of Energy (2010) are intended to show the promise of a technology versus a set of performance and cost targets. No knowledge of the actual battery chemistry is needed. Thus, the evaluation of cells concentrates on their performance and life and how their life is affected by SOC, time, temperature, and type of test. Calendar and cycle life tests were performed to determine the aging characteristics of the blended cathode material under PHEV testing conditions (US Department of Energy, 2010). It should be noted that the calendar tests in the PHEV and HEV (Department of Energy, 2003) manuals are essentially the same. The cycle life test is more complex, depending on what is to be learned. Instead of just a single cycling mode of operation, charge sustaining (CS), as in an HEV, the PHEV operates in charge

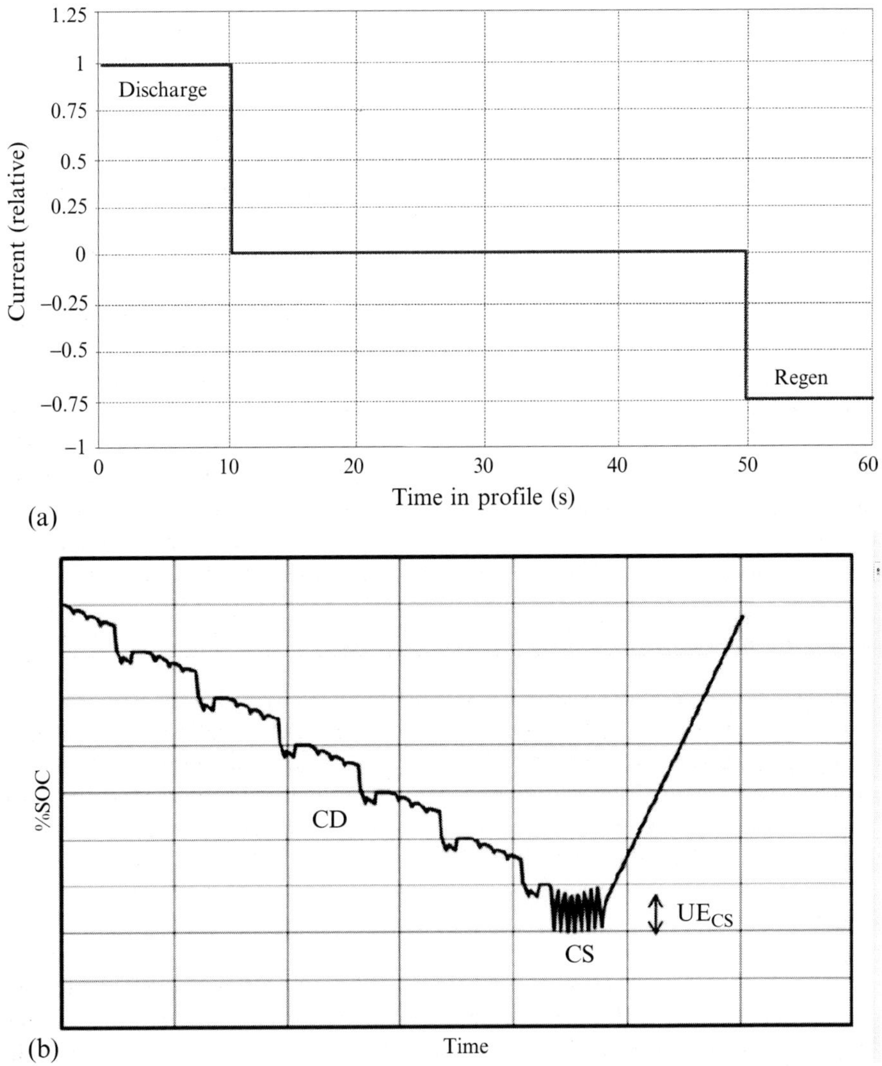

Figure 3.19 PHEV battery charging-regen recommendations and operating modes: (a) test profile and (b) operating modes indicating charge depletion (CD) and charge sustaining (CS) modes.

depleting (CD) as well as CS modes, as shown schematically in Figure 3.19(b). In the CD mode of operation, the battery powers the vehicle directly; the internal-combustion engine is not used at all. After a while, the battery energy becomes exhausted. At this low SOC, the PHEV will operate in CS mode, similar to that of an HEV.

Given the above summary of operational details of a PHEV driving cycle and BMS requirements, and the required longer calendar life of a vehicle battery pack, designers have to look into battery behavior in terms of capacity degradation and parameter deterioration. As detailed by Belt et al. (2011), Figure 3.20(a) depicts the hybrid-pulse power characterization (HPPC) of a 10 kW capable pack at 30 °C based on the test profile as per Figure 3.19(a). Figure 3.20(b) and (c) indicates the typical test profiles for CS- and CD-type operations as detailed in US Department of Energy (2010).

Li-ion test cells aimed at PHEV-type applications were tested as per details in Belt et al. (2011), and overall results indicate that in Li-ion chemistry the cell resistance keeps increasing with age and the number of cycles, while the deliverable capacity reduces with age and number of cycles. This is shown in Figure 3.21. Dotted lines in Figure 3.21 show the case for CS cycle.

Comparing these results with those obtained from a calendar life test as detailed by Belt et al. (2011) shows that most of the resistance increases (Figure 3.21(a)) are due to calendar life effects at all temperatures. At 30 and 40 °C, there were no significant differences between the calendar and CS cycle curves. Only at 50 and 60 °C were significant differences observed; the cells from the CS cycle test displayed about 9-19% additional resistance increase. Figure 3.21(b) compares the capacity fade results in a calendar life test to those found in a CS cycling imposed on the cells.

In all cases, CS cycling increased capacity fade and, in most cases, CS cycling changed the kinetic behaviors. Capacity fade in the calendar life test tended to follow Arrhenius-like, linear-kinetics in the temperature range of 30-50 °C; those from the CS cycle test exhibited $t^{1/2}$ kinetics. Only at 60 °C in the calendar test were parabolic kinetics observed. A detailed discussion on these degradation aspects is found in Belt et al. (2011) and Bloom et al. (2006).

Figure 3.22 presents a lumped parameter battery equivalent circuit suitable for depicting the impedance rise in high power Li-ion cells as discussed by Bloom et al. (2002). In this equivalent circuit, a large capacitor 1/OCV$'$ is also used to depict the changes in diffusion process in the positive electrode. This lumped parameter model (LPM) described in PNGV documents attempts to linearize the behavior and response of a battery during a simple repeatable current profile such as CS in US Department of Energy (2010).

The equivalent circuit used in the LPM is in Figure 3.22, where OCV is the ideal OCV; R_o is the internal resistance due to ohmic, interfacial, and kinetic impedances in the cell; R_p is the concentration polarization of the electrolyte and electrodes; C is a capacitance included to simulate the time constant during the polarization of the electrolyte and electrodes; I_L is the load current; $I_p(t)$ is the current through the polarization resistance; V_L is the load voltage; R_L is the load resistance; t is the time; and 1/OCV$'$ is related to the changes in the diffusion coefficient of Li+in the electrodes with time. The relationship between V_L and other parameters could be described by

$$V_L = \text{OCV} - \text{OCV}'\left[\int I_L \, dt\right] - R_o[I_L] - R_p\left[I_p(t)\right] \tag{3.12}$$

More details on combining this relationship with linear regression function (LINEST) in Microsoft Excel to calculate the parameters are given in Bloom et al. (2002).

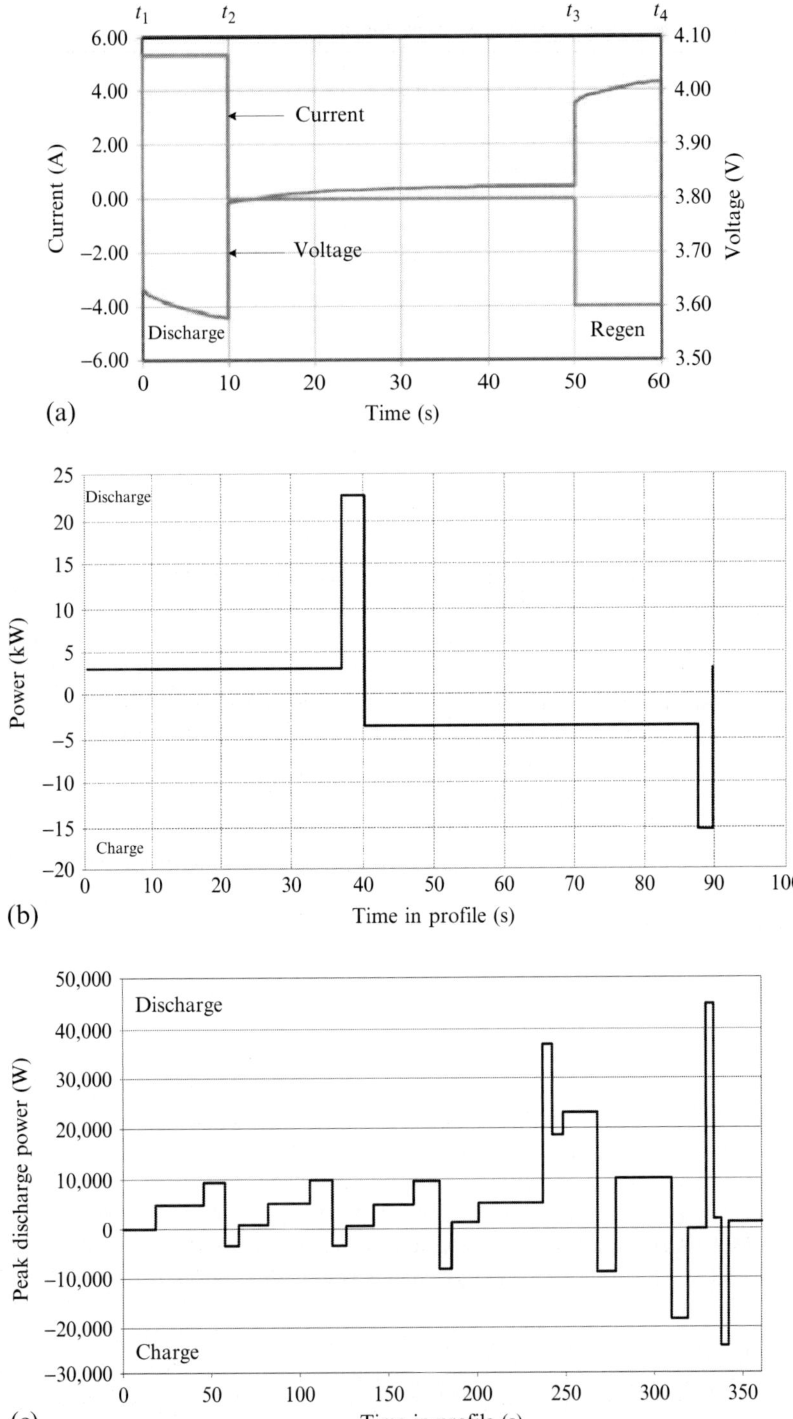

Figure 3.20 Hybrid pulse power characterization (HPPC) testing of a pack of Li-ion test cells and PNGV specified test profiles (Belt et al., 2011) for PHEV battery packs: (a) HPPC profile and voltage response of a typical Li-ion cell; (b) typical recommended test profile for charge sustaining (CS) mode; and (c) typical recommended test profile for charge depleting (CD) mode. *Source:* (a and c) Belt et al., 2011.

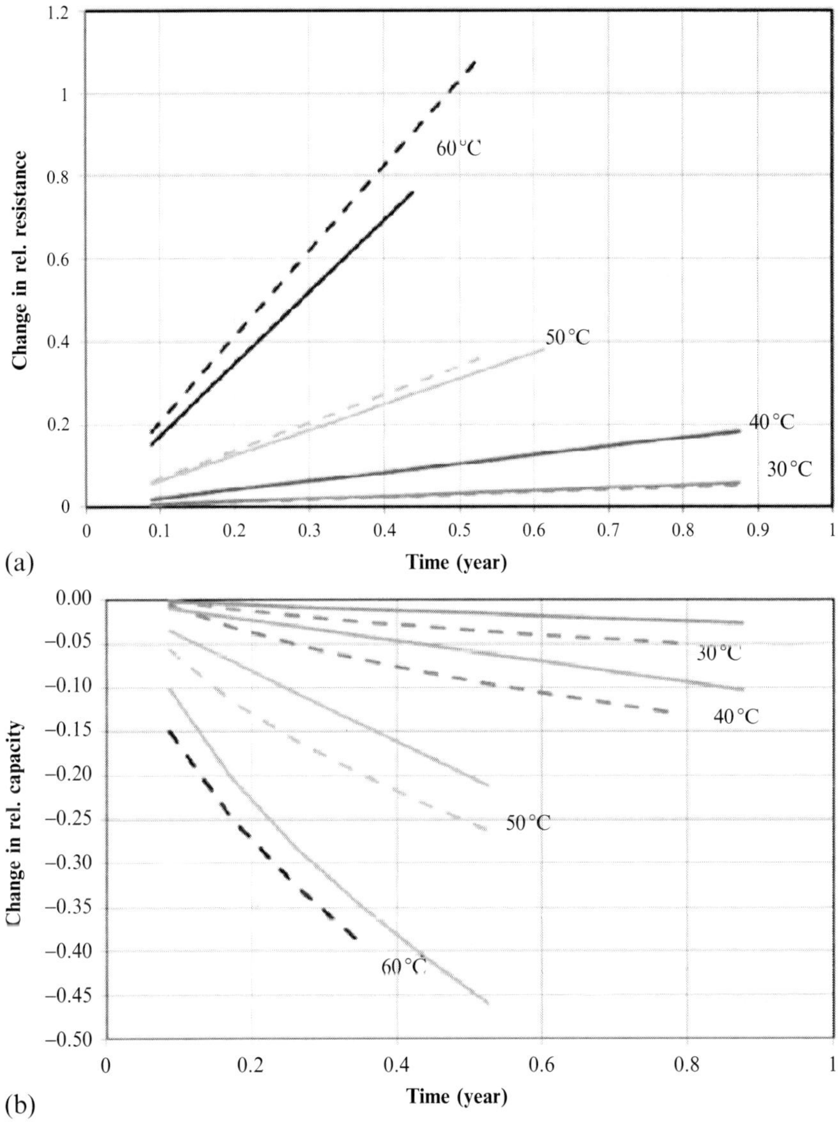

Figure 3.21 Battery performance versus time: (a) cell resistance increase with time and (b) cell capacity drop with time [Dotted lines show CS test and continuous lines show calendar test]. *Source:* Belt et al., 2011.

Pattipati et al. (2011) also discuss an estimation procedure for SOC/SOH of a battery pack used for EVs and PHEVs based on the same model as in Figure 3.22. More discussion on aging of Li-ion is available in Bloom et al. (2001). In Urbin et al. (2010), a model similar to the case in Bloom et al. (2002) is used for energetical model based on a distributed parameter model similar to the model of a transmission line.

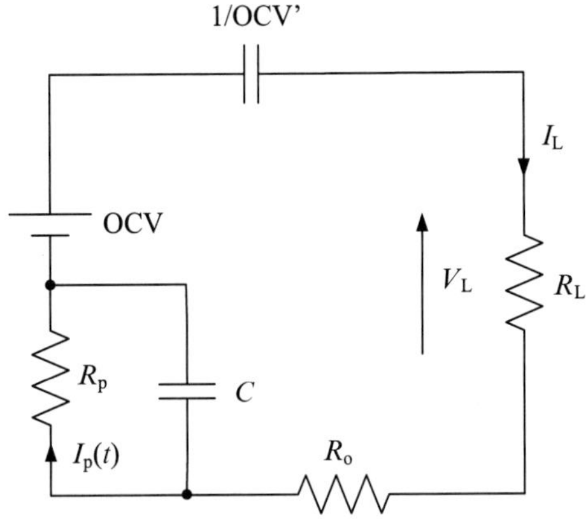

Figure 3.22 Li-ion battery lumped parameter model (LPM).
Source: Bloom et al., 2002.

3.5.7 Lithium ferro phosphate batteries

With the automotive world getting inclined toward electric drives, engineering community sees the prime barrier—a long-lasting (cycle-life based), safe, and energy-dense battery packs for the energy storage. On another front, green energy world with renewable energy resources getting integrated with the grid, energy storage systems require similar capability battery packs with longer calendar life. In order to meet this demand, Li-based rechargeable chemistries such as lithium ferro phosphate ($LiFePO_4$, also known as lithium iron phosphate) and lithium titanate (LTO) have been developed.

$LiMO_2/Li_4Ti_5O_{12}$-based anode in LTO or Li_xFePO_4-based cathode in lithium ferro phosphate (LFP) batteries offers long lifetime, high charge/discharge rates, high intrinsic safety for automotive applications and power tools, etc. Table 3.1 provides a comparison of different Li-ion battery chemistries based on cathode and anode material options (Świerczyński et al., 2014).

Figure 3.23 shows the schematic description of the model of an LFP cell. As pictorially indicated in Figure 3.23, negative active particle radius is about 140 times bigger than the positive active particle radius (Marcicki et al., 2013). Electrochemistry-based modeling to predict the performance of this chemistry can be found in Wang et al. (2011), Forgez et al. (2010), Kassem and Delacourt (2013), and Kassem et al. (2012).

Similar to the models developed based on electrochemistry, thermodynamic, or otherwise for other chemistries, for LFP cells EIS-based techniques can be developed to predict the SOC/SOH of the battery. Figure 3.24 depicts the equivalent circuit model based on EIS data and the associated Nyquist plot (Greenleaf et al., 2013) for a commercially available Powerizer LFP-RCR 123A LFP battery. This Powerizer cell is comprised of a Li_xFePO_4 cathode and a graphite anode with a nominal voltage of 3.2 and a rated capacity of 450 mAh as per manufacturer's datasheet.

Table 3.1 Technical and economic comparison of Li-ion chemistries

	Cathode					Anode
Performance parameter	**LCO (lithium cobalt oxide)**	**NMC (lithium nickel manganese cobalt oxide)**	**NCA (lithium nickel cobalt aluminum oxide)**	**LMO (lithium manganese oxide)**	**LFP (lithium iron phosphate)**	**LTO (lithium titanate)**
Lifetime	Low	Good	Very good	Low	Very good	Very good
Cost per cycle	Average	Average	High	Average	Low	Low
Fast response	Very good	Very good	Very good	Very good	Very good	Very good
Performance	Good	Good	Good	Low	Good	Very good
Safety	Low	Good	Low	Good	Very good	Very good
Self-discharge	Very good	Very good	Very good	Very good	Good	Very good

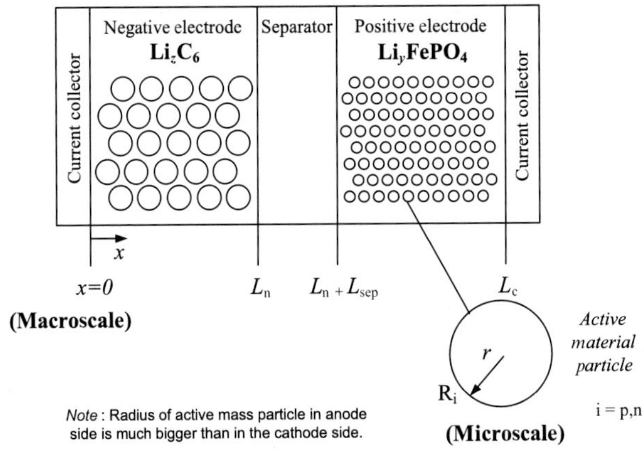

Figure 3.23 Schematic representation of the LFP cell.

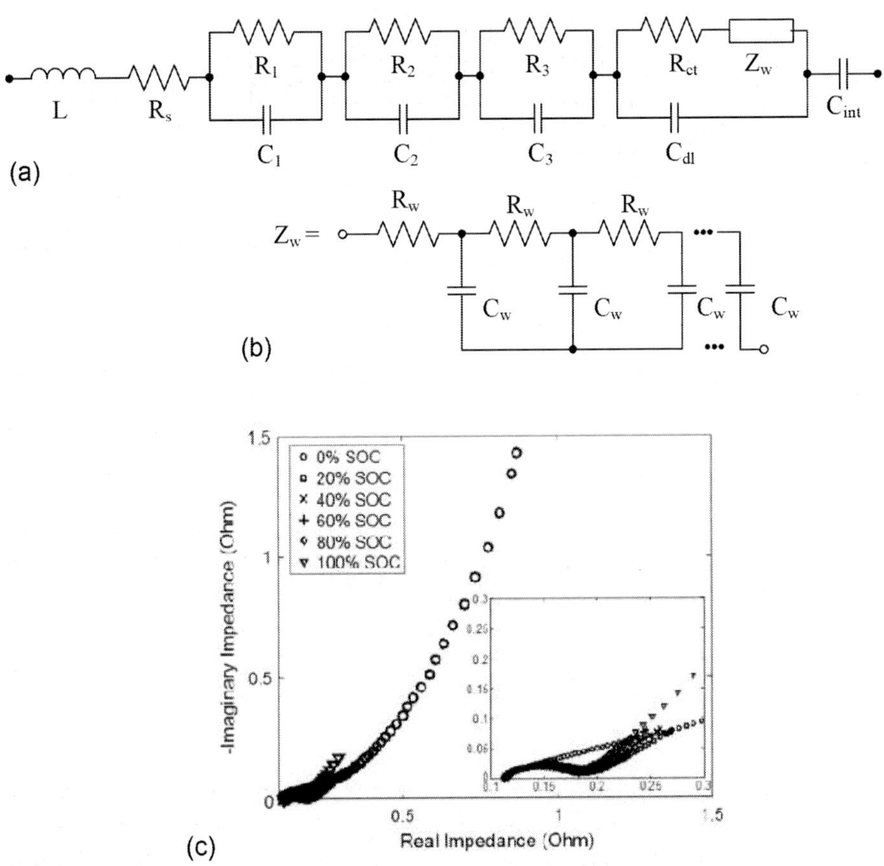

Figure 3.24 Powerizer LFP battery equivalent circuit and Nyquist plot: (a) total equivalent circuit (b) expanded Warburg impedance (Z_W) (c) Nyquist plot from 10 mHz to 2 kHz. *Source:* Greenleaf et al., 2013.

Table 3.2 List of parameter values for LFP based on EIS measurements at different SOC values.

Parameter	Description	SOC percentage		
		0	40	100
L (μH)		0.78	0.86	1.17
R_s (mΩ)	Ohmic resistance of solution, separator, and contacts	118	116	115
R_1 (mΩ)	Components representing multilayer surface film created during initial cycling and further thickened by successive cycles (SEI)	53	4	6
C_1 (F)		1.39	100	153
R_2 (mΩ)		1540	11	24
C_2 (F)		20.7	2.0	12.6
R_3 (mΩ)		43.5	28.4	24.1
C_3 (F)		0.192	0.126	0.440
R_{ct} (mΩ)	Charge-transfer kinetics at electrolyte/electrode boundary	301	30	36
C_{dl} (F)		0.024	0.019	0.023
C_{int} (F)	Intercalation/deintercalation of Li-ion from active mass	700	675	156.5
R_w (Ω)	Warburg impedance components. T_w is the time constant of the branch circuit	1.07	0.25	0.15
$T_w = R_w * C_w$ (s)		23.4	127.8	125
OCV (V)	Open-circuit voltage	2.53	3.31	3.41

Adapted from Greenleaf et al., 2013.

Table 3.2 shows the component values based on EIS measurements at different SOC values.

It is interesting to see that the value of almost all resistive components keeps increasing as the SOC goes down. This tells us the reason for a badly depleted cell showing very high internal resistance in general and this is true for almost all battery chemistries. Frequency-dependent Warburg impedance components that describe the diffusion of Li-ion through the porous electrode can be expressed at a particular measurement frequency as

$$Z_w(\omega) = R_w \frac{\tanh\sqrt{(j\omega T_w)}}{\sqrt{(j\omega T_w)}} \tag{3.13}$$

which is the same as the relationships in Equations (3.6) and (3.7).

The premise of the model shown in Figure 3.24 is intended to relate the electrochemical and phenomenological processes observed in the energy storage devices to their analogous electronic mechanisms, thereby providing a valuable set of data to quantify the electrical behavior of the cell. In order to describe the battery's performance in time domain, the frequency domain circuit components shown in Figure 3.24 can be transformed using Fourier transform techniques and the result will

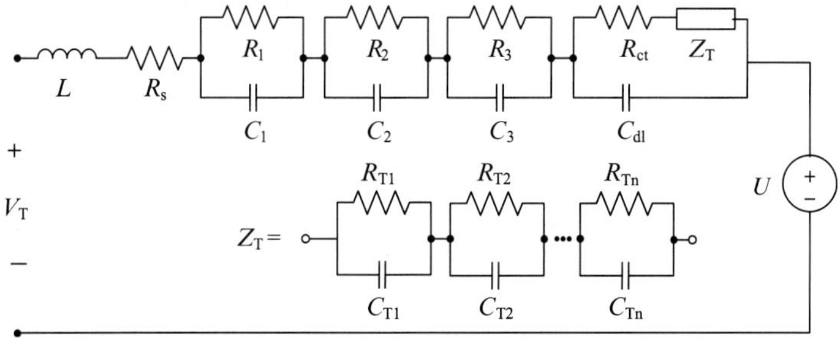

Figure 3.25 Time domain transformation of frequency domain circuit model shown in Figure 3.24.

be the circuit shown in Figure 3.25. Component values in Figure 3.25 and details of the transformation process are detailed in Greenleaf et al. (2013). A similar discussion will be found later in Chapter 5 related to supercapacitors, which has similar processes such as diffusion.

R_{Tn} and C_{Tn} values in Figure 3.25 can be obtained from a simple conversion into a transmission line model (Greenleaf et al., 2013) as

$$R_{Tn} = \frac{8R_w}{(2n-1)^2\pi^2} \tag{3.14}$$

$$C_{Tn} = C_w/2 \tag{3.15}$$

Albertus et al. (2008) provide a discussion on capacity versus electrode combinations possible for electric vehicle battery packs, based on some electrode material options indicated in Table 3.1, including an electrode pair based on LTO and LFP materials. A detailed discussion on these is beyond the scope of this book, since these subjects require significant knowledge components on electrochemistry of batteries. Most references provided in this chapter (Albertus et al., 2008; Belt et al., 2011; Bloom et al., 2001, 2002, 2005, 2006, 2010; Buller et al., 2003, 2005; Bundy et al., 1998; Department of Energy, 2003; Fang et al., 2012; Forgez et al., 2010; Gao and Manthiram, 2009; Greenleaf et al., 2013; Grimnes and Martinsen, 2008; Gould et al., 2009; Gu et al., 1999; Huet, 1998; Jossen, 2006; Karden et al., 2000; Kassem and Delacourt, 2013; Kassem et al., 2012; Khun et al., 2006; Kularatna, 2008; Lai and Rose, 1992; Li et al., 2014; Lindahl et al., 2012; Marcicki et al., 2013; Mauracher and Karden, 1997; Orazem and Tribollet, 2008; Pan et al., 2002; Pattipati et al., 2011; Ramadass et al., 2004; Randles, 1947; Ruetschi, 2004; Singh et al., 2004; Stevanatto et al., 2014; Stiaszny et al., 2014; Świerczyński et al., 2014; Tan and Baras, 2005; Tang et al., 2008; Thele et al., 2005; US Department of Energy, 2010; Valoen et al., 1997; Wang et al., 2011; Warburg, 1899; Wright et al., 2002; Wu et al., 2001; Urbin et al., 2010; http://www.ultraglobalprt.com/?page_id=24) are work done by many electrochemists and physicists and therefore electronic engineers probably require a basic course on electrochemistry to appreciate these discussions. In the sections 3.6 to 3.8, a summary of battery equivalent circuits and BMSs in an electronic engineers viewpoint is provided.

3.6 Battery management in practical applications

About three decades ago, battery management was having a reliable, fast, and safe charging methodology to be selected for a battery bank, together with the monitoring facilities for detecting the discharged condition of the battery pack. With modern battery technologies emerging, the demands from the cost-sensitive portable product market, as well as the medium power range products such as UPS and telecom power units and the automotive sector (in EV, HEV, and PHEV sectors), attributes of a modern BMS may include:

- Practical battery modeling suitable for electronic systems design.
- Battery charging methods and charge control.
- Determination of SOC/SOH/RUL and the end of discharge (EOD).
- Gas gaging.
- Monitoring battery health issues.
- Communication with the host system/or power management subsystems.
- Battery safety.

The Sections 3.6 and 3.7 provide a discussion on some concepts and techniques related to simplified and practical modeling of batteries for best performance extraction, managing a given chemistry for best runtime and the longest life, safety of battery packs, and prognostics for health management of battery packs.

3.6.1 *Practical modeling of a cell to reflect its electrochemistry—an electronic engineer's viewpoint*

As engineers we tend to work with device datasheets, leaving the device design and its behavior to be decided by the device expert, and a team of electrochemists/physicists to give us a mass producible cell. This electrochemical cell is a multidisciplinary systems which involves complex chemical, electrical, and thermodynamic processes. As we discussed in Section 3.5, while modeling a battery requires lot of details, as electronic engineers in an oversimplified equivalent circuit, we tend to oversimplify a battery to a voltage source (with OCV) and a series resistance. However, practical voltage profiles of batteries during discharge and charge do not exactly reflect the accuracy of this simplified assumption. Figure 3.26 indicates the overly simplified chemical process within a battery as it discharges its stored chemical energy to a load.

As we have seen in Figure 2.1, battery capacity decreases as the discharge rate increases. Also as per Figure 2.5, we see that a cell can have high pulse discharge, if we allow resting between pulses. These kinds of behavior are simply explained by Figure 3.26. Figure 3.26(a) shows the case of fully charged cell with maximum concentration of active species. When a load is connected, the load current causes this active species to be consumed at the electrode surface (Figure 3.26(b)) and

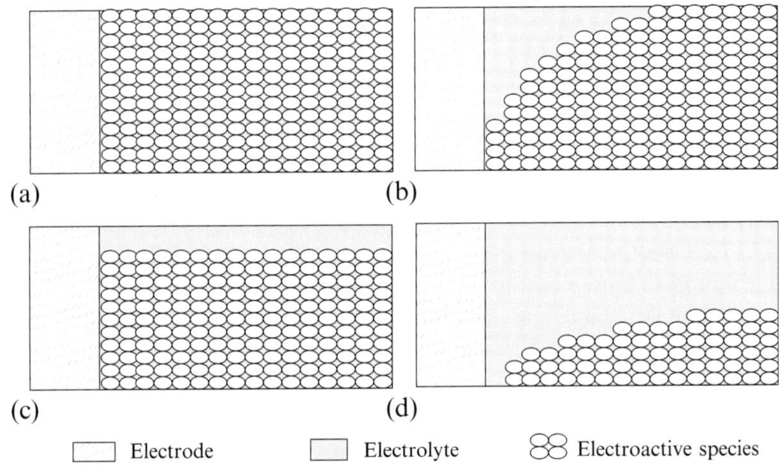

Figure 3.26 Behavior of a symmetric electrochemical cell: (a) fully charged state; (b) before recovery; (c) after recovery; and (d) discharged state.

replenished by the diffusion process from the bulk of the electrolyte. A higher load current causes a higher concentration gradient, and thus a lower concentration of active species near the electrode surface. When this concentration falls below a particular threshold, a voltage cutoff occurs. However, the unused charge is not physically lost and it is simply unavailable at the electrode surface due to lag between reaction and diffusion rates. Decreasing the discharge rate effectively reduces this lag as well as the concentration gradient in the vicinity of the electrode. When the recovery occurs (Figure 3.26(c)), a high current discharge is possible until the active species are fully consumed as in Figure 3.26(d). For more details and related modeling of batteries, article by Rao et al. (2003) is suggested.

This simplified discussion on the process of charge-transfer and diffusion process, etc. leads to the curiosity toward more complex chemical behavior of a cell, which was summarized in Section 3.5 for different chemistries. Studying such complex chemical processes inside a cell leads us to develop a more accurate circuit model suitable for modeling a battery based on simplified engineering information extracted from a datasheet. NiMH or Li-ion cells can be modeled using modern finite element methods (Gao et al., 2002; Renhart et al., 2008; Schweighofer et al., 2003) for temperature behavior, while automated test systems can be used to extract information for simplified models (Schweighofer et al., 2003) to reflect the cell behavior. For much more simplified requirements, extracting data from cell manufacturer datasheets is also possible (Gao et al., Sept, 2002).

As discussed in the literature cited in Sections 3.4 and 3.5, based on different approaches there are several possible battery models as listed below:

(i) Physical models.
(ii) Empirical models.
(iii) Abstract models.
(iv) Mixed models.

In general, microscopic-level models are important for a battery designer, while macroscopic-level models are useful for the BMS designer or the battery user. A user-level model should capture only the gross characteristics of the battery behavior.

Physical models are also known as electrochemical models (Rao et al., 2003) and they are very accurate, but require detailed consideration of electrochemical processes, thermodynamic process, physical considerations for both charging and discharging characteristics. These are computationally expensive and practical application of them in engineering of BMS is limited.

Empirical models are easy to configure and are represented by mathematical expressions with less number of parameters. However, they fail to provide accurate estimation of battery SOC/SOH, etc. under varying load conditions.

Abstract models are simplified equivalent circuit representations such as PSPICE models or based on stochastic processes (Chiasserini and Rao, 2001).

Mixed models are based on high-level abstraction, avoiding excessive details of physical laws (i.e., electrochemical process) governing the battery characteristics which leads to the derivation of simplified analytical expression. Sometimes these are called analytical models (Rong and Pedram, 2006a).

An example of an empirical lead-acid battery model for charge/discharge and lifetime estimation is given in Agrawal et al. (2010). This empirical model can be extended to other chemistries, and in this work Rakhmatov and Virdhula model (Rakhmatov et al., 2003) is used as the base. This analytical model, which is useful for portable product BMS area, allows for a tradeoff between the accuracy and the amount of computation performed. A fast prediction model for SOC/SOH estimation with about 5% accuracy for Li-ions is presented by Rong and Pedram (2003).

3.6.2 Application-specific approaches to battery modeling

Dominating battery chemistries used in current systems are lead-acid, NiMH, and Li-based chemistries. With the rapid growth of electric vehicles, larger capacity batteries are developed for HEV, PHEV, and EV (or sometimes called battery electric vehicles) applications and fractional Ah to few Ah packs are developed for portable devices. Table 3.3 depicts a summary of typical battery characteristics expected in HEV, EV, and portable electronics with the expected accuracies in SOC/SOH estimations.

In BMS systems, development designers prefer to use electrical analog battery models such as the ones discussed by Li et al. (2012). For off-line estimation of battery SOC/SOH, these analog battery models with simplified equivalent circuits are commonly used (Chen and Rincon-Mora, 2006; Gao et al., Sept, 2002; Husseni et al., 2011; Li et al., 2012; Zhang et al., 2010). As discussed by Li et al. (2012), an analog battery model such as in Figure 3.27 can be developed, where the impedance Z shown in Figure 3.27(a) is a combination of RC sections and the series resistance we discussed in Figures 3.24 and 3.25. Figure 3.27(b) shows the dynamic response of a battery such as Ultralife UBBL 10 Li-ion pack with 6.8 Ah capacity used in EV-type applications. Series resistance in Figures 3.24 and 3.25 is responsible for the instantaneous drop, and RC sections are responsible for the longer duration transient response. More details are in Li et al. (2012).

Table 3.3 Typical characteristics envisaged in three different battery environments

Characteristic/parameter	Hybrid electric vehicle (HEV)	Electric vehicle (EV)	Portable electronics (PE)
Maximum charge or discharge rate (C)	±20	±5	±3
Rate profile	Very dynamic	Moderate	Piecewise constant
SOC estimation required	Very precise	Precise	Crude
Predict available power	Yes	Yes	No
SOH estimation	Required	Required	Not essential
Cell balancing	Continuous	Continuous or on charge mode only	On charge only
Lifetime expected (years)	10-15	10-15	<5

Source: Plett, 2004a.

3.6.2.1 Battery equivalent circuits and practical ways to estimate parameter values

Battery equivalent circuits are useful for modeling the behavior of a battery pack, to predict its short-term behavior as well as the long-term performance with regard to discharge rates, temperature changes, and life cycle prediction, as well as pulse discharge performance. There are many different models available ranging from a simple voltage source with internal resistance to equivalent circuits modeling the effects of charge state, temperature, pressure, aging, and many more aspects of the battery pack.

To predict the performance of a battery in the short term, simplified equivalents such as shown in Figure 3.28(a) is possible. This model is valid for time scales ranging from milliseconds up to a few seconds at a constant SOC/discharge and temperature.

In Figure 3.28(a)

- U_B and I_B are the terminal voltage and current.
- U_o is the OCV.
- R_i is the constant part of the internal resistance (connector, electrodes, electrolyte).
- R_D and C_D are the effects on the surface of the electrode (double-layer capacitance effect).
- R_k and C_k reflect the effect of diffusion process in the electrolyte.

To deal with longer prediction times or different temperature situations, the values of components have to be adapted according to the changes inside the battery (Bernardi and Carpenter, 1995; Gu et al., 1995; Landfords et al., 1995). Schweighofer et al. (2003) discuss the details of a test setup that can be used to extract the parameters for the equivalent circuit model shown in Figure 3.20(a). Figure 3.20(b) shows some details of such a test setup. Table 3.4 indicates the component values related to a 9 Ah NiMH battery with 270 A current capacity. Given the time constants for the surface effects of the electrodes, R_D and C_D, and the time constants related to the diffusion process, R_k and C_k, one should be able to predict the battery behavior to match the

(a)

(b)

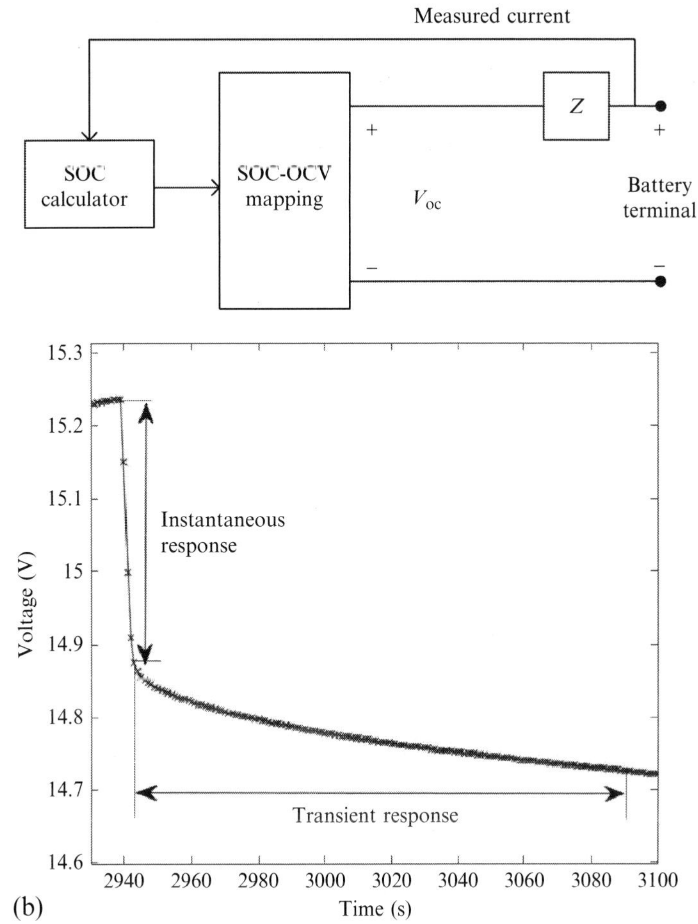

Figure 3.27 Analog battery model useful in electric vehicle battery pack developments: (a) model and (b) dynamic response of Ultralife UBBL 10 Li-ion battery module. *Source:* Li et al., 2012.

experimental data available in datasheets. For details, article by Schweighofer et al. (2003) is suggested.

In case a much simplified model is used, public information available on battery datasheets can be used to arrive at a model (Gao et al., Sept, 2002) shown in Figure 3.29. In this case, the battery voltage curves based on different discharge rates, measured in multiples or fractional C rates, can be used using simplified graphical and calculative approaches. This three-component model is based on

- OCV (equilibrium potential), E.
- An internal resistance R_{int} having two components R_1 and R_2.
- An effective capacitance that characterizes the transient response of charge double layers in porous electrodes.

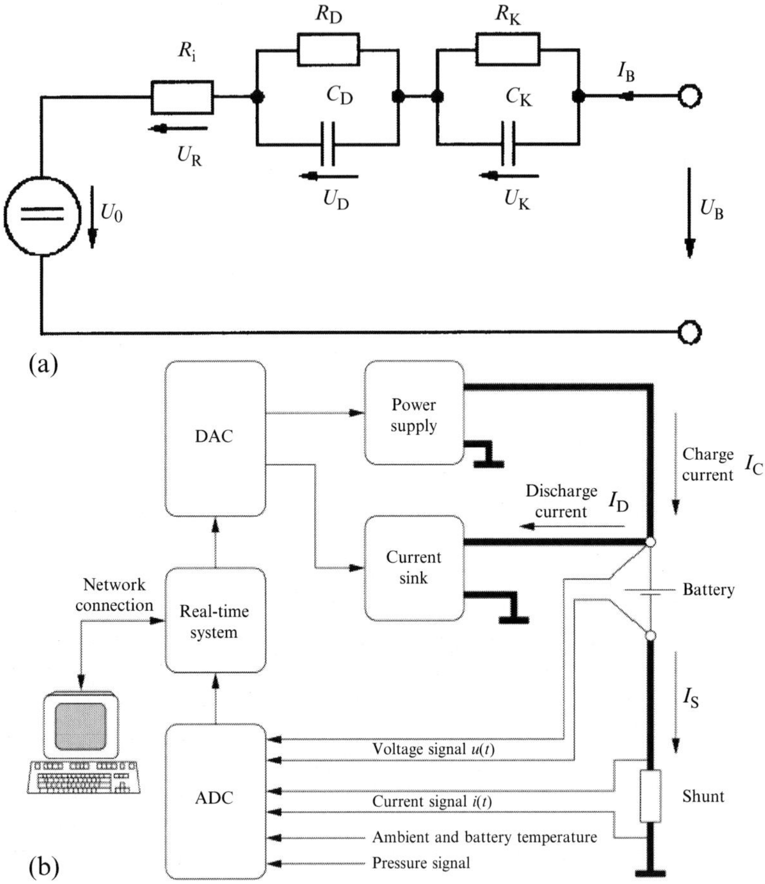

Figure 3.28 Battery equivalent circuit models and parameter extraction: (a) for short-term behavior and thermal performance and (b) test setup for extracting parameters of model. *Source:* Schweighofer et al., 2003. Reproduced with the permission of IEEE, USA.

Table 3.4 **Model parameters for a typical 9 Ah NiMH cell**

Components of cell internal resistances (mΩ)		**Capacitances**	
R_i	1		
R_D	0.35	C_D	171 F
R_k	1.6	C_k	16,000 F

Adapted from Schweighofer et al., 2003.

More details related to this procedure are discussed by Gao et al. (2002). For design of battery packs for portable systems, an accurate prediction of battery models for run-time and *V-I* performance Cadence simulation can be practically used. More details and various equivalent circuits suitable for this purpose are discussed by Chen and

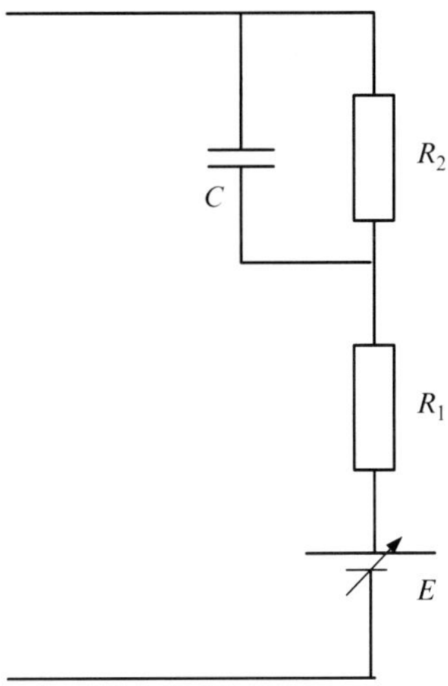

Figure 3.29 A simplified equivalent circuit model for a Li-ion cell usable for datasheet parameter extraction.

Rincon-Mora (2006), in a design engineer's viewpoint. A set of battery models suitable for Cadence-type simulation work as proposed by Chen and Rincon-Mora (2006) is shown in Figure 3.30. This uses two sets of parameters where Figure 3.30(a) indicates a capacitor- and a current-controlled current source inherited from run-time-based models. RC network-based model in Figure 3.30(b) based on Thevenin models simulates the transient response (Chen and Rincon-Mora, 2006).

Another important condition in a battery is the internal battery temperature and the ambient temperature of the battery pack. By using finite element models (FEM), it is possible to analyze and predict the thermal behavior of a battery pack or a cell using

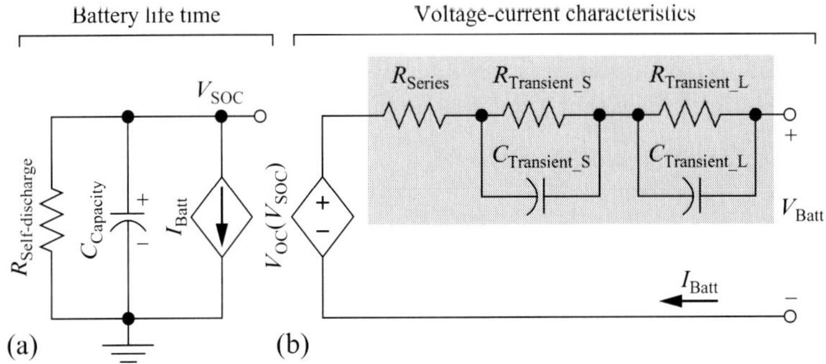

Figure 3.30 A battery model suitable for Cadence simulation: (a) model suitable for capacity, SOC, and runtime estimation and (b) model suitable for transient response estimation. *Source:* Chen and Rincon-Mora, 2006. Reproduced with the permission of IEEE, USA.

modern FEM software combined with thermal models of the battery (Renhart et al., 2008).

For standby power system requirements such as UPS and telecommunication power supplies, where VRLA-type lead-acid cells are used, reserve life estimation is very useful. In such situations, reserve life estimation is based on SOH measurements on the battery pack. To estimate the SOH in such situations, several parameters and measurements can be utilized:

(a) Float voltage and float current measurement.
(b) Temperature rise in cells in a battery bank.
(c) Number of charge-discharge cycles.
(d) Impedance, conductance, and resistance measurements.
(e) Coup de fouet techniques.

While (a)-(d) are simple enough to understand, coup de fouet (translated as "crack of the whip") requires a brief discussion. In general, start-of-discharge of a VRLA battery is dominated by two transient voltage responses: response associated with batteries internal series resistance and the inductance; and, a more complex phenomenon that is attributed to the electrochemical reactions within the battery known as coup de fouet. As shown in the example of Figure 3.31 for a Hawker 2H1275-type cell (Pascoe and Anbuky, 2002, 2005), coup de fouet region for a lead-acid battery occurs whenever a fully charged cell is discharged, where the terminal voltage drops to a minimum, or trough, and then recovers back to the plateau value in a relatively short time. A detailed discussion on this subject is found in Pascoe and Anbuky (2002), and Pascoe and Anbuky (2005) provide an overview on how the above parameters or battery behaviors (a)-(e) could be used to estimate the reserve life or the RUL of a lead-acid battery pack.

3.6.2.2 On-line parameter estimation and techniques

With battery-powered systems such as electric vehicles and portable products introduced into the market keep growing, in many situations designers of BMS wish to estimate the SOC/SOH dynamically while the battery pack is powering the electronics or electrical systems. In these cases, parameters such as SOC and SOH should be estimated online and few mathematically intensive techniques are currently considered to achieve this target. Few well-documented techniques are

(i) Kalman filter (KF) based
(ii) Neural network (NN) based
(iii) Markov chain based
(iv) Fuzzy logic (FL) based

In all techniques, the modeling of a battery pack based on either electrochemical models or equivalent circuit models is used with intensive mathematical approaches. With high power, high discharge capability battery packs used in different types of electric vehicles, where the driver should be kept informed of battery pack SOC/SOH and other important parameters, the above mathematically complex approaches are considered as possible solutions, for online estimation of the parameters.

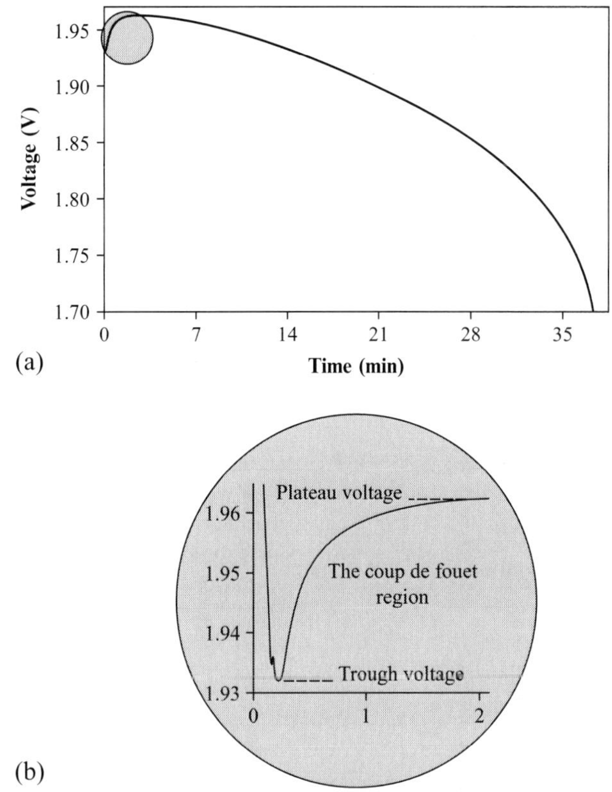

Figure 3.31 Coup de fouet phenomenon in lead-acid VRLA battery—an example from Hawker 2H1275 battery after a full charge: (a) voltage versus time with highlighted coup-de-fouet region and (b) expanded coup-de-fouet region.
Source: Pascoe and Anbuky, 2002.

3.6.2.2.1 KF-based approach

KFs are an intelligent—and sometimes optimal—means for estimating the present value of the time-varying "state" of a dynamic system. An additional benefit of the KF is that it automatically provides dynamic estimation error bounds on these estimates as well, and this fact was exploited to give online performance from a battery pack, without fear of causing damage by overcharge or overdischarge. By modeling a battery system to include the wanted unknown quantities in its state description, in a three part series of papers, Plett (2004a,b,c) details the use of KF to estimate their values. A brief tutorial to assist applying KFs in BMS is found in Plett (2004a), with a specific case of extended Kalman filtering (EKF) to take care of the nonlinear nature of the battery parameters. The second part of the papers (Plett, 2004b) describes some mathematical cell models that may be used with this method. It also gives an overview of other modeling methods in the literature and shows how an EKF may be used to adaptively identify unknown parameters in a cell model, in real time, given the cell input-output data. The third part of the paper (Plett, 2004c) covers the parameter estimation problem; namely, how to dynamically estimate SOC, power fade, capacity fade, and so forth with an EKF used in conjunction with the cell model. The cell model may be fixed or may itself have adaptable parameters so that the model tracks cell aging effects.

Figure 3.32 shows a simple block diagram for the algorithm function, and the HEV BMS performs many tasks, including communicate with the vehicle controller, measure cell physical quantities of interest (e.g., cell voltage, current, and temperature), and manage cell balancing. Here, we are only interested in the algorithmic considerations as motivated by the requirements imposed by the environment and the vehicle. In relation to Figure 3.32, following processes should happen in the algorithm, based on a regular sample collection of battery terminal voltage V_k, temperature T_k, and the battery current I_k.

- *Initialization.* When the vehicle is turned on, the algorithms must be initialized. The predominant dynamics of a cell while at rest is simply "self-discharge." If the level of self-discharge is too high, the SOH should be flagged as a warning or fault condition.
- *SOC update.* Once in every measurement interval, the voltages, temperatures, and module current are measured. The cell/pack SOC estimate must be updated based on these measurements.
- *SOH update.* Battery capacity and other parameters change over the lifetime of the pack. These must be continuously estimated in order to maintain safety and to obtain maximum performance from the pack.
- *Maximum available power.* Based on the SOC estimate and its uncertainty, and a dynamic cell model, the BMS must be able to estimate the maximum dis/charge power available at any time that will not cause voltage, SOC, or other design limits to be violated.
- *Equalization.* Series strings of cells with unequal capacities (as all are) will become unbalanced. That is, even if the SOC of all cells start with the same value, they will drift slowly apart as the system operates. The BMS must determine which cells must have their levels of charge altered to keep the pack balanced.

Many of the algorithmic requirements just described prescribe estimating parameters of a battery pack that may not be directly measured. It is well known that Kalman filtering provides an elegant and powerful solution to situations such as this. Kalman filtering is an established technology for dynamic system state estimation that is in common use in many fields including: target tracking, global positioning, dynamic systems control, navigation, and communication, but is not widely known in the field of BMS until the work done during the last decade. The KF comprises a set of recursive equations that are repeatedly evaluated as the system operates. The reader is

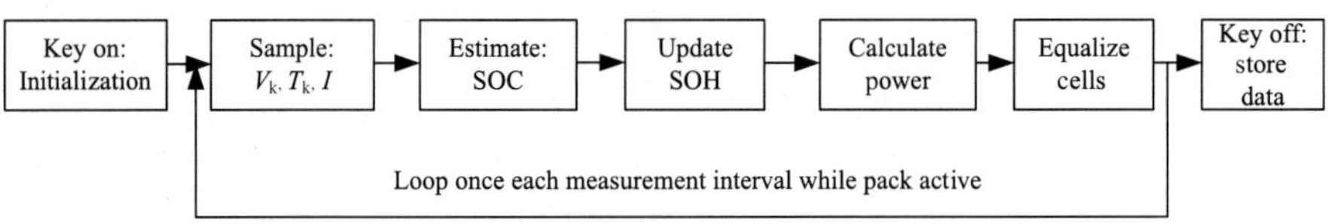

Figure 3.32 Algorithms in an HEV battery management system.
Source: Plett, 2004a.

referred to Kalman's original paper and several textbooks listed under bibliography for further derivation details together with the content of the three-part paper (Plett, 2004a,b,c) cited here.

3.6.2.2.2 NN-based techniques

The inspiration for developing NN came from examination of central nervous systems. In an ANN, simple processing nodes, called "neurons," "neurodes," "processing elements," or "units," are connected together to form a network which mimics a biological NN. NNs are frequently employed to classify patterns by learning from examples. Pattern statistics are determined from a set of training samples, and then new patterns are classified based on these statistics. In a variation called probabilistic NNs (PNN), heuristic approaches are often used to discover the underlying class statistics. They require long computation times for training. Incremental adaptation approaches are susceptible to convergence of an error function to the local minima. The network structure of PNN is similar to that of back propagation, but the activation function is replaced by one of the class of functions, such as exponential. Training PNN requires only a single pass. It is based on the Bayesian decision boundaries as the number of training samples grows. After the PNN is trained, all training samples are stored and then used in classifying new patterns. The computation time of PNN is much faster than that of back propagation types. PNN can be programmed easily into processors or chips to achieve even higher speeds through the use of parallel hardware. A discussion on how this is applied in battery SOC/SOH estimations is available in Lin et al. (2013), Shen et al. (2005), and Shahriari and Farrokhi (2013).

3.6.2.3 Markov chain and FL-based approaches

Recently, Markov chain and FL-based approaches have been attempted by the researcher community involved with the BMS. Micea, et al. (2011), Moura et al. (2013), Gholizadeh and Salmasi (2014), Lin et al. (2012), and Hsieh et al. (2001) provide some insight into these approaches.

3.7 Prognostics in battery health management

Prognostics attempt to estimate the remaining component lifetime when an abnormal condition has been detected. The key to useful prognostics is not only an accurate remaining lifetime estimate, but also an assessment of confidence of the uncertainty estimate. The phrase "battery health monitoring" has a wide variety of connotations, ranging from estimating the approximate voltage and other parameters to fully automated online monitoring of various measurements and estimated battery parameters. Prognostics of battery health could be very critical in electric vehicle, aerospace, and space exploration applications. For example, a catastrophic battery failure has occurred in NASA's Mars Global Surveyor, which stopped operating in November 2006. It was revealed later that when the spacecraft was commanded to go into safe mode, radiator of the batteries was positioned toward the sun, which created a temperature issue on the battery system (Goebel et al., 2008).

Prognostics and health management (PHM) in batteries for EVs and HEVs is a growing topic (Gao et al., Sept, 2002; Goebel et al., 2008). PHM dynamic models use sophisticated reasoning schemes applied with the goal of estimating the SOC, SOH, and the state of life (SOL). However, it remains a difficult task to predict the EOL for a battery pack (Goebel et al., 2008). In estimating battery SOC, SOH, SOL, etc., monitoring the variation in battery impedance becomes very useful, as this parameter has a significant variation as the battery packages.

3.7.1 Battery impedance and its time variation estimation as a prognostic parameter

Impedance of a battery is one of the key parameters vary as the battery ages. Due to various electrochemical reasons within a cell, as the cell ages impedance keeps increasing in general. Figure 3.33 indicates this situation for NiMH and Li-ion cells. Figure 3.33(a) is a simple DC resistance variation for a well-used versus a badly used (overcharged repeatedly) NiMH cell. Figure 3.33(b) indicates the results of measuring the impedance of an 18,650-type Li-ion cell in a PHM exercise (Goebel et al., 2008), and Figure 3.33(c) indicates the variation in the Li-ion cell impedance versus frequency after different numbers of discharge cycles (Karfthoefer, 2005). In Figure 3.33(c), it clearly shows that at lower frequencies below 10 Hz, as the discharge cycles are increased the impedance gradually increases. There are new battery management ICs that monitor this behavior in gas gaging (discussed later) (Karfthoefer, 2005).

In large standby telecommunication battery banks or UPS installations, large numbers of series and parallel connected VRLA cells (sealed lead acid (SLA) cells) are used and any single or multiple misbehaving cells could in the longer run create expensive replacement issues. In these cases of large and expensive battery banks, battery impedance testing is used as a prognostic technique. By automatic or manual measurement of individual cell voltages over the lifetime of the bank, test engineers can easily predict the potential failure of individual cells. In these situations, typical VRLA cells behave as per Figure 3.34 in general (Langnan, 2000). A detailed discussion on the impedance variations and terminal voltages is available in Langnan (2000).

3.7.2 Energy-aware battery modeling concepts for best runtime

With the availability of techniques for computationally feasible mathematical models of cells, "energy-aware systems design" is another new concept currently developing, to get the best out of a battery pack, with multiple cells or batteries connected to a given electronic system. Given the chemical structure and the processes inside a battery, as depicted in Figure 3.26, if a battery pack can contain multiple cells, where each one of them can be connected to the load selectively depending on the best utilization model for the total energy within the pack, runtime of a portable product can be improved significantly (Rao et al., 2003). This is particularly the case with systems based on CMOS logic where the power supply source voltage V_{DD} and threshold voltage, V_{th} of the FETs are the key variables for best power consumption. In case, these

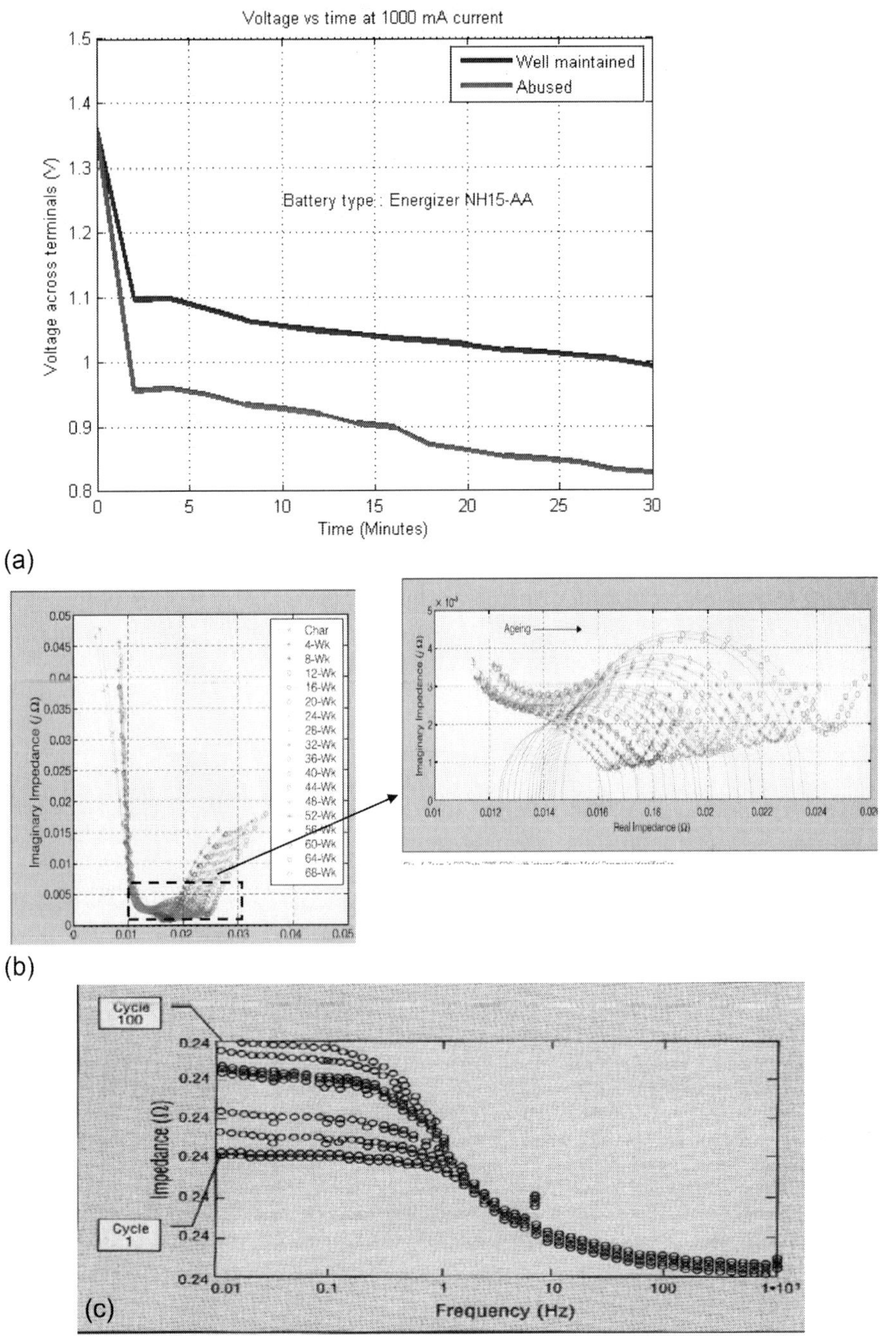

Figure 3.33 Battery impedance variation with age and abuse: (a) DC impedance comparison of a well-used NiMH cell and an abused one; (b) electrochemical impedance spectrometry (EIS) data for a 18,650-type Li-ion cell with aging (Goebel et al., 2008); and (c) typical variation of Li-ion cells with life cycle at different frequencies (Karfthoefer, 2005).

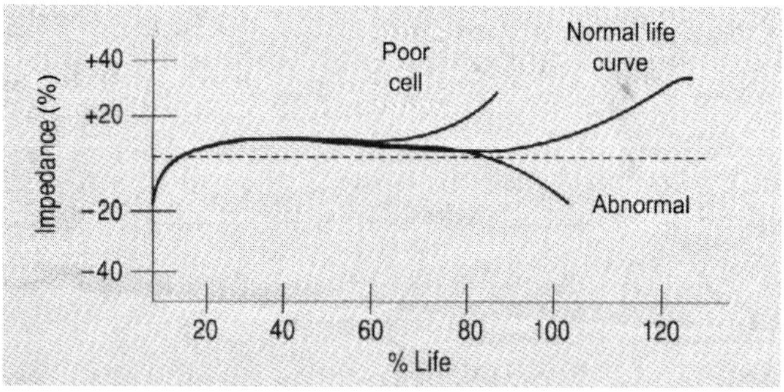

Figure 3.34 VRLA battery impedance behavior over lifetime.
Adapted from Langnan, 2000.

parameters can be optimized with a battery efficiency model, an interleaved power supply system (with multiple batteries selectively supplying the load conditions) can provide the best runtime, given the case that a battery pack has only a limited energy after an optimal charging process. This concept is shown in a simplified case of Figure 3.35, with two batteries that can be selectively connected to a suitable DC-DC converter, which supplies the VLSI circuit.

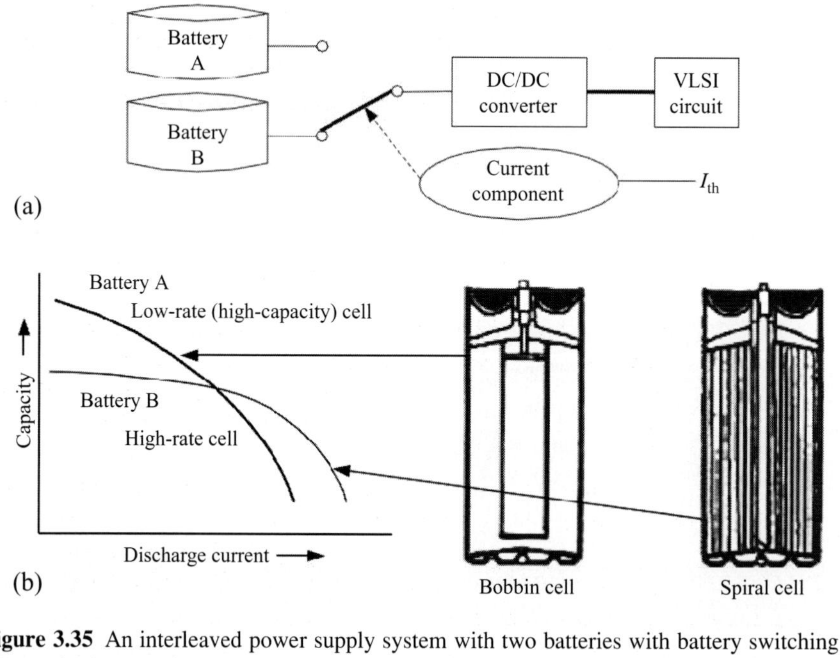

Figure 3.35 An interleaved power supply system with two batteries with battery switching scheme for best runtime: (a) basic concept and (b) an example of implementation with a high-rate spiral cell and a low-rate bobbin cell.
Source: Rao et al., 2003, p. 86. Reproduced with the permission of IEEE, USA.

In this example, where an interleaved dual battery concept is used as in Figure 3.35 (b), working principle is simple. When the discharge current is lower than a threshold value, I_{th}, the system uses battery A as the energy source and vice versa with battery B. Battery A is low-rate-high capacity cell, while the other one is a high-rate-low capacity cell. In this process of switching based on suitable algorithms, a suitable battery model is taken into account for optimum discharge behavior (Wu et al., 2000). This kind of dynamic power management of battery powered electronics can be used for battery life optimization as well (Benini et al., 2003). Energy efficient communication is another application of these concepts with electrochemical battery models (Chiasserini and Rao, 2001).

3.8 Fast charging of batteries

With electric vehicles and portable products proliferating fast charging of rechargeable batteries have become a valuable topic for R & D. However, in fast charging, if the charger cannot enter into a cutoff mode, battery pack is subjected to a significant overcharge and the life of the battery pack could be reduced, or the pack may even fail.

While all major battery families can accept either a standard (16-24 h) or a fast (2-4 h) charge, most of the discussion here is limited to fast charging methods. Slower charging schemes tend to be found in simpler, price-sensitive applications, which do not need (or cannot afford) much beyond a simple charger and a low-battery indicator. The objective of fast charging a battery is to cram as much energy as it takes to bring the battery back to fully charged state in the shortest possible time without damaging the battery or permanently affecting its long-term performance. Since current is proportional to charge divided by time, the charging current should be as high as the battery systems will reasonably allow.

Nickel-based chemistries prefer constant-current-type charging, while lead-acid and lithium cells prefer constant voltage (current limited) charging. For the constant-current cells (NiCd and NiMH), a 1 C charge rate will typically return more than 90% of the battery's useable discharge capacity within the first hour of charging. The constant-potential (CP) cells (lead acid and lithium ion) are a bit slower to reach the 90% mark, but can generally be completely recharged within 5 h. See Table 3.5 for representative and typical charging recommendations for different chemistries. Most appropriate method should be selected in consultation with the manufacturer.

Fast charging has compelling benefits, but places certain demands upon the battery system. A properly performed fast charge, coordinated to the specifications of a battery rated for such charging, will deliver a long cycle life. The high charging rates involved, however, cause rapid electrochemical reactions within the cells of the battery. After the battery goes into overcharge, these reactions cause a sharp increase in internal cell pressure and temperature. Uncontrolled high-rate overcharge quickly cause irreversible battery damage. Thus, as the battery approaches full charge, the charging current must be reduced to a lower "top-off" level, or curtailed entirely.

Table 3.5 Representative charging recommendations for different batteries

	NiCd or NiMH	Sealed lead acid	Li-ion
Charging current (C)	1	1.5	1
Voltage per cell (V)	1.80	2.5	4.20 ± 0.05
Charging time (h)	~3	~3	2.5-5.0
Method for optimum fast-charge termination point	See Table 3.6	Current cutoff	Typically a timer[a]
Back-up charge termination method	See Table 3.6	Timer	–
"Top off" rate	0.1 C	0.002 (trickle)	
Temperature range (°C)	10-40 (NiCd) 15-30 (NiMH)	0-30	0-40

[a]Depending on manufacturers recommendation.

3.8.1 Charge termination methods

If a rapid charge is applied to a battery pack, it is necessary to select a reliable method to terminate charging at the fully charged position. Two practical approaches for charge termination are *temperature termination method* and *voltage termination method*.

3.8.1.1 Voltage termination methods

Four commonly available voltage termination methods are maximum voltage (V_{max}), negative delta voltage ($-\Delta V$), zero slope, and inflection point (d^2V/dt^2). The maximum voltage method senses the increase in battery voltage as the battery approaches full charge. However, this is accurate only on a highly individualized basis. It is necessary to know the exact value at the voltage peak, otherwise the batteries may be over or undercharged.

Temperature compensation is also required because of the negative temperature coefficient of battery voltage. The maximum voltage will increase if the batteries are cold, causing an undercharge because the charging voltage will reach the maximum voltage trip point early. If the batteries are hot, the maximum voltage may never be reached and the batteries will be destroyed due to overheating. Therefore, the V_{max} method is generally not recommended for fast-charge rates.

The negative delta voltage is the most popular of the fast-charge termination schemes. It relies on the characteristic drop in cell voltage that occurs when a battery enters overcharge, as shown in Figure 3.36. With most NiCd cells, the voltage drop is a very

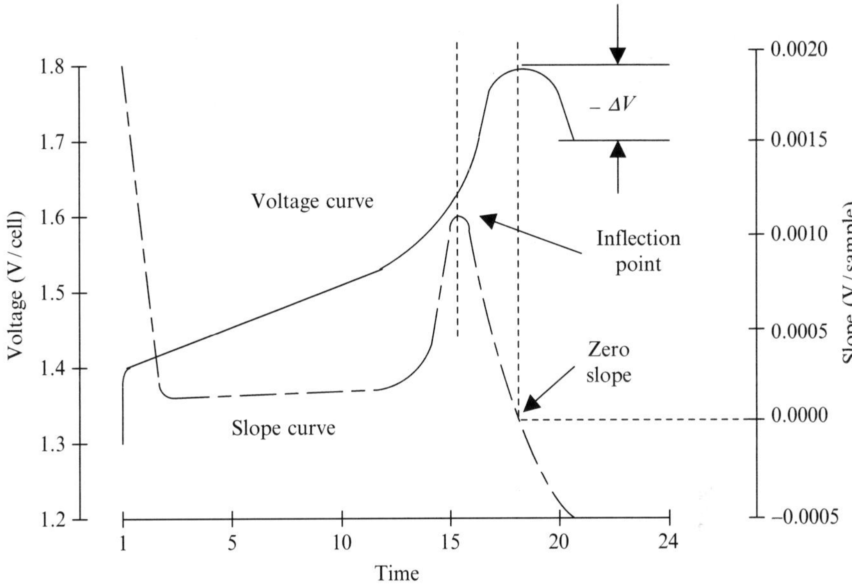

Figure 3.36 Termination methods based on changes in voltage-and-voltage slope.

consistent indicator, and the $-\Delta V$ method is fine for charge rates up to 1 C. An inherent problem with this method is that the batteries must be driven into the overcharge region to cause the voltage decrease. Pressures and temperatures rise very rapidly at fast-charge rates beyond 1 C. In cyclic applications, the battery must be able to endure that continual abuse.

Another concern is that cells such as NiMH types do not always have the characteristic decrease in voltage, compared to NiCds as shown in Figure 2.10. This creates a problem of forward compatibility when moving from NiCd to NiMH cells. Most manufacturers of NiMH cells do not advocate the -ΔV method of charge termination.

The zero slope method monitors the point where the slope of the battery voltage reaches zero. This method is reliable for rapid charge rates up to 4 C, and is less susceptible to noise on the voltage sense lines. However, a few types of batteries such as the NiCd button cells may have a voltage slope that never quite reaches zero. Therefore, the zero slope method is better suited as a backup method.

In the inflection point (d^2V/dt^2) method, the system monitors the change in voltage over time, and is the most sensitive indicator for preventing overcharge. The inflection point method relies on the changes in the voltage slope, shown in Figure 3.36, which occurs during charge and is an excellent primary termination method for up to 4 C charge rates.

The change in the voltage slope is an extremely reliable and repeatable indication of charge. It does not rely on the decrease in voltage that may not always occur. Instead, this method looks for the flattening of the voltage profile as the battery reaches full charge. By monitoring the relative change in the steepness of the voltage slope, this method avoids having to use absolute numbers.

3.8.1.2 Temperature termination methods

Temperature is the main cause of failure in a rechargeable cell, so it makes sense to monitor the cell temperature to determine when to shut off charge to a battery. Three methods of charge termination, based on temperature, are common: maximum temperature cutoff (MTC), temperature difference (DT), and temperature slope (dT/dt). The MTC system is the easiest and cheapest to implement, but is the least reliable. Using a bimetallic thermal switch or a positive temperature coefficient thermistor, a simple, low cost circuit can shut down a charging current at an appropriate temperature.

The DT method measures the difference in ambient and cell temperatures to compensate for a cool environment. It requires monitoring two temperature sensors, one for the battery temperature and one for the ambient. This method may be unsuitable if the difference between cell and ambient temperature is very large.

The DT method can become unreliable with a quickly changing ambient temperature unless an equal thermal mass is attached to the ambient sensor. This means that the DT method is suitable for a primary charge termination at lower charge rates up to $C/5$ if the ambient temperature is not going to change often. The DT method also provides an excellent backup charge termination scheme.

The dT/dt method, a more sophisticated temperature termination scheme, measures the change in temperature over time. This method uses the slope of the battery temperature, and therefore is less dependent on changes in ambient temperatures or in large differences between ambient and battery temperatures. Accurately adjusted to a particular pack and with careful attention paid to the type and placement of the temperature sensor, the dT/dt method works very well. This dT/dt method is suitable for charge rates up to 1 C and provides an excellent backup method.

3.8.2 NiCd and NiMH fast-charge methods

Nickel-based batteries such as NiCd and NiMH types are mature chemistries. Although it is not correct to consider the NiCd and NiMH electrochemistries or charging regimens as being interchangeable, they are similar enough that they can be discussed together.

There is no one best way to fast charge a NiCd or NiMH battery. Variables introduced by the allowable cost and size of the end application, the choice of charge termination method(s), and the specific battery vendor's recommendations will all influence the final choice of charging technique.

Figure 3.37(a) shows the voltage, pressure, and temperature characteristics of a NiCd cell being charged at 1 C rate. Figure 3.37(b) shows similar data for NiMH cell.

These curves illustrate the need for a reliable termination of the high-current portion of the charge cycle, and assist in understanding the various fast-charge termination methods outlined in Table 3.5. For both electrochemistries, the ideal fast-charge termination point is at 100-110% of returned charge. The charging current is then reduced to the top-off value for 1-2 h, to bring the cell into a state of slight overcharge. This compensates for the inefficiencies of the charging process (e.g., heat generation).

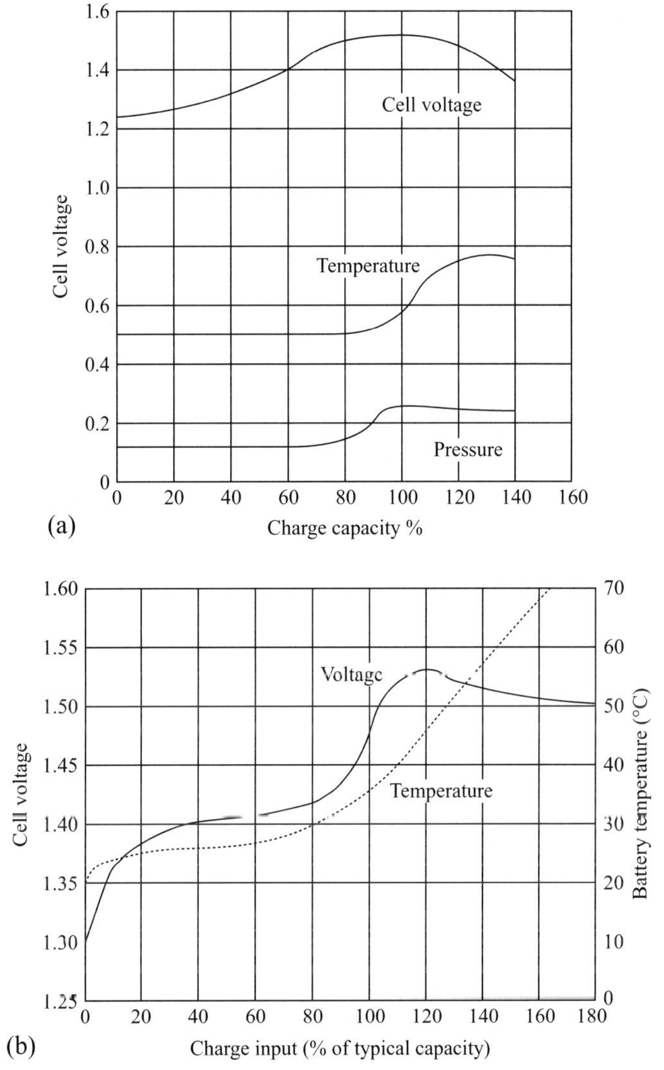

Figure 3.37 Charging indications for nickel-based batteries at 1 C charge rate; (a) cell voltage, temperature, and pressure for a typical NiCd cell at 1 C rate charging and (b) NiMH voltage and temperature characteristics.

If the specific application will have the battery on standby for more than several weeks, or at high temperatures, the top-off charge is followed by a continuous, low-level "trickle" charge to counter the self-discharge characteristics of NiCd and NiMH cells.

Under certain conditions, particularly following intervals of storage, a NiMH battery may give an erroneous voltage peak as charging commences. For this reason, the charger should deliberately disable any voltage-based charge termination technique for the first 5 min of the charging interval. Table 3.6 summarizes the fast-charge termination methods for NiCd and NiMH cells.

3.8.3 Charging SLA batteries

Unlike nickel-based batteries, SLA batteries are charged using a "CP" regimen. CP charging employs a voltage source with a deliberately imposed current limit (a current-limited voltage regulator). A significantly discharged battery undergoing CP charging will initially attempt to draw a high current from the charger. The current-limiting function of the CP regimen serves to keep the peak charging current within the battery's ratings.

Table 3.6 Fast-charge termination methods for NiCd and NiMH batteries

Charging technique	Description
Negative $\Delta V(-\Delta V)$	Looks for the downward slope in cell voltage which a cell exhibits ($\gg$30-50 mV for NiCd, 5-15 mV for NiMH) upon entering overcharge. Very common in NiCd applications due to its simplicity and reliability
Zero ΔV	Waits for the time when the voltage of the cell under charge stops rising, and is "at the top of the curve" prior to the downslope seen in overcharge. Sometimes preferred over-ΔV for NiMH, due to NiMH's relatively small downward voltage slope
Voltage slope (dV/dt)	Looks for an increasing slope in cell voltage (positive dV/dt) which occurs somewhat before the cell reaches 100% returned charge (prior to the Zero ΔV point)
Inflection point cutoff (d^2V/dt^2, IPCO)	As a cell approaches full charge, the rate of its voltage rise begins to level off. This method looks for a zero or, more commonly, slightly negative value of the second derivative of cell voltage with respect to time
Absolute temperature cutoff (TCO)	Uses the cell's case temperature (which will undergo a rapid rise as the cell enters high-rate overcharge) to determine when to terminate high-rate charging. A good backup method, but too susceptible to variations under ambient temperature conditions to make a reliable primary cutoff technique
Incremental temperature cutoff (ΔTCO)	Uses a specified increase in the cell's case temperature, relative to the ambient temperature, to determine when to terminate high-rate charging. A popular, relatively inexpensive, and reliable cutoff method
Delta temperature/delta time ($\Delta T/\Delta t$)	Uses the rate of increase of a cell's case temperature to determine the point at which to terminate the high-rate charge. This technique is inexpensive and reliable once the cell and its housing have been properly characterized

Following the current-limited phase of the charging profile, an SLA battery exposed to a constant voltage will exhibit a tapering current profile as shown in Figure 3.38. When returned charge reaches 110-115% of rated capacity, allowing a dischargeable capacity of 100% of nominal, the charge cycle is complete.

The specifics of fast-charging SLA batteries are more vendor-dependent than those of NiCd or NiMH units. The information in Table 3.5 uses data from GS battery (USA) Inc. The primary termination method, "current cutoff," looks at the absolute value of the average charging current flowing into the battery.

When that current drops below 0.01 C, the battery is fully charged. If it will be in standby for a month or more, a trickle current of 0.002 C should be maintained. The backup termination method, according to the vendor's recommendations, should be a 180-min time-out on the charging cycle (Schwartz, 1995).

To satisfy more stringent charge control recommendations, where the battery temperature, voltage, and current need be sampled, many dedicated charge controller ICs are available in the market. Bq2031 lead-acid fast-charge IC from Texas/Benchmarq and UC3906 (SLA charger) from Texas/Unitrode Integrated Circuits are examples. The UC3906 battery charge controller contains all of the necessary circuitry to optimally control the charge and hold cycle for SLA batteries. These integrated circuits monitor and control both the output voltage and current of the charger through three separate charge states; a high-current bulk-charge state, a controlled overcharge, and a precision float-charge, or standby state. Sacarisen and Parvereshi (1995) and Application Note U-104 (Unitrode Inc., n.d.) provide details about SLA charge control using UC3906.

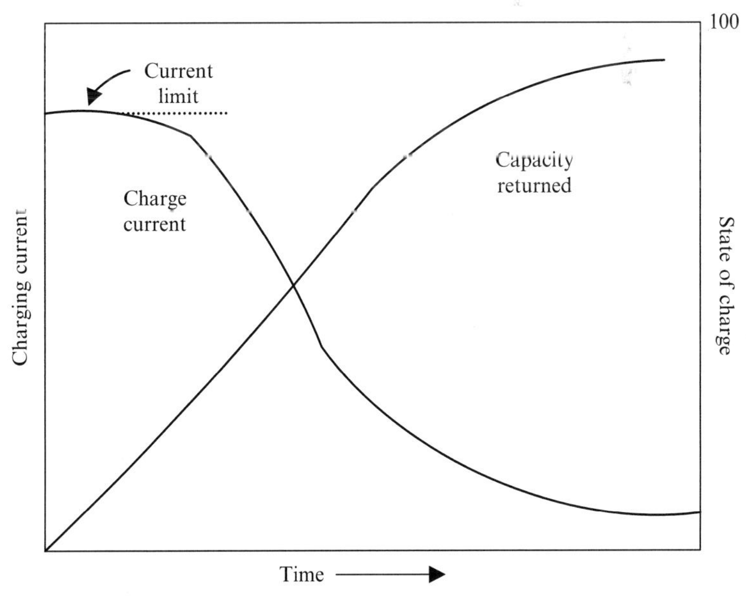

Figure 3.38 Typical current and capacity returned *Vs* charge time for CP charging in sealed lead-acid batteries.

3.8.4 Li-ion chargers

Lithium-Ion batteries require a constant potential charging regimen, very similar to that used for lead-acid batteries. Because the voltage of Li-ion cell under constant-current charge could continue to rise to the point of cell destruction, absolute voltage limits are required (Bentley and Heacock, 1996). Typical recommendations for Li-ion fast charge are indicated in Table 3.7. As with lead-acid batteries, a Li-ion cell under charge will reduce its current draw as it approaches full charge. Figure 3.39 indicates this condition. However, if the charger has poor regulation

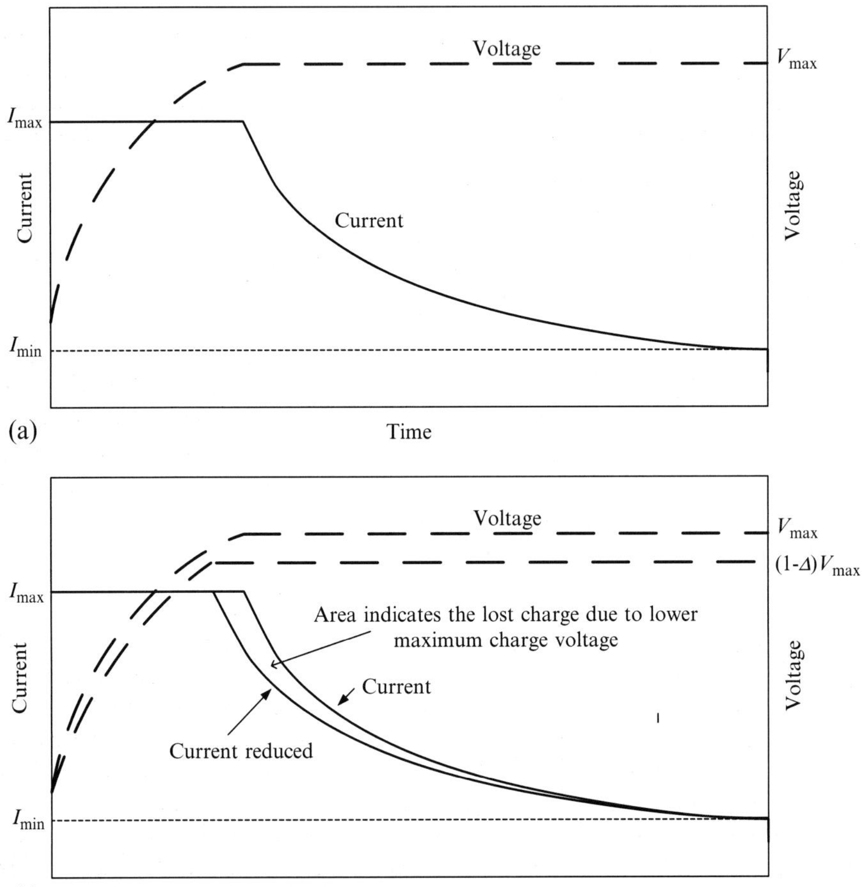

Figure 3.39 Voltage-regulated charging profile for Li batteries: (a) current-limited voltage-regulated charging profile and (b) charge dependence on voltage regulation.
Adapted from Bentley and Heacock, 1996.

on the output voltage and creates a lower output voltage, it will decrease the capacity as depicted in Figure 3.39(b). Area between the graphs of current indicates the lost charge.

If the cell vendor's recommendation for charging voltage (generally 4.20 V $\pm$ 50 mV at 23 °C) is followed, the cells will be able to completely recharge from any "normal" level of discharge within 5 h. At the end of that time, the charging voltage should be removed. Trickle current is not recommended. If the voltage on a Li-ion cell falls below 1.0 V (or to an abnormal level), recharging of that cell should not be attempted. If the voltage is between 1.0 V and the manufacturer's nominal minimum voltage (typically 2.5-2.7 V), it may be possible to salvage the cell by charging it with a 0.1 C current limit until the voltage across the cell reaches the nominal minimum, followed by a fast charge.

Due to special characteristics of Li-ion batteries, most Li-ion manufacturers incorporate custom protection circuits into their battery packs to monitor the voltage across each cell within the battery and to provide protection against overcharge, battery reversal, and other major faults. These circuits are not to be confused with charging circuits. For example, MC 33347 protection circuit is such an IC from Motorola (Alberkrack, 1996).

3.8.5 Portable chargers and comparison of recharge requirements for different chemistries

With the availability of charge controllers, protectors, and other battery management ICs today, a designer can find many commercial ICs, and supporting reference designs (Israelsohn, 2001). Choosing the best chemistry requires a deliberative comparison between battery attributes and the required power source specifications in general. As summarized above, the four common chemistries require different recharge algorithms and give different indications when they have completed the discharge cycle. Table 3.7 provides a summary, as adapted from source (Israelsohn, 2001).

Most secondary batteries tolerate trickle charging for long periods. Simplest charging strategy, therefore, uses a simple linear regulator IC or a pulse width modulator IC in conjunction with a series transistor and a currents sense resistor. Circuits such as these are available from many vendors and with a range of auxiliary features for charging single cells or nickel chemistry stacks (Israelsohn, 2001). They sometimes use an adaptive method that adjusts the charger's behavior according to the battery's SOC. A charger of this type may start by testing the battery for deep discharge, which is determined by comparing the battery's terminal voltage with a threshold value. If the battery is deeply discharged, the charger enters a precharge mode in which it limits its current to some fraction, K, of resistor-programmed maximum charge current, I_{PGM}. Figure 3.40 gives representative options of charging algorithms. Figure 3.40(a) is a constant-current/constant-voltage charging algorithm, while Figure 3.40(b) is a current foldback mode.

Table 3.7 Comparison of recharge requirements of different chemistries

Parameter	SLA	NiCd	NiMH	Li-ion
Standard Charge				
Current (C rate)	0.25	0.1	0.1	0.1
Voltage per cell (V)	2.27	1.5	1.5	4.1 ± 50 mV
Time (hours)	24	16	16	16
Temperature range (°C)	0-45	5-40	5-40	5-40
Termination	None	None	Timer	None
Fast Charge				
Current (C rate)	1.5	1	1	1
Voltage per cell (V)	2.45	1.5	1.5	4.1 ± 50 mV
Time (hours)	1.5	3	3	2.5
Temperature range (°C)	0-30	15-40	15-40	10-40
Primary termination method	I_{min}[1], ΔTCO	dT/dt, $-\Delta V$	Zero dv/dt, $-\Delta V$, d^2V/dt^2, ΔTCO	I_{min} + timer, dT/dt, dT/dt
Secondary termination method	Timer, ΔTCO	TCO, timer	TCO, timer	TCO, timer

[1]Minimum current termination threshold.
Adapted from Israelsohn, 2001.

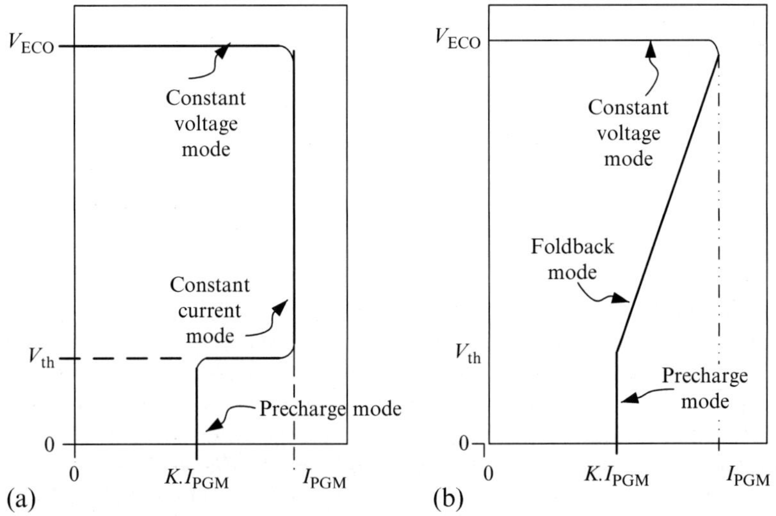

Figure 3.40 Charger voltage-current algorithms: (a) constant-current/constant-voltage case and (b) current foldback case.
Source: Israelsohn, 2001.

3.8.6 EOD determination

Determination of the point at which a battery has delivered all of its usefully dischargeable energy is important to the longevity of the cells which form that battery. Discharging a single cell too far will often cause irreversible physical damage inside the cell.

If multiple cells are placed in series, unavoidable imbalances in their capacities can cause the phenomenon known as "cell reversal," in which the higher-capacity cells force a backward current through the lowest-capacity cell. Knowing the EOD point provides a "zero capacity" reference for coulometric gas gaging.

The actual determination of the EOD point is typically done by monitoring the cell voltage. For the most accurate determination of EOD when the load is varying, correction factors for load current and the battery's SOC should be applied, especially to SLA and Li-ion batteries. The essentially flat discharge profiles of NiCd and NiMH make these corrections a matter of user discretion for most load profiles. Table 3.8 shows voltages commonly used to indicate the EOD point for the four battery types.

3.8.7 Gas gaging

The "gas gaging or fuel gaging" concept discussed here does not refer to the gases which may be evolved by the battery reactions, but rather to the concept of the battery as a fuel tank powering the product. Gas gaging, therefore, involves real-time determination of a battery's SOC, relative to the battery's nominal capacity when fully charged.

It is possible to make an inexpensive and moderately useful SOC measurement from a simple voltage reading, if the battery being used has a sloping voltage profile. Hence, Li-ion batteries and, to a lesser extent, SLA batteries, should be amenable to such an approach. In practice, the results are less than optimal: cell voltages are dependent upon loading, internal impedance, cell temperature, and other variables. This reduces the attractiveness of the simple voltage method of gas gaging; the fact that it is not suitable for NiCd or NiMH batteries, due to their essentially flat voltage

Table 3.8 Typical end of discharge voltages

Cell type	EODV (V)	Comments
Lead acid	1.35-1.9 (1.8 V typical)	Depend upon loading, state of charge, cell construction, and manufacturer
NiCd	0.9	Essentially constant
NiMH	0.9	Essentially constant within recommended range of discharge rates
Li-ion	2.50-2.70	Depend upon manufacturer, loading, and state of charge

profile, makes it commercially untenable. A clever and effective alternative is the "coulometric" method (or earlier discussions referred to as coulomb counting).

Coulometric gas gaging, as its name implies, measures the actual charge ($\int I\mathrm{d}t$) going into and out of the battery. By integrating the difference of current in and current out, it is possible to determine the charge status of the battery at any given time. There are of course real-world details which must be observed in the actual implementation of such a gas gage; some of the most important of these are:

(a) It is necessary to have an accurate starting point for the integrator, corresponding to a known SOC in the battery. This is often resolved by zeroing the integrator when the battery reaches its EODV.

(b) Compensation for the temperature. The actual capacity of lead-acid batteries increases with temperature, and that of nickel-based batteries decreases as battery temperature rises.

(c) Appropriate conversion factors should be applied for the particular charge regimen and discharge profile used. Under conditions of highly variable battery loading, dynamic compensation may be advisable.

3.8.8 Battery health in a practical viewpoint

SOC is a widely used concept to represent the remaining capacity of a battery in a charge/discharge cycle. However, due to cycle-aging phenomenon, using SOC alone may result in large errors because the full charge capacity (FCC) may be significantly less than the design capacity (DC) that denotes the FCC of a new battery.

Considering the aging effect, SOH of a battery is defined as the FCC of cycle-aged battery to its DC. Analytical modeling of batteries can be used to get theoretically accurate values for SOC and SOH (Rong and Pedram, 2006b). With battery health defined as a battery's actual capacity relative to its rated capacity, the health of the battery can be determined and maintained in three steps:

(i) Discharge the battery to the EOD point, preferably into a known load.

(ii) Execute a complete charge cycle, while gas gaging the battery.

(iii) Compare the battery's measured capacity to its rated capacity.

This sequence will simultaneously "condition" the battery (which means to overcome the so-called "memory effect" of capacity of NiCd batteries, e.g.) and will indicate the capacity of the battery after conditioning. The information so obtained can be used to ascertain whether the battery is in good shape, or is approaching the end of its useful life. Online SOC estimation with monitoring system is discussed in Kutluay et al. (2005).

3.9 Battery communication and related standards

In the early 1990s, power management and battery management were rarely used in portable products. With more advanced processors introduced around the mid-1990s, with very high power consumption and relatively high load currents of tens of amperes, placed a burden on the available rechargeable battery chemistries due to rapid increase of discharge current. To meet system run times expected by the users, innovative power management concepts had to be employed. For this reason, around mid-1990s new industry standards were proposed to standardize the battery and power management subsystems within portable products, where concepts such as Advanced Configuration and Power Interface by Intel were utilized. Following were the standards proposed:

(i) System Management Bus [SMBus] Specification.
(ii) Smart Battery Data Specification.
(iii) Smart Battery Charger Specification.
(iv) Smart Battery Selector Specification.

The above specifications form the smart battery systems (SBS) specification. This specification presents a solution for many of the issues related to batteries used in portable equipment such as laptop computer systems, cellular telephones, or video cameras. Fundamental to the SBS system is the concept that the battery contains all of the necessary components to determine the battery's SOC, predict time to full and time to empty, specify the charging voltage and current, and determine when the battery is fully charged to fully discharged.

Concept of a SBS is shown in Figure 3.41(a). Figure 3.41(b) indicates details. It consists of an AC-DC converter (unregulated), power switch, system power supply, smart battery charger, smart battery selector, all of which communicate with the system host and the system elements themselves via SMBus. In this case, smart battery A powers the system while battery B is getting conditioned and/or charged.

The System Management Bus (SMBus) is a two-wire interface through which simple power-related chips can communicate with the rest of the system. It uses I^2C as its backbone (Phillips Semiconductors, 2003). A system using SMB passes messages to and from devices instead of tripping individual control lines. Removing the individual control lines reduces pin count. Accepting messages ensures future expandability. With SMBus, a device can provide manufacturer information, tell the system what its mode/part number is, save its state for a suspended event, report different types of errors, accept control parameters, and return its status. The SMBus may share the same host device and physical bus, provided that an appropriate electrical bridge is provided between the respective devices. For details, articles by Benchmarq Microelectronics Inc. et al. (1996 a,b) and Heacock (1998) are suggested.

3.10 Battery safety

With the introduction of Li-based batteries, with high gravimetric energy density, safety of the battery powered portable products has become a concern during the last decade. In early 2000s, some laptop fires scared the public and promoted widespread recalls of Li-ion cells. During the year 2013, new aircrafts such as the Boeing Dreamliner was grounded due to battery pack issues. Since these events, individual companies have taken many steps to ensure that such disasters do not recur, and the industry has joined forces to unify the standards. By 2004, IEEE has published two standards

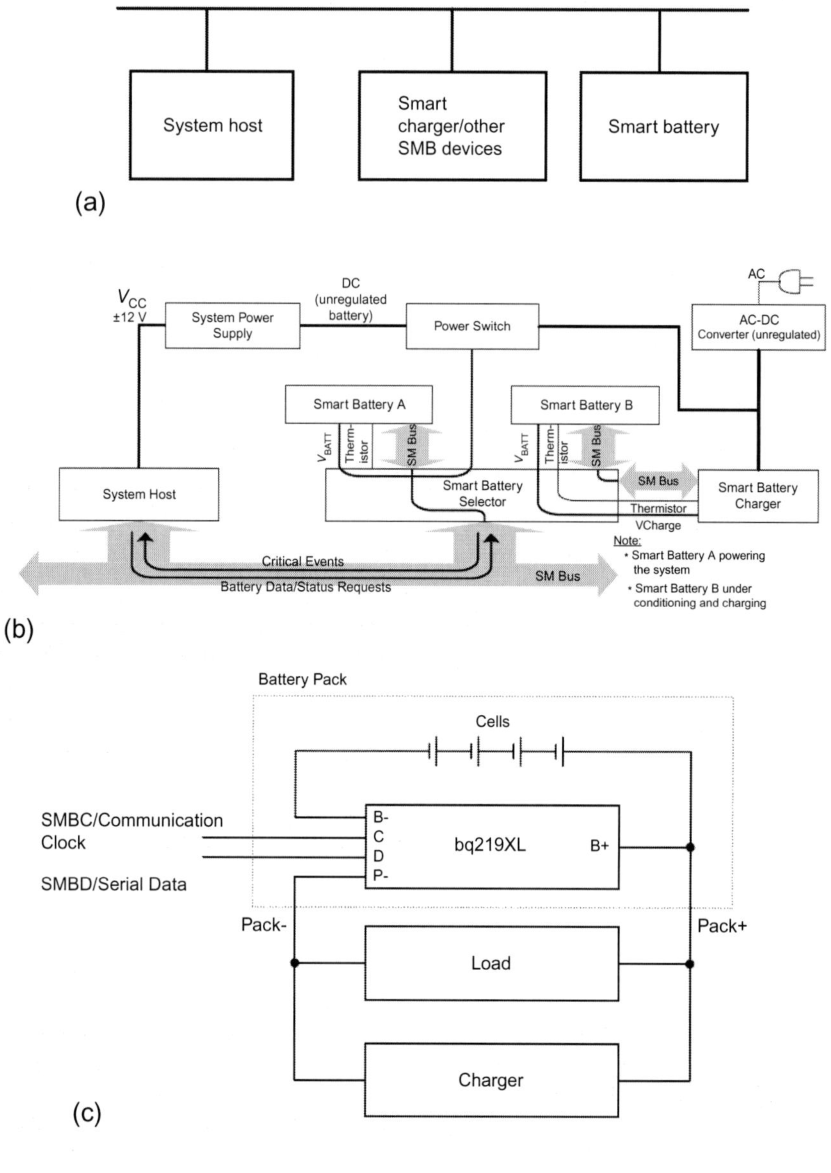

Figure 3.41 Smart battery system and a SMBus module: (a) concept; (b) a typical multiple smart battery system; and (c) typical control IC (bq 219XL) and its connections.

relevant to the design and manufacture of portable systems: The IEEE 1625 Standard for Rechargeable Batteries for Portable Computing, and a related IEEE 1725 for Rechargeable Batteries for Cellular Phones.

The standard covers design approaches that ensure reliable operation and that minimize the occurrence of faults leading to hazards in portable computing devices and other rechargeable battery-powered systems. The standard guides the system and subsystem design through five major areas:

- systems integration,
- cell,
- pack,
- host device, and
- total system reliability.

Also covered are the critical operational parameters and how they change with time and environment, the effect of extremes in temperature, and the management of component failure. Figure 3.42 indicates the case of a portable computing device broken into subsystems and within the environment.

In general portable equipment battery packs usually contain an analog front end (AFE) which is monitoring the conditions of the cells, and it usually provides a first level of protection to the pack. A battery management unit BMU (similar to the example in Figure 3.42(c)) usually communicates with the cell stack to monitor and communicate with the host system via SMBus. Referring to Figure 3.42(b), in the 1625 standard, it proposes to move the second level of protection into the AFE, which communicates directly with the BMU (Elder, 2004). An industry-wide methodology such as design failure modes and effects analysis (DFMEA), which can be used to highlight and prioritize the possible roots of faults and hazards, is to be used with the 1625 conditions in order to get the benefits proposed in the standard. Figure 3.42(c) indicates the block diagram of a battery pack compliant with the SBS 1.1with 4 series Li-ion cells (Bernardi and Carpenter, 1995). This change in architecture helps clarify the DFMEA analysis. Compared to the older architecture, in the new architecture (Figure 3.42(b)), BMU and AFE monitor each other as well as the cell stack. Theoretically, this allows the second-level protection to be integrated with the first-level protection device.

As in Figure 3.42(c), the new architecture consists of a BMU, an AFE-integrated circuit, current sense resistor, safety FET, and a chemical fuse. BMU could be a 3.3-V microcontroller with measurement capabilities and a programmable flash. The AFE provides a high-voltage interface to the cell stack for voltage measurements, along with cell-balancing control, over current protection, and a low dropout regulator to power the BMU. Based on possible problem levels, for discrete components, BMU and AFE interaction DFMEA tables can be prepared similar to the case in Table 3.9 and evaluate each item and score according to the following criteria:

- Severity of the fault (SEV).
- Probability of occurrence (OCC).
- Difficulty of detection (DET).

Typically, evaluation and scoring are based on experience and statistical reliability data. However, these data may not be always available and guidance in 1626 standard may be helpful. For more details, article by Elder (2004) is suggested.

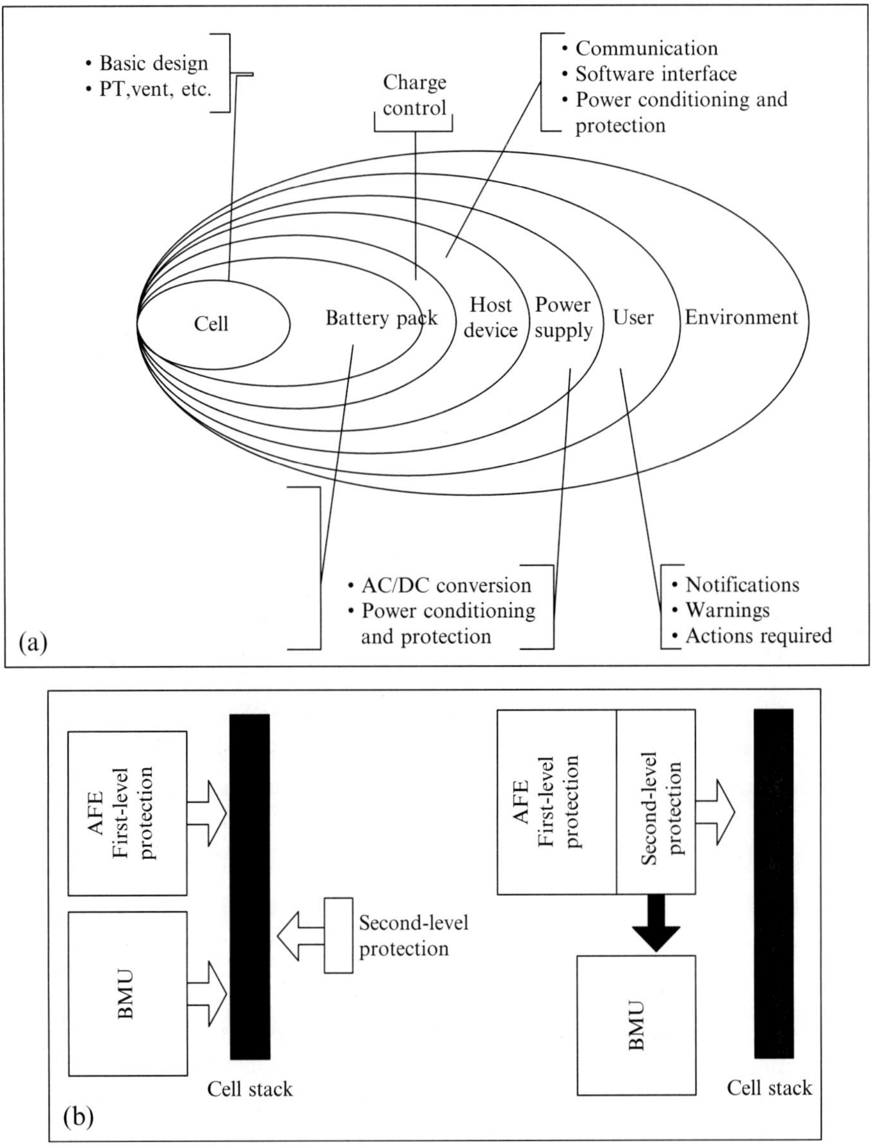

Figure 3.42 A portable computing device and its environment within the case of IEEE 1625: (a) portable computing device broken out into subsystems; (b) proposed level of safety compared with older battery packs; and

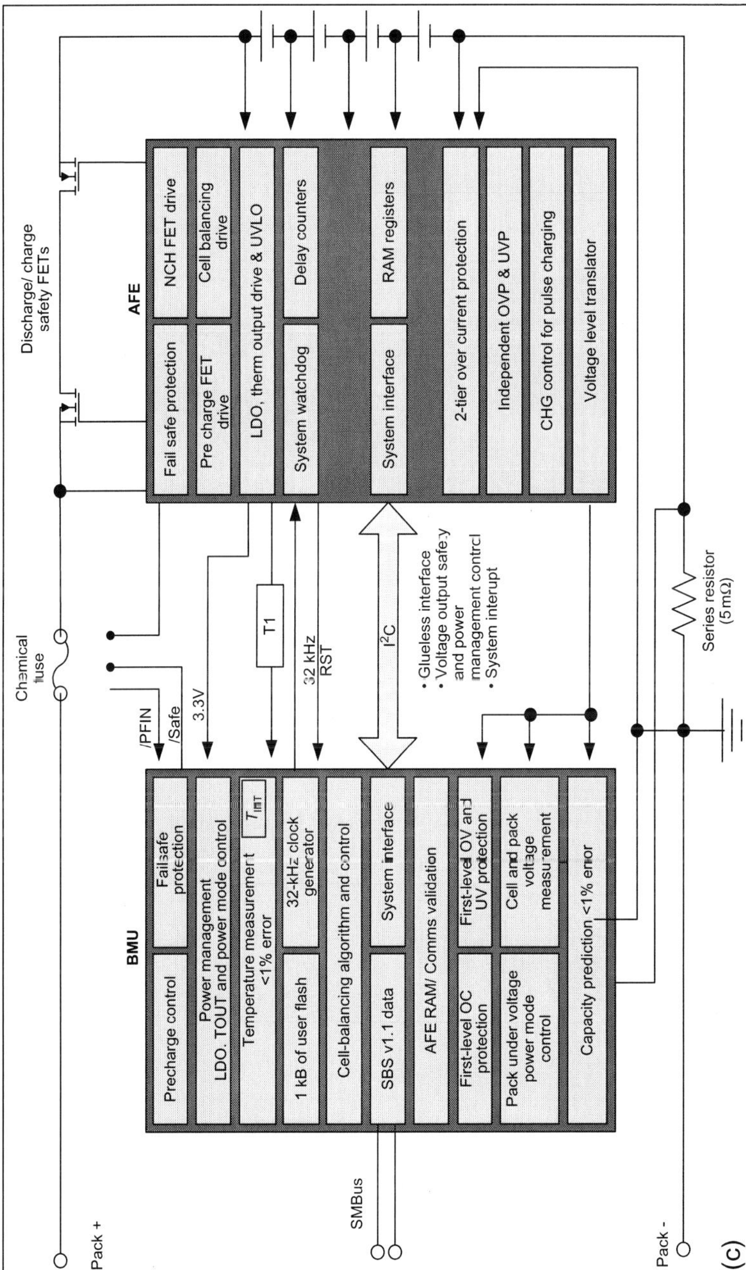

Figure 3.42—Cont'd (c) implementation of IEEE-1625 concepts.

Table 3.9 **An example of a DFMEA table for discrete components**

Problem	Problem level	SEV	Issue	OCC	Protection features	DET	APN
BMU to AFE I^2C latch up	Low safety	6	BMU voltage measurements, RAM verification, and AFE control nonfunctional	2	BMU will detect and count I^2C failures and optionally blow the fuse. Fail counter needs to be able to be periodically cleared based on normal operating conditions. First-level current protection still available	1	12
AFE to BMU VCC < VCC (min)			See Table 3.3				
AFE RST output latched high	Severe customer experience	4	BMU is held in reset	1	BMU no longer functions, so the WDI input will stop and the AFE turn off the FETs and optionally blow the fuse	1	4
AFE RST output latched low	Severe customer experience	5	BMU will never be reset on POR or AFE watchdog fault	2	BMU will not be reset on watchdog fault, but watchdog will still turn off FETs and optionally blow the fuse	5	50
AFE TOUT output latched high	Low customer experience	1	Thermistor will always be powered	1	AFE will consume extra current, but there is no safety issue	7	7
AFE TOUT output latched low	Severe safety	9	Thermistor will never be powered	1	BMU will measure out-of-range temperature, which will cause the fuse to blow. All current and voltage protection methods are still fully functional	1	9
BMU to AFE CLK < CLK (min)	Severe customer experience	5	AFE will not function correctly	1	The AFE watchdog will timeout, causing a BMU reset, the FETs to turn off, and optionally blow the fuse	2	10

BMU sense of CELL latched high	Moderate safety	7	BMU voltage measurements will be 0 V	2	First-level voltage protection is not valid, but second-level overvoltage is a completely separate circuit; so, it is fully operational. All current- and temperature-based protections are fully functional. Additional measurement validation checks can allow the BMU to blow the fuse	2	28
BMU sense of CELL latched low	Severe customer experience	5	BMU voltage measurements will be full scale (high)	2	The BMU will read all cells at overvoltage; so, pack is not able to be charged but can be discharged. Current and temperature protections still functioning. Additional measurement validation checks can allow the BMU to blow the fuse	2	20
AFE XALERT latched high	No issues	1	XALERT will never activate	2	BMU polls STATUS for updates and checks XALERT, so no issues	2	4
AFE XALERT latched low	Low customer experience	2	XALERT will always be activated	2	BMU will always try to check status and clear XALERT, so will consume more power. All protection functions are functioning	2	8

Source: Elder, 2004. Reproduced with permission from Power Electronics Technology Magazine.

Future With the consumer electronics, EVs, and the portable product markets are growing, newer battery technologies may enter the market. One recent indication in the battery market research reports (www.sdle.co.il) is that high-voltage Li-ion chemistries with better anode and cathode materials will come into the market with higher gravimetric energy densities. In addition, new technologies to combine batteries and supercapacitors may be available.

Major application growth areas for batteries are the transportation systems. Battery packs can challenge the designer because they are no longer simple configuration of cells, but could contain so much safety features, intelligence, energy-aware models and selective batteries feeding the host product, and serial data communication and even recycling suggestions.

In all these situations, a simple and useful concept that should be remembered by the designers is that "batteries are like human beings," and they need care and useful intelligence.

References

Agrawal, V., Uthaichana, K., DeCarlo, R.A., Tsoukalas, L.H., 2010. Development and validation of a battery model useful for discharging and charging power control and lifetime estimation. IEEE Trans. Energy Conv. 25 (3), 821–835.

Alberkrack, J., 1996. A programmable in-pack rechargeable lithium cell protection circuit. In: HFPC Conference Proceedings, September 1996, pp. 230–237.

Albertus, P., Couts, J., Srinivasan, V., Newman, J., 2008. A combined model for determining capacity usage and battery size for hybrid and plug-in hybrid vehicles. J. Power Sources 183, 771–782.

Belt, J., Utgikar, V., Bloom, I., 2011. Calendar and PHEV cycle life aging of high energy, li-ion cells containing blended spinel and layered-oxide cathodes. J. Power Sources 196, 10213–10221.

Benchmarq Microelectronics Inc., Duracell Inc., Energizer Power Systems, Intel Corporation, et al., Smart battery system specifications—system management bus specifications, Revision 1.0, February 15, 1995a.

Benchmarq Microelectronics Inc., Duracell Inc., Energizer Power Systems, Intel Corporation, et al., Smart battery system specifications—smart battery data specifications, Revision 1.0, February 15, 1995b.

Benchmarq Microelectronics Inc., Duracell Inc., Energizer Power Systems, Intel Corporation, et al., Smart battery system specifications—smart battery charger specifications, Revision 1.0, June 27, 1996a.

Benchmarq Microelectronics Inc., Duracell Inc., Energizer Power Systems, Intel Corporation, et al., Smart battery system specifications—smart battery selector specifications, Revision 1.0, September 05, 1996b.

Benini, L., Bruni, D., Macii, A., Macii, E., Poncino, M., 2003. Discharge current steering for battery life optimization. IEEE Trans. Comput. 92 (8), 985–995.

Bentley, W.F., Heacock, D.K., 1996. Battery management considerations for multichemistry systems. In: Proceedings of the 11th Annual Battery Conference, Long Beach, CA, January 1996.

Bernardi, D.M., Carpenter, M.K., 1995. A mathematical model of the oxygen recombination lead acid cell. J. Electrochem. Soc. 142 (8), 2631–2642.

Bloom, I., Cole, B.W., Sohn, J.J., Jones, S.A., Polzin, E.G., Battaglia, V.S., Henriksen, G.L., Motloch, C., Richardson, R., Unkelhaeuser, T., Ingersoll, D., Case, H.L., 2001. An accelerated calendar and cycle life study of Li-ion cells. J. Power Sources 101, 238–247.

Bloom, I., Jones, S.A., Polzin, E.G., Battaglia, V.S., Henriksen, G.L., Motloch, C.G., Wright, R. B., Jungst, R.G., Case, H.L., Doughty, D.H., 2002. Mechanisms of impedance rise in high-power, lithium-ion cells. J. Power Sources 111, 152–159.

Bloom, I., Jansen, A.N., Abraham, D.P., Knuth, J., Jones, S.A., Battaglia, V.S., Henriksen, G.L., 2005. Differential voltage analyses of high-power, lithium-ion cells: 1. Technique and application. J. Power Sources 139, 295–303.

Bloom, I., Potter, B.G., Johnson, C.S., Gering, K.L., Christophersen, J.P., 2006. Effect of cathode composition on impedance rise in high-power lithium-ion cells: long-term aging results. J. Power Sources 155, 415–419.

Bloom, I., Walker, L.K., Basco, J.K., Abraham, D.P., Christophersen, J.P., Ho, C.D., 2010. Differential voltage analyses of high-power lithium-ion cells. 4. Cells containing NMC. J. Power Sources 195, 877–882.

Buller, S., Thele, M., Karden, E., Doncker, R.W., 2003. Impedance based non-linear dynamic battery modeling for automotive applications. J. Power Sources 113, 422–430.

Buller, S., Thele, M., Doncker, R.W., Karden, E., 2005. Impedance based simulation models of supercapacitors and Li-ion batteries for power electronic applications. IEEE Trans. Industry Appl. 41 (3), 742–747.

Bundy, K., Karlsson, M., Lindbergh, G., Lundqvist, A., 1998. An electrochemical impedance spectroscopy method for prediction of the state of the charge of a nickel-metal hydride battery at open circuit and during discharge. J. Power Sources 72, 118–125.

Chen, M., Rincon-Mora, G.A., 2006. Accurate electrical battery model capable of predicting run time and I-V performance. IEEE Trans. Energy Conv. 21 (2), 504–511.

Chiasserini, C.B., Rao, R.R., 2001. Energy efficient battery management. IEEE J. Selected Areas Commun. 19 (7), 1235–1245.

Department of Energy, Freedom CAR Battery Test Manual for Power-Assist Hybrid Electric Vehicles, DOE/ID-11069, October 2003, 14 pages.

Elder, G., 2004. IEEE 1625 helps promote safety and reliability. Power Electron. Technol. April, 34–43.

Fang, k., Mu, D., Chen, S., Wu, B., Wu, F., 2012. A prediction model based on artificial neural network for surface temperature simulation of nickel metal hydride battery during charging. J. Power Sources 208, 378–382.

Forgez, C., Do, D.V., Friedrich, G., Morcrette, M., Delacourt, C., 2010. Thermal modeling of a cylindrical LiFePO4/graphite lithium-ion battery. J. Power Sources 195 (9), 2961–2968.

Gao, J., Manthiram, A., 2009. Eliminating the irreversisble capacity loss nof high capacity layered Li $[Li_{0.2}Mn_{0.54}Ni_{0.13}Co_{0.13}]O_2$ cathode by blending with other lithium insertion hosts. J. Power Sources 191, 644–647.

Gao, L., Liu, S., Dougal, R.A., 2002. Dynamic lithium-ion battery model for system simulation. IEEE Trans. Components Packaging Technol. 25 (3), 495–505.

Gholizadeh, M., Salmasi, F.R., 2014. Estimation of state of charge, unknown nonlinearities, and state of health of a Li-ion battery based on a comprehensive unobservable model. IEEE Trans. Ind. Electron. 61 (3), 1335–1344.

Goebel, K., Saha, B., Saxena, A., Celaya, J.R., Christopherson, J.P., 2008. Prognostics in battery health management. IEEE I & M Mag. 11 (4), 33–40.

Gould, C.R., Bingham, C.M., Stone, A., Bentley, P., 2009. New battery model and state-of-health determination through subspace parameter estimation and state observer techniques. IEEE Trans. Veh. Technol. 58 (8), 3905–3916.

Greenleaf, M., Li, H., Zheng, J.P., October 2013. Modeling of LIXFEPO4cathode Li-ion batteries using linear electrical circuit model. IEEE Trans. Sustainable Energy 4 (4), 1065–1070.

Grimnes, S., Martinsen, O.G., 2008. Bioimpedance and Bioelectricity Basics. Elsevier, Academic Press, Amsterdam, 471 pages.

Gu, W.B., Wang, C.Y., Liaw, B.Y., 1995. Numerical modelling of coupled electrochemical and transport processes in lead-acid batteries. J. Electrochem. Soc. 144 (6), 2053–2061.

Gu, W.B., Wang, C.Y., Li, S.M., Geng, M.M., Liaw, B.Y., 1999. Modeling discharge and charge characteristics of nickel-metal hydride batteries. Electrochim. Acta 44, 4525–4541.

Heacock, D., 1998. Enabling smart batteries for portable devices. In: Proceedings of Global Forum on Mobile Handsets (London), June 1998.

Hsieh, G.-C., Chen, L.-R., Huang, K.-S., 2001. Fuzzy-controlled Li-ion battery charge system with active state-of-charge controller. IEEE Trans. Ind. Electron. 48 (3), 585–593.

Huet, F., 1998. A review of impeadance measurements for determination of the state-of-charge or state-of-health of secondary batteries. J. Power Sources 70, 59–69.

Husseni, A., Kutkut, N., Batarseh, I., 2011. A hysteresis model for a lithium battery cell with improved transient response. In: Proceedings of 26th IEEE APEC 2011 Conference, pp. 1790–1794.

Israelsohn, J., Battery management included, EDN, January 18, 2001, pp. 65-74.

Jossen, A., 2006. Fundamentals of battery dynamics. J. Power Sources 154, 530–538.

Karden, E., Buller, S., Doncker, R.W., 2000. A method for measurement and interpretation of impedance spectra for industrial batteries. J. Power Sources 85, 72–78.

Karfthoefer, B., 2005. Measure battery capacity precisely in medical design. Power Electron. Technol. January, 30–38.

Kassem, M., Delacourt, C., 2013. Postmortem analysis of calendar-aged graphite/LiFePO4 cells. J. Power Sources 235, 159–171.

Kassem, M., Bernard, J., Revel, R., Pélissier, S., Duclaud, F., Delacourt, C., 2012. Calendar aging of a graphite/LiFePO4 cell. J. Power Sources 208, 296–305.

Khun, E., Forgez, C., Lagonotte, P., Friedrich, G., 2006. Modeling NiMH battery using Cauer and Foster structures. J. Power Sources 158, 1490–1497.

Kularatna, N., 2008. Digital and Analogue Instrumentation. IET, London Chapter 4, pp. 113-162.

Kutluay, K., Cadirici, Y., Ozkazanc, Y.S., Cadirici, I., 2005. A new on line state of charge estimation and monitoring system for sealed lead-acid batteries in telecommunication power supplies. IEEE Tarns. Ind. Electron. 52 (5), 1315–1327.

Lai, J.S., Rose, M.F., April 1992. High energy density double layer capacitors for energy storage applications. IEEE AES Mag. 7 (4), 14–19.

Landfords, J., Simonsson, D., Sokirko, A., 1995. Mathematical modelling of a lead/acid cell with immobilized electrolyte. J. Power Sources 55, 217–230.

Langnan, P.E., 2000. VRLA battery impedance analysis: cell evaluation via changes in the slope of the impedance curve. Power Qual. Assur. September/October, 24–29.

Li, J., Mazzola, M., Gafford, J., Younan, N., 2012. A new parameter estimation algorithm for an electrical analogue battery model. In: Proceedings of IEEE APEC 2012 Conference, pp. 427–433.

Li, S.E., Wang, B., Peng, H., Hu, X., 2014. An electrochemistry-based impedance model for lithium-ion batteries. J. Power Sources 258, 9–18.

Lin, F.J., Huang, M.S., Yeh, P.Y., Tsai, H.N., Kuan, C.H., 2012. DSP-based probabilistic fuzzy neural network control for Li-ion battery charger. IEEE Trans. Power Electron. 27 (8), 3782–3794.

Lin, H.-T., Liang, T.-J., Chen, S.-M., 2013. Estimation of battery state of health using probabilistic neural network. IEEE Trans. Ind. Informatics 9 (2), 679–685.

Lindahl, P.A., Cornachione, a., Shaw, S.R., 2012. A time-domain least squares approach to electrochemical impedance spectroscopy. IEEE Trans. Instrum. Meas. 61 (12), 3303–3311.

Marcicki, J., Canova, M., Conlisk, A.T., Rizzoni, G., 2013. Design and parametrization analysis of a reduced-order electrochemical model of graphite/LiFePO4 cells for SOC/SOH estimation. J. Power Sources 237, 310–324.

Mauracher, P., Karden, E., 1997. Dynamic modeling of lead/acid batteries using impedance spectroscopy for parameter identification. J. Power Sources 67, 69–84.

Micea,, M.V., Ungurean, L., Carstoiu, G.N., Groza, V., 2011. Online state of health assessment for battery management systems. IEEE Trans. Instrum. Meas. 60 (6), 1997–2006.

Moura, S.J., Stein, J.F., Fathy, H.K., 2013. Battery Health conscious power management in plug-in hybrid electrical vehicles via electrochemical modeling and stochastic control. IEEE Trans. Control Syst. Technol. 21 (3), 679–694.

Nelson, P., Bloom, I., Amine, K., Henriksen, G., 2002. Design modeling of lithium-ion battery performance. J. Power Sources 110, 437–444.

Orazem, M.E., Tribollet, B., 2008. Electrochemical Impedance Spectroscopy. John Wiley, New Jersey 523 pages.

Pan, Y.H., Srinivasan, V., Wang, C.Y., 2002. An experimental and modeling study of isothermal charge/discharge behavior of commercial Ni-MH cells. J. Power Sources 112, 298–306.

Pascoe, P.E., Anbuky, A.H., 2002. The behavior of the coup de fouet of valve-regulated lead-acid batteries. J. Power Sources 111, 304–319.

Pascoe, P.E., Anbuky, A.H., 2005. Standby power system VRLA battery reserve life estimation scheme. IEEE Trans. Energy Conv. 20 (4), 887–895.

Pattipati, B., Sankavaram, C., Pattipati, K.R., 2011. System identification and estimation framework for pivotal automotive battery management system characteristics. IEEE Trans. Sys. Man. Cybern. 41 (6), 869–884.

Phillips Semiconductors, 2003. The I2C bus and how to use it; Document # 98-8080-575-01.

Plett, G.L., 2004a. Extended Kalman filtering for battery management systems of LiPB-based HEV battery packs: part 1. Background. J. Power Sources 134, 252–261.

Plett, G.L., 2004b. Extended Kalman filtering for battery management systems of LiPB-based HEV battery packs: part 2. Modeling and identification. J. Power Sources 134, 262–276.

Plett, G.L., 2004c. Extended Kalman filtering for battery management systems of LiPB-based HEV battery packs: part 3. Sate and parameter estimation. J. Power Sources 134, 277–292.

Rakhmatov, D., Vrudhula, S., Wallach, D.A., 2003. A model for a battery lifetime analysis for organizing applications on a pocket computer. IEEE Trans. VLSI Syst. 11 (6), 1019–1030.

Ramadass, P., Haran, B., Gomadam, P.M., White, R., Popov, N., 2004. Development of first principles capacity fade model for Li-ion cells. J. Electrochem. Soc. 151, A196–A203.

Randles, J.E.B., 1947. Kinetics of rapid electrode reactions. Discuss. Faraday Soc. 1, 11–19.

Rao, R., Vrudhala, S., Rakhmatov, D.N., 2003. Battery modeling for energy aware system design. IEEE Comput. 36 (12), 77–87.

Renhart, W., Magele, c., Schweighofer, B., 2008. FEM based thermal analysis of NiMH batteries for hybrid vehicles. IEEE Tans. Magnetics 44 (6), 802–805.

Rong, P., Pedram, M., 2003. An analytical model for predicting the remaining battery capacity of Li-ion batteries. In: Proceedings of IEEE Design, Automation and Test Conference, Europe, 2003, pp. 1148–1149.

Rong, P., Pedram, M., 2006a. An analytical model for predicting the remaining battery capacity of Li-ion batteries. IEEE Trans. VLSI Syst. 14 (5), 441–451.

Rong, P., Pedram, M., 2006b. An analytical model for predicting the remaining capacity of lithium-ion batteries. IEEE Trans. VLSI Syst. 14 (5), 441–451.

Ruetschi, P., 2004. Aging mechanisms and service life of lead acid batteries. J. Power Sources 127, 33–44.

Sacarisen, P.S., Parvereshi, J., 1995. Lead acid fast charge controller with improved battery management techniques. In: South Conference, March 1995.

Schwartz, P., 1995. Battery management. In: Portable by Design Conference, pp. 525–547.

Schweighofer, B., Raab, K.M., Brasseur, G., August 2003. Modeling of high power automotive batteries by the use of an automated test system. IEEE Trans. Instrum. Meas. 52 (4), 1087–1091.

Shahriari, M., Farrokhi, M., 2013. Online state-of-health estimation of VRLA batteries using state of charge. IEEE Trans. Ind. Electron. 60 (1), 191–202.

Shen, W.X., Chau, K.T., Chan, C.C., Lo, E.W.C., 2005. Neural network based residual capacity indicator for nickel metal hydride batteries in electric vehicles. IEEE Trans. Veh. Technol. 54 (4), 1705–1712.

Singh, P., Fennie, C., Reisner, D., 2004. Fuzzy logic modeling of state-of-charge and available capacity of nickel/metal hydride batteries. J. Power Sources 136, 322–333.

Song, L., Evans, J.W., 2000. Electrochemical-thermal model of lithium polymer batteries. J. Electrochem. Soc. 147 (6), 2086–2095.

Stevanatto, L.C., Brusamarello, V.J., Tairov, S., April 2014. Parameter identification and analysis of uncertainty measurements of lead-acid batteries. IEEE Trans. Instrum. Meas. 63 (4), 761–768.

Stiaszny, B., Ziegler, J.C., Krauζ, E.E., Schmidt, J.P., Ivers-Tiffée, E., 2014. Electrchemical characterization and post-mortem analysis of aged $LiMnO_4 \cdot Li(Ni_{0.5}Mn_{0.3}Co_{0.2})O_2$/graphite li-ion batteries, part 1. J. Power Sources 251, 439–450.

Świerczyński, M., Store, D.I., Stan, A., Teodorescu, R., Sauer, D.U., 2014. Selection and performance-degradation modeling of $LiMO_2/LI_4$ TI $_5O_{12}$ and Li FE PO_4/C battery cells as suitable energy storage systems for grid integration with wind power plants: an example for the primary frequency regulation service. IEEE Trans. Sustainable Energy 5 (1), 90–101.

Tan, X., Baras, J., 2005. Adaptive identification and control of hysteresis in smart materials. IEEE Trans. Automatic Control 50 (6), 827–839.

Tang, X., Zhang, X., Koch, B., Frisch, D., 2008. Modeling and estimation of nickel metal hydride battery hysteresis for SOC estimation. In: Proceedings of IEEE International Conference on Prognostics and Health Management, pp. 1–12.

Thele, M., Buller, S., Sauer, D.U., Doncker, R.W., Karden, E., 2005. Hybrid modeling of lead acid batteries in frequency and time domain. J. Power Sources 144, 461–466.

Unitrode Inc. Improved Charging Methods for Lead-Acid Batteries Using the UC 3906, Application Note U-104.

Urbin, M., Hinaje, M., RaëS, Davat, B., Desperez, P., 2010. Energetical modeling of lithium-ion batteries including porosity effects. IEEE Trans. Energy Conv. 25 (3), 862–872.

US Department of Energy, Battery Test Manual for Plug_in Hybrid Electric Vehicles, Rev 2, INL/EXT-07-12536, December 2010, 56 pages.

Valoen, L.O., Sunde, S., Tunold, R., 1997. An impedance model for electrode processes in metal hydride electrodes. J. Alloys Compounds 253 (254), 656–659.

Wang, J., Liu, P., Hicks-Garner, J., Sherman, E., Soukiazian, S., Verbrugge, M., Tataria, H., Musser, J., Finamore, P., 2011. Cycle-life model for graphite-LiFePO4 cells. J. Power Sources 196, 3942–3948.

Warburg, E., 1899. Über das Verhalten sogenannte unipolarisierbare Electroden gegen Wechselstrom. Ann. Phys. Chem. 67, 493–499.

Wright, R.B., Motloch, C.G., Belt, J.R., Christophersen, J.P., Ho, C.D., Richardson, R.A., Bloom, I., Jones, S.A., Battaglia, V.S., Henriksen, G.L., Unkelhaeuser, T., Ingersoll, D., Case, H.L., Rogers, S.A., Sutula, R.A., 2002. Calendar- and cycle-life studies of advanced technology development program generation 1 lithium-ion batteries. J. Power Sources 110, 445–470.

Wu, Q., Qiu, Q., Pedram, M., 2000. An interleaved dual-battery power supply for battery operated electronics. In: Proceedings: 2000 Conference, Asia and South Pacific Design Automation, pp. 387–390.

Wu, B., Mohammed, M., Brigham, D., Elder, R., White, R.E., 2001. A non-isothermal model of a nickel-metal hydride cell. J. Power Sources 101, 149–157.

Zhang, J., Ci, S., Sharif, H., Alahmad, M., 2010. An enhanced circuit-based model for single-cell battery. In: Proceedings of 25th IEEE APEC 2010 Conference, pp. 672–675.

Bibliography

Kalman, R.E., 1960. A new approach to linear filtering and prediction problems. Trans. ASME—J. Basic Eng., Ser. D 82, 35–45.

The Seminal Kalman Filter Paper. http://www.cs.unc.edu/~welch/kalman/kalmanPaper.html, 1960 (accessed 20.05.03).

Haykin, S., 1996. Adaptive Filter Theory, third ed. Prentice-Hall, Upper Saddle River, NJ.

Haykin, S., 2001. Kalman filters. In: Haykin, S. (Ed.), Kalman Filtering and Neural Networks. Wiley/Interscience, New York, pp. 1–22.

Burl, J., 1999. Linear Optimal Control: H2 and H∞ Methods. Addison Wesley, Menlo Park, CA.

Kiehne, H.A., 2003. Battery Technology Handbook. Marcel Dekker Inc, New York.

Capacitors as energy storage devices—simple basics to current commercial families

4

Kosala Gunawardane

4.1 Capacitor fundamentals

A capacitor is a device that stores electrical charge and releases them when it is required by the circuit. The arrangement shown in Figure 4.1 illustrates the simplest configuration of a capacitor: two parallel conducting plates separated by a dielectric material (electric insulator) where parallel plates have an area A and separation d.

The charge on either one of the parallel conductors in the capacitor is zero in the uncharged state. When a DC voltage, V, is applied to the capacitor electrodes, an electrical charge builds up on the capacitor plates with electrons producing a positive charge ($+Q$) on one plate and an equal and opposite negative charge ($-Q$) on the other plate. Current, I, flow continues until the potential difference between the two electrodes becomes equal to the supply voltage, in a fully-charged capacitor.

The ratio $\frac{Q}{V}$ is defined as the capacitance,

$$C = \frac{Q}{V} \tag{4.1}$$

with capacitance units of farad (F).

The behavior of a capacitor is described by the equation,

$$i(t) = \frac{\mathrm{d}q}{\mathrm{d}t} = C\frac{\mathrm{d}v}{\mathrm{d}t} \tag{4.2}$$

where t is the time, i is the current through the capacitor at time t, and v is the voltage across the capacitor leads at time t. Integrating Eqn (4.2), capacitor voltage can be obtained as,

$$v = \frac{1}{C}\int_0^t i(t)\,\mathrm{d}t + v(0) \tag{4.3}$$

where $v(0)$ is the initial voltage of the capacitor at the beginning of the charging process.

As shown in Figure 4.1, the simplest form of a capacitor consists of two conducting plates of area, A, which are parallel to each other and separated by distance d by a dielectric material such as ceramic, polymer, or aluminum oxide. In such a

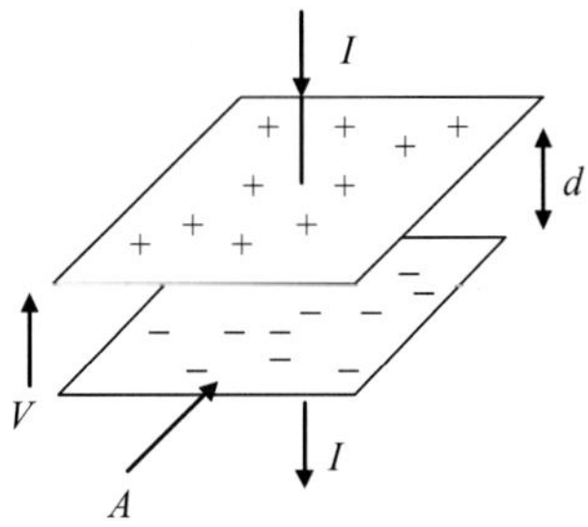

Figure 4.1 Construction of a parallel-plate capacitor.

parallel-plate capacitor, the electric energy is stored statically in the electric field between two electrodes.

The capacitance formula for a parallel-plate capacitors is given by,

$$C = \frac{\varepsilon A}{d} \tag{4.4}$$

Thus capacitance, increases with electrode area (A) and increases with permittivity, ε, of the dielectric material, but decreases with the electrode separation, d. Conventional capacitors yield capacitance values in the range of 5 pF to 1 F with a voltage range of 10 to over 400 V. Commonly available large capacitors go up to about 100,000 μF in general. Larger a capacitor and its voltage rating, they are more expensive.

4.1.1 Capacitor charging

A typical charging circuit of a capacitor is shown in Figure 4.2 using a constant DC voltage V_{in} and a series resistance R. This resistor represents the equivalent series resistance (ESR) of the capacitor, resistance of external leads and connections, and any deliberately introduced resistance. Assume that the initial voltage of the capacitor, $V(0)$, is zero.

Using Kirchoff's voltage law,

$$V_{\text{in}} = i(t)R + \frac{1}{C}\int_0^t i(t)'\,\mathrm{d}t' \tag{4.5}$$

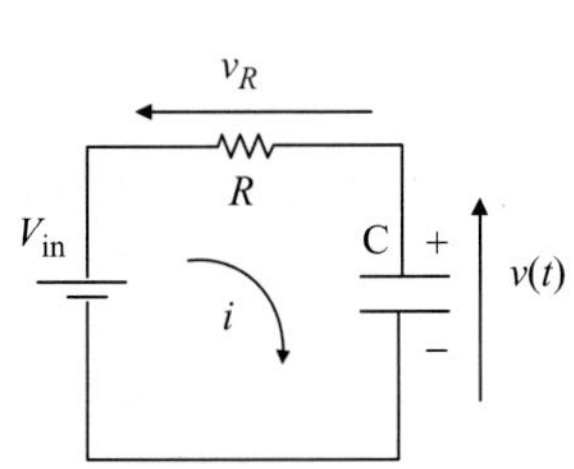

Figure 4.2 Charging a capacitor from a voltage source.

Solving Eqn (4.5) for current we obtain,

$$i(t) = \frac{V_{\text{in}}}{R} \exp\left[\frac{-t}{RC}\right] \tag{4.6}$$

Using Eqn (4.6), the capacitor voltage at a given time during the charging process is,

$$v(t) = V_{\text{in}}\left[1 - \exp\left[\frac{-t}{RC}\right]\right] \tag{4.7}$$

Figure 4.3(a) shows the voltage and current behavior of the capacitor during the charging process.

The rate of charge of the capacitor depends on the product of R and C. This is usually called the time constant, τ,

$$\tau = RC \tag{4.8}$$

It is also important to note that when a capacitor is charged from 0 V to its rated voltage, while the capacitor stores an energy content of $\frac{1}{2}QV$, the resistive elements in the charging loop dissipates the same amount of energy, irrespective of the value of the total loop resistance. Compared to this case, when the same charge is transferred into an electrochemical battery with an electrochemical voltage of V volts across the anode-cathode pair (assumed to be constant), it stores an energy of QV during the charging process (assuming that no losses inside the battery). This means that an ideal battery can be charged with theoretically 100% efficiency, a fully discharged capacitor charging can achieve only a 50% charging efficiency. However, if the charging of a capacitor occurs from a non-zero initial value, the charging loop losses are less. This principle is practically used in AC-DC converters where a large capacitor is used to smooth out the ripple, but it discharges only a small percentage of the total stored energy in a cycle of charging and discharging.

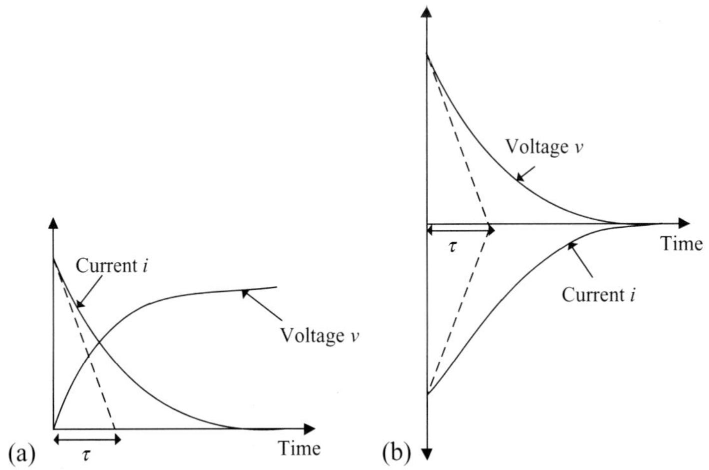

Figure 4.3 Capacitor voltage-current behavior: (a) charging process; (b) discharging process.

4.1.2 Capacitor discharging

If the DC supply is replaced with a short circuit, the capacitor will discharge through the load resistance R. When discharging, the current behaves the same as for charging, but flows in the opposite direction. Voltage across the capacitor will decay exponentially to zero. Equations for both voltage and current discharge are given by,

$$v(t) = V_f \exp\left[\frac{-t}{RC}\right] \tag{4.9}$$

$$i(t) = \frac{V_f}{R} \exp\left[\frac{-t}{RC}\right] \tag{4.10}$$

where V_f is the voltage at the starting of capacitor discharge (see Figure 4.3(b)).

4.1.3 Capacitor energy storage

The total energy stored in a capacitor, E, can be shown as the product of the capacitance, C, and the square of the voltage, V.

$$E = \frac{1}{2} C V^2 \tag{4.11}$$

According to the law of conservation of charge, if two pre-charged capacitors are connected together as shown in Figure 4.4, the total charge on the parallel combination is equal to the sum of the original charges on the capacitors.

The two capacitors shown in Figure 4.4, C_1 and C_2, are charged to voltages V_1 and V_2, respectively. When the switch is closed, an impulse of current flows and the charge is redistributed between the two capacitors. The total charge at the beginning of the parallel combination of the two capacitors is,

$$Q_T = C_1 V_1 + C_2 V_2 \tag{4.12}$$

This charge is distributed between the two capacitors, so that the total charge across the two capacitors remains the same.

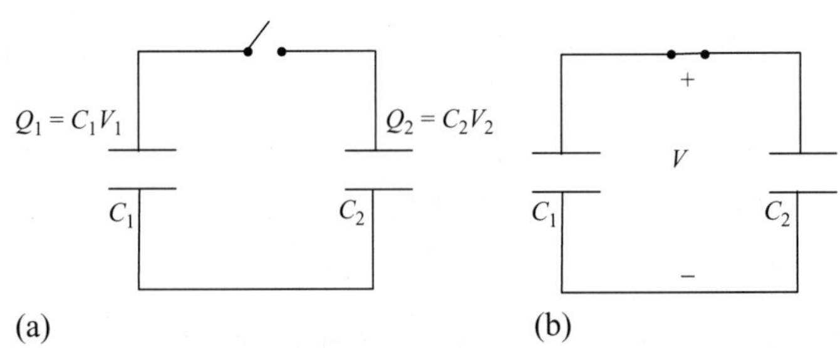

Figure 4.4 Charge distribution between capacitors.

$$Q_T = C_1V + C_2V = (C_1 + C_2)V \tag{4.13}$$

where V is the new voltage across the parallel combination.

$$V = \frac{C_1V_1 + C_2V_2}{C_1 + C_2} \tag{4.14}$$

The total energy loss E_L due to paralleling is,

$$E_L = \left[\frac{1}{2}C_1V_1^2 + \frac{1}{2}C_2V_2^2\right] - \frac{1}{2}(C_1 + C_2)V^2 \tag{4.15}$$

which gives,

$$E_L = \frac{1}{2}\frac{C_1C_2}{C_1 + C_2}(V_1 - V_2)^2 \tag{4.16}$$

If C_2 is very much smaller than C_1 the relationship in Eqn (4.16) can be simplified as,

$$E_L \approx \frac{1}{2}C_2(V_1 - V_2)^2 \tag{4.17}$$

4.1.4 Capacitor models

In practical circumstances, the actual capacitor model is more complex as in Figure 4.5 (a). Figure 4.5(b) shows a simplified version of this equivalent circuit model.

ESR and equivalent series inductance (ESL) indicate the ESR (composed of the lead resistances, electrodes, and terminating resistances) and equivalent series inductance respectively. These two parameters are important design parameters when the capacitor is used at higher frequencies or in high-voltage situations. R_{ac} and R_{dc} represent the equivalent resistance due to AC dielectric losses and leakage resistance due to DC dielectric losses, respectively (Kularatna, 2008). In certain circuits such as precision integrators or sample/hold circuits, the dielectric absorption of the capacitor is

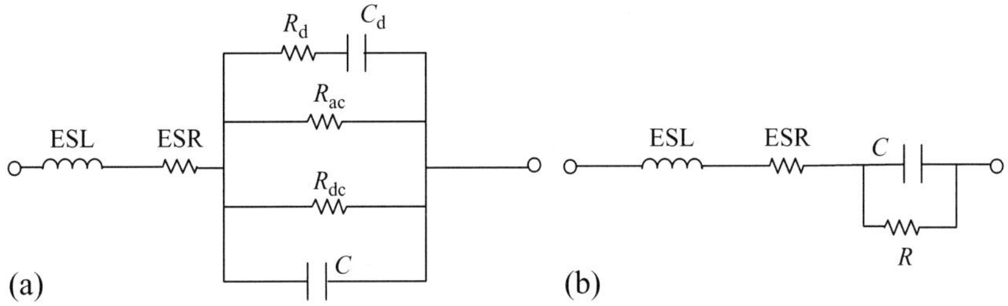

Figure 4.5 Capacitor models: (a) a comprehensive equivalent circuit; (b) a simplified version (Kularatna, 2008).

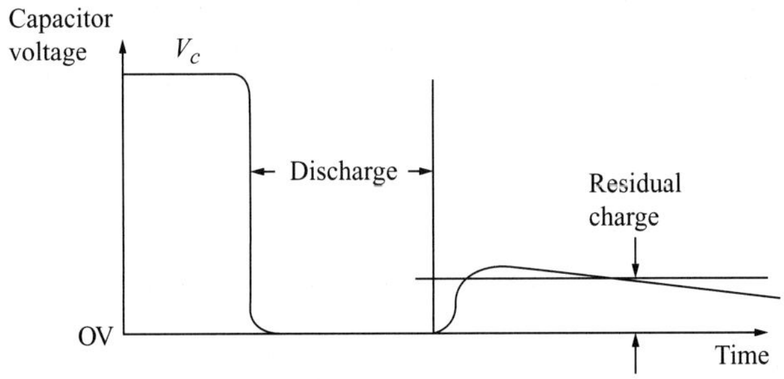

Figure 4.6 Effect of dielectric absorption (Kularatna, 2008).

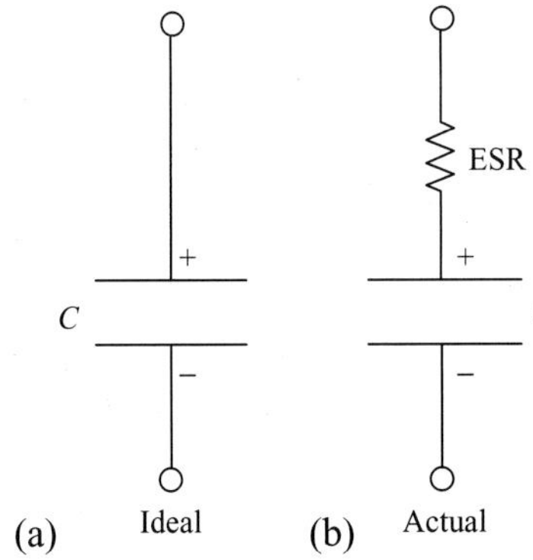

Figure 4.7 (a) Ideal capacitor; (b) simplest first-order model.

an important parameter. Dielectric absorption is a property, in which the dielectric material does not become polarized instantly. This causes a memory effect and is represented by R_d and C_d in Figure 4.5(a). Dielectric absorption property prevents a capacitor from totally discharging, even when short-circuited for a short period of time (http://en.wikipedia.org/wiki/Dielectric_absorption). In such a situation, if the capacitor is charged and then discharged, and left open, it will recover some of its charge which will gradually reappear as a small amount of DC voltage at the terminals of the capacitor due to the parasitic effect of R_d and C_d. Figure 4.6 shows the effect of dielectric absorption when a capacitor is fully charged and then discharged by a short circuit (Kularatna, 2008; http://en.wikipedia.org/wiki/Dielectric_absorption).

In ideal conditions, the capacitor is simply represented as the capacitance, C, as shown in Figure 4.7(a). However a simplified first-order model of the capacitor which has capacitance and ESR as shown in Figure 4.7(b) is used in most DC applications.

4.2 Capacitor types and their properties

The capacitor is one of the most critical passive components used in electronic circuits to provide smoothing, buffering, filtering, bypassing, DC blocking, etc. Capacitors have the capability of storing electrical charges and releasing them when required by the circuit. There is a wide variety of capacitor types, based on the capacitor construction and the technology used. The most commonly used capacitor types are,

1. Film capacitors
2. Ceramic capacitors
3. Electrolytic capacitors

Film capacitors are general-purpose capacitors that are used in applications requiring tight capacitance tolerances and very low leakage currents. They have very low ESR and ESL values and much more stable electrical parameters. Film capacitors have relatively large sizes and weights, but have much higher surge or pulse load capabilities (http://en.wikipedia.org/wiki/Film_capacitor). Film capacitors are not polarized, so they can be used in AC voltage applications (http://www.cornell-dubilier.com [Aluminum electrolytic capacitors]).

Ceramic capacitors are also fixed value capacitors that are suited for AC applications as they are non-polarized (Aluminum electrolytic capacitors). Electrical behavior of the capacitor is defined by the composition of the ceramic material. The applications of the ceramic capacitors are divided into two stability classes (http://en.wikipedia.org/wiki/Types_of_capacitor):

- Class 1 ceramic capacitors have high stability and low losses for resonant circuit application.
- Class 2 ceramic capacitors have high volumetric efficiency for buffer, by-pass, and coupling applications.

Electrolytic capacitors are available in several types as aluminum, tantalum, and niobium versions.

Aluminum electrolytic capacitors are available in the range of less than 1 μF to 1 F with working voltages up to several hundred DC volts, resulting in more capacitance and energy storage per unit volume (Aluminum electrolytic capacitors; Eliasson and Daly, 2003; http://en.wikipedia.org/wiki/Types_of_capacitor). Aluminum electrolytic capacitors are polar devices having distinct positive and negative terminals. They are capable of providing a high ripple current capability together with a high reliability.

Tantalum capacitors are generally available in low-voltage versions up to several hundred microfarads. Because of their smaller size for a given capacitance and lower impedance at high frequencies they are popular in miniature applications (http://en.wikipedia.org/wiki/Types_of_capacitor). This capacitor category has a lower energy density and is produced to tighter tolerances than aluminum electrolytics (http://en.wikipedia.org/wiki/Types_of_capacitor).

Niobium capacitor were first introduced as cheaper alternative to tantalum capacitors. But, the materials suppliers could not make a capacitor-grade niobium powder that matched the capacitance/voltage (*CV*) values required to compete with tantalum. Limitations of niobium capacitors include lower volumetric efficiency (*CV* in a given capacitor size), lower voltage range and higher DC leakage compared to tantalum capacitors (Roos, 2012).

Each capacitor type is suitable for some applications but not adequate for others. A summary of essential features that is given in Table 4.1 (http://www.hitachiaic.com/docs/technical/1-Capacitors/Basics_of_capacitors.pdf).

Table 4.1 **Characteristics of various capacitor types (http://www.hitachiaic.com/docs/technical/1-Capacitors/Basics_of_capacitors.pdf)**

	Aluminum	Tantalum	Niobium	Ceramic	Film
Dielectric	Aluminum oxide	Tantalum pentoxide	Niobium pentoxide	Based on barium titanate, etc.	Polyester, polypropylene, etc.
Shape	Screw terminal type, snap-in type, lead terminal type, chip type	Chip type, Dip type	Chip type	Chip type, Dip type	Dip type, case type for SMD
Advantages	Cheap, small-size and large-capacity	Small size and comparatively large capacitance, semi-permanent service life	Small size and comparatively large capacitance, semi-permanent service life	Small-size, no polarity	Good characteristics, can be made for low- to high-voltage applications, high reliability
Disadv.	Short service life in hot environment, large tolerance, polarized	Can be used with some voltage leeway, polarity	Can be used with some voltage leeway, polarity	Large changes in capacitance due to changes in temperature and DC voltage	Large outside dimensions

Figure 4.8 shows some common capacitor types. Figure 4.9 depicts a summary of ranges of capacitance and rated voltage of various types of capacitors (Basics of Capacitors).

Some typical capacitor applications in the electronics industry are (www.holystonecaps.com):

- *DC-blocking capacitor*: When fully-charged, the capacitor blocks the flow of DC current between two parts of a circuit.
- *Capacitor as an energy (charge) buffer*: Capacitors are used as a charging unit and the energy is stored and released when required by the circuit.
- *Bypass capacitor*: The reactance of a capacitor decreases as the frequency increases. In certain applications this property is utilized by placing a capacitor in parallel with other components to bypass a specified frequency.
- *Capacitor as a filter*: In filter circuits such as, low-pass, high-pass, band-pass capacitors are used as the main filter elements.
- *Coupling capacitor*: The ability of a capacitor to pass AC signal, allows it to couple sections of an electronic circuit that require DC isolation.

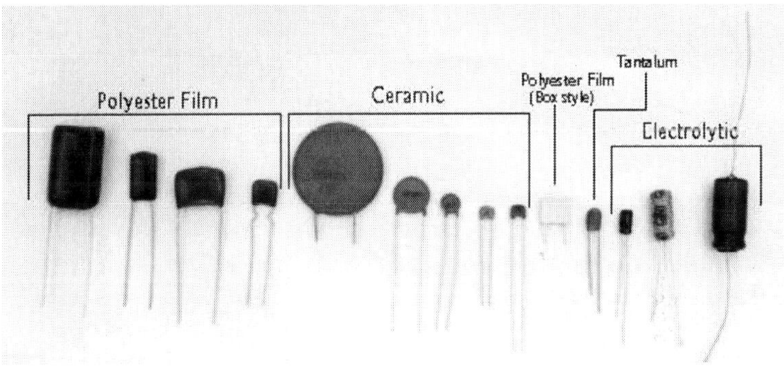

Figure 4.8 A sample of some commercially available capacitors.

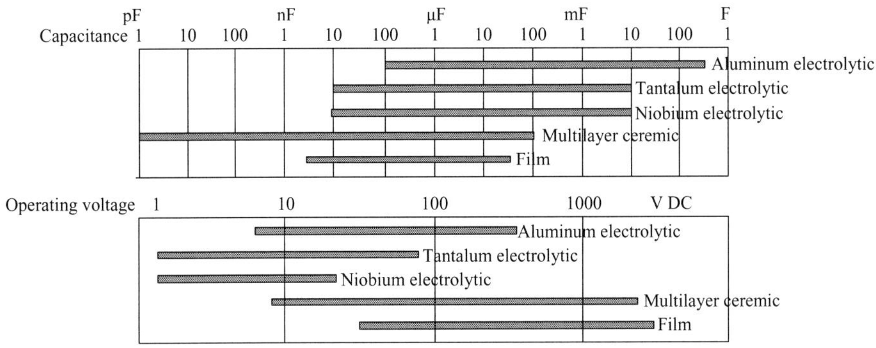

Figure 4.9 Ranges of capacitance and operating voltage of various capacitors (http://www.hitachiaic.com/docs/technical/1-Capacitors/Basics_of_capacitors.pdf).

- *Decoupling capacitor*: A capacitor is used to minimize noise and disturbances to the logic signal. Typically in such applications, the capacitor is placed very close to the IC output and serves as a local energy source to provide some extra current.
- *Snubber capacitor*: A snubber capacitor is used to limit the high voltage transient across the circuit, which is frequently used in power electronic circuits.

In real-time applications, capacitors do not show their ideal characteristics due to the effect of factors such as parasitic resistance, inductance and capacitance variations over temperature and voltage. Choosing the wrong capacitor can lead to unpredictable circuit behavior, circuit instability, excessive power dissipation or noise (Morita, 2010). A capacitor has a rated capacitance, and a tolerance which is expressed as a percentage decrease or increase of the rated capacitance. The standard capacitance tolerance is $\pm 20\%$. In general, capacitors can have a tolerance in the range of -20% to as high as $+80\%$ (http://www.cornell-dubilier.com; http://www.electronics-tutorials.ws/capacitor/cap 3.html; http://en.wikipedia.org/wiki/Types_of_capacitor#Leakage_current).

The capacitance of a capacitor varies with temperature as the changes in temperature affects the dielectric properties. Furthermore, this variation is dependent to a small degree on the rated voltage and capacitor size. The normal working range for most capacitors is -30 °C to $+125$ °C with nominal voltage ratings (http://www.cornell-dubilier.com; http://www.electronics-tutorials.ws/capacitor/cap 3.html). ESR of a capacitor also increases with temperature.

The effective capacitance decreases as frequency increases. Impedance of an ideal capacitor is expected to drop with the inverse of the frequency, based on the relationship given in Eqn (4.18). However, the impedance of a non-ideal capacitor model takes a dip at self-resonance, f_0 and then begins to increase with frequency as illustrated in Figure 4.10. The inductive component (ESL) takes over at frequencies greater than f_0 (Kularatna, 2008; http://www.maximintegrated.com/app-notes/index.mvp/id/3660).

$$X_c = \frac{1}{2\pi f C} \tag{4.18}$$

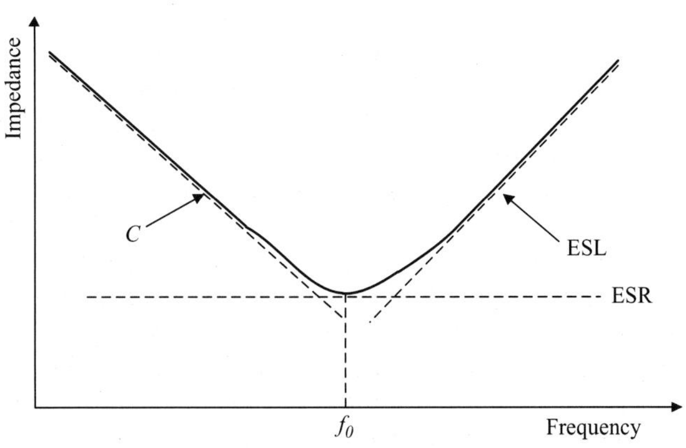

Figure 4.10 Frequency dependency of a non-ideal capacitor (Kularatna, 2008).

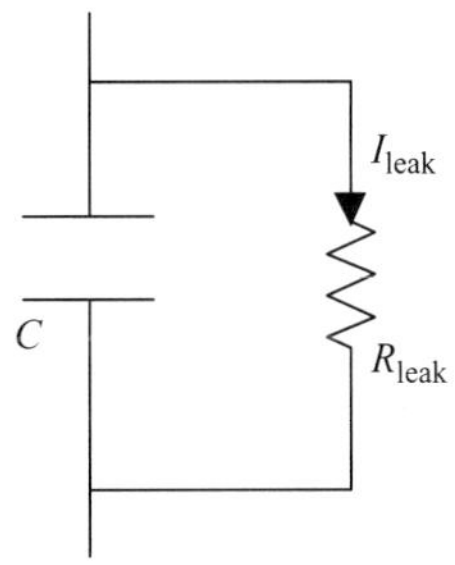

Figure 4.11 Leakage model of a capacitor.

The dielectric material of the capacitor is not an ideal insulator, as it results in a very small current flowing through the dielectric due to the influence of the powerful electric fields built up by the charge on the plates (http://www.electronics-tutorials.ws/capacitor/cap 3.htm). This small DC current flow, in the range of nanoamperes, is called the leakage current. However, electrolytic capacitors may have very high leakage currents typically in the order of about 5-20 μA (http://www.electronics-tutorials.ws/capacitor/cap 3.html). Leakage can be represented as equivalent to a resistor in parallel with the capacitor as shown in Figure 4.11. Furthermore, the capacitor leakage current is also dependent upon the charging period, applied voltage, and temperature.

4.3 Ragone plot

Energy storage devices (ESD) are characterized by the energy and the available power for a load (Christen and Carlen, 2000). The Ragone plot is a chart used for performance comparison of various energy storing devices where the values of energy density (in Wh/kg) are plotted versus power density (in W/kg) (http://en.wikipedia.org/wiki/Ragone_chart). ESDs such as batteries, capacitors, supercapacitors, flywheels, and magnetic ESDs are located in characteristic regions in the power-energy plane in the Ragone plot. Both axes are usually presented in logarithmic scale, which allows comparing performance of very different devices. As shown in Figure 4.12, this plot compares the performances of various ESDs like batteries, supercapacitors, and conventional capacitors (Schneuwly and Gallay, 2000). The choice of the ESD depends particularly on the power density required by the application.

The sloping lines (isotherms of characteristic time constant) on the Ragone plot indicate the relative time taken to get the charge in or out of the device (Christen and Carlen, 2000). At one extreme, capacitor can pump power in and out within microseconds; whereas at the other extreme, devices which have a poor dynamic performance take a few hours to generate and deliver their energy. Supercapacitors are somewhere in between and provide a reasonable compromise between the two.

Portable devices have a wide range of energy-storage needs. Some portable devices require high-power, high-energy for hours of run-time, whereas others require only a

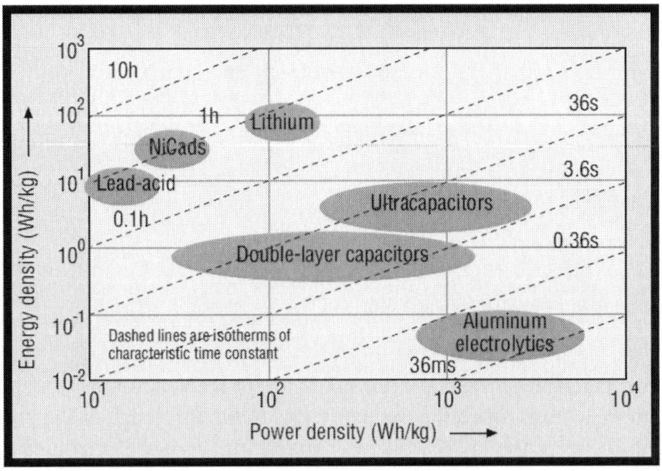

Figure 4.12 The Ragone plot showing energy density vs. power density (Kim, 2003).

few seconds of operation before recharging (Dispennette, 2005). As illustrated in the Ragone plot, batteries have high energy densities but low power densities making them suitable for applications which need longer runtime. Batteries suffer from several drawbacks, such as limited lifetime and expensive maintenance. In contrast, capacitors have rather high power densities but, low energy densities with much longer lifetimes and cycle life (Christen and Carlen, 2000).

References

Aluminum electrolytic capacitors. Cornell Dubilier. Available: http://www.cornell-dubilier.com

Basics of Capacitors. Hitachi AIC Inc. Available: http://www.hitachiaic.com/docs/technical/1-Capacitors/Basics_of_capacitors.pdf

Capacitor applications. Holystone, Available: www.holystonecaps.com.

Capacitor characteristics. Available: http://www.electronics-tutorials.ws/capacitor/cap_3.html.

Christen, T., Carlen, M.W., 2000. Theory of Ragone plots. J. Power Sources 91, 210–216.

Dispennette, J., 2005. Ultracapacitors Bring Portability to Power. Power Electronics Technology.

Eliasson, L., Daly, J., 2003. Optimize Electrolytic Capacitor Selection. Power Electronics Technology.

Kim, Y., 2003. Ultracapacitor Technology Powers Electronics Circuits. Power Electonics Technology.

Kularatna, N., 2008. Electronic Circuit Design: From Concept to Implementation. CRC Press.

Morita, G., 2010. Capacitor selection guidelines for analog devices, Inc. LDOS, Analog Devices.

PCB layout techniques to achieve RF immunity for audio amplifiers. Maxim Integrated, July, 2006, Available: http://www.maximintegrated.com/app-notes/index.mvp/id/3660.

Roos, G., 2012. Niobium capacitors slow to take hold. Digi-Key Corporation. Available: http://www.digikey.com/supply-chain-hq/us/en/articles/.

Schneuwly, A., Gallay, R., 2000. Properties and applications of supercapacitors from the state-of-the-art to future trends. Proceeding of PCIM.

Electrical double-layer capacitors: fundamentals, characteristics, and equivalent circuits

5

Co-Author - Jayathu Fernando

5.1 Introduction

Previous chapter provides a practical overview of common capacitor families used in electronic circuits and a revision of simple concepts related to capacitors. In general, common capacitor families do not have devices beyond several hundred thousand microfarads, but these devices come with a wide range of DC voltage ratings from 5 V to several kilovolts. In general, these devices are used in well-known circuit applications such as DC blocking, filtering, temporary energy storage, and building tuned circuits.

However, when we consider energy storage requirements similar to batteries, relatively a newer family of capacitors called electrochemical (EC) capacitors has entered the market from about late 1980s, and within the last decade many manufacturers have introduced many variations in very high capacitance, low voltage devices. These EC devices, variously referred to by manufacturers as supercapacitors, ultracapacitors, or electrochemical double-layer capacitors (EDLCs), store charge in an electrical double layer at the interface between a high surface area electrode and an electrolyte. Current commercial devices are relatively low voltage single-cell devices with DC voltage ratings up to about 16 V or series-connected modules with voltage ratings up to and over 100 V DC. As we have indicated in Chapter 1, supercapacitors are very much capable to provide short-duration power bursts, and they are having life cycles of half a million to several millions, compared to batteries. As a general guideline, supercapacitors are over 10 times more capable in terms of specific power (W/kg) rating and about 10 times smaller compared to rechargeable batteries in terms of gravimetric energy density (Wh/kg) rating.

As at mid-2014, there are over 70 supercapacitor manufacturers in the world according to market search reports. Market is expected to achieve a compound average growth rate of 30% and the overall market will be close to 6 billion US Dollars.

This chapter provides a general discussion on EDLC principles, modeling, equivalent circuits, device properties, and characterization.

5.2 Historical background

Hermann von Helmholtz, a German physicist first described the concept of double-layer capacitance in 1853. General Electric Company in 1957 first patented (Rightmire, 1966) the electrochemical capacitor based on the double-layer

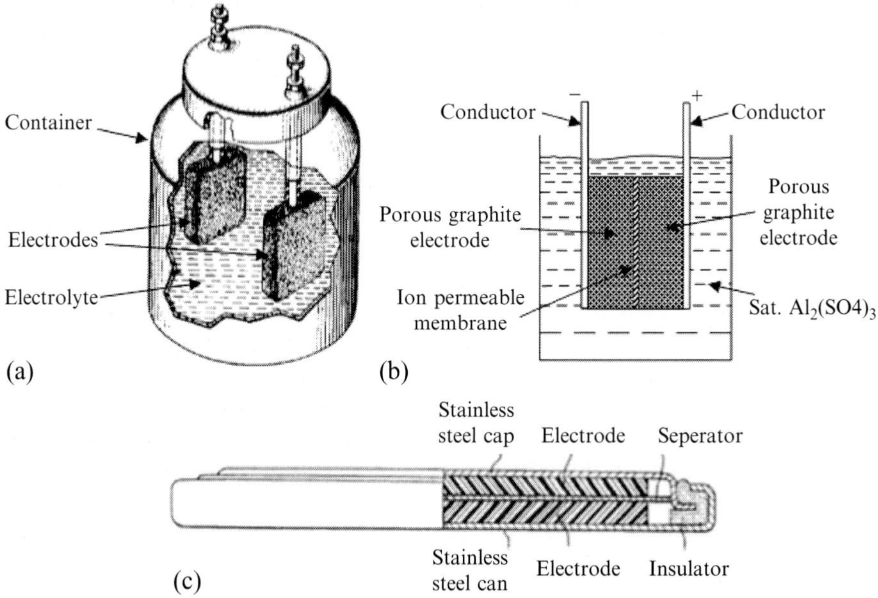

Figure 5.1 Early supercapacitors (a) from 1957 patent, (b) from 1966 patent, and (c) early button cell supercapacitor.
Source: (a) Becker (1957), (b) Rightmire (1966), and (c) Endo et al. (2001).

capacitance structure (Figure 5.1(a)), and this capacitor consisted of porous carbon electrodes with a double-layer capacitance. Commercial double-layer capacitors originated in the Standard Oil of Ohio (SOHIO) Research Center (Cleveland, USA) in 1961 and 1962 and a patent was granted to Rightmire titled "electrical energy storage apparatus" (Becker, 1957). Figure 5.1(b) depicts the concept. Development work continued and the first advertising brochure was released in 1969. SOHIO also utilized the double-layer capacitance of high surface area carbon materials, but in a nonaqueous solvent containing a dissolved tetra-alkyl-ammonium salt electrolyte. At this time, SOHIO acknowledged that the *double-layer* at the interface behaves like a capacitor of relatively high specific capacity. SOHIO went onto patent a disc-shaped capacitor in 1970 utilizing a carbon paste soaked in an electrolyte. Due to lack of sales, in 1971, SOHIO halted their development efforts and later licensed the technology to Nippon Electric Company (NEC) who developed and marketed commercial double-layer products. NEC of Japan licensed the technology from SOHIO and introduced the first EC products to the marketplace as memory backup devices.

Another group, Carborundum Company of Niagara Falls, New York (formerly SOHIO) started selling double-layer capacitor products under the name MAXCAP and these devices are still sold by a company named Kanthal Globar. Several series with capacitance values from about 0.047 to 1 F with ESR values from fractional ohms to few ohms, high temperature range from −25 to 85 °C, small diameter or low height packages are offered. Product brochures describing the MAXCAP products show a similar construction to the NEC capacitors and are another source of early products demonstrating the success of the double-layer effect (Endo et al., 2001). These early devices, due to high ESR compared to modern day supercapacitors, are only useful for limited applications such as memory backup or similar applications where high ESR could be tolerated at low load currents.

During the 1980s, Matsushita Electric Industrial Co., Japan, now known as Panasonic, patented methods for producing double-layer capacitors containing improved electrodes (Endo et al., 2001). One, made from activated carbon (AC) fibers woven into a fabric, was the initial basis for the line of electric double-layer capacitor called Gold capacitors. Figure 5.1(c) shows the cross-sectional view of practical button cell using an AC fiber as a polarizable electrode (Endo et al., 2001). During the last decade, the market of electric double-layer capacitor has been drastically developing due to the increasing demand from devices such as mobile phones and laptop computers, and larger systems such as electric vehicles (EVs). Moreover, these environmentally friendly devices (compared to batteries) have become attractive for clean energy technology such as renewable sources with the increased demand for environmentally friendly solutions.

5.3 Electrical double-layer effect and device construction

Supercapacitors are based on the same simple principles applicable to electrostatic capacitors where larger area plates and shorter distance between plates could give a higher effective capacitance. In supercapacitors, electrical double layer formed next to a large area electrode and an electrolyte is effectively used and hence these devices are technically called electric double-layer capacitors (EDLCs).

At this stage, it is worth summarizing the difference between electrochemical (EC) cells and electrochemical capacitors. In EC capacitors, actual positive and negative electrostatic charges physically reside at some two-dimensional plates set. In an EC cell, charges are stored indirectly and can be admitted or withdrawn from the electrodes based on Faradaic processes to transfer charges across the interface, involving oxidation and vice versa. Two types of processes therefore are distinguished as non-Faradaic processes and Faradaic processes in capacitors and batteries, respectively.

5.3.1 Electric double-layer effect

Figure 5.2 depicts the basic idea of an EDLC, where two electrodes are inserted into an electrolyte which can act as an EDLC. As shown in Figure 5.2(a), when a power source is applied to the electrodes made up of a large surface area such as AC, at the collectors attached to electrodes, positive and negative charge layers built up with oppositely charged layers are formed within the electrolyte, next to the electrode surfaces. The use of a porous carbon, such as AC, is motivated by its high electrical conductivity, low cost, and high chemical stability with a large specific surface area (SSA) (or large surface area/mm^2). The large surface area is due to the distribution

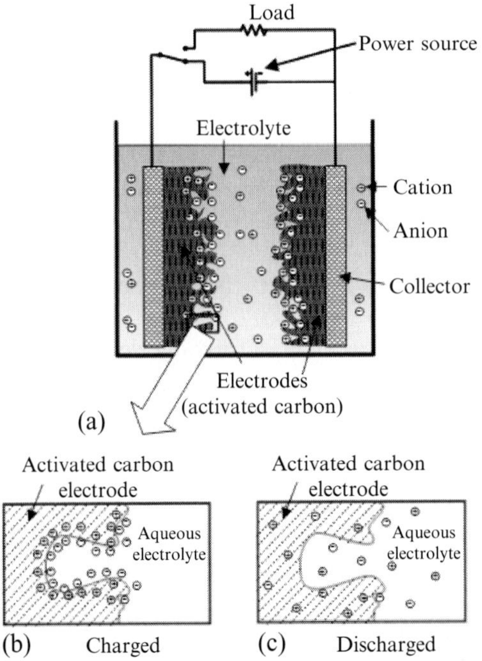

Figure 5.2 Idea of an EDLC capacitor formed by the double layers at each electrode. (a) Conceptual electrochemical capacitor, (b) charge conditions, and (c) enlarged view of porous carbon under discharged condition.

of pores, creating large areas as shown in Figure 5.2(b), for the case of each electrode-electrolyte surface in Figure 5.2(a).

Unlike batteries where energy storage is achieved via reduction and oxidation (redox) reactions creating electron transfers between chemical species, EDLCs are based on the charge separation occurring at an electrode-electrolyte interface. Figure 5.3 depicts the simplified case of an EDLC construction to explain the behavior of porous electrode. In a given particle of the porous material, different pore sizes are mixed and in each of these randomly distributed pores, an electrical double layer is formed giving the simple case of a capacitor plates with very high surface area.

Figure 5.4 indicates an expanded view of the cross section of the case of a porous electrode, where electrode surface is composed of multiple diameter pores with different pore size distribution (PSD). Depending on the diameter of the pores, it is categorized into subsets such as macropore, mesopore, and micropore as in Figure 5.3. This indicates that a porous electrode like AC can have a very a large effective surface area, to create a large capacitor at each electrode. In order to mathematically predict the behavior of the porous electrode, Levie (1963) developed analytical models based on circular cylindrical surfaces of fixed radius, which lead to the transmission line models (TLMs) usable for porous electrode behavior. Figure 5.4(a) depicts this approximation in a pictorial manner, as adapted from this early publication. More details on this can be found in Itagaki et al. (2007) and Levie (1963). In order to achieve high energy density and power density in commercial devices, electrode materials must possess a large SSA available for ion electroadsorption, PSD optimized

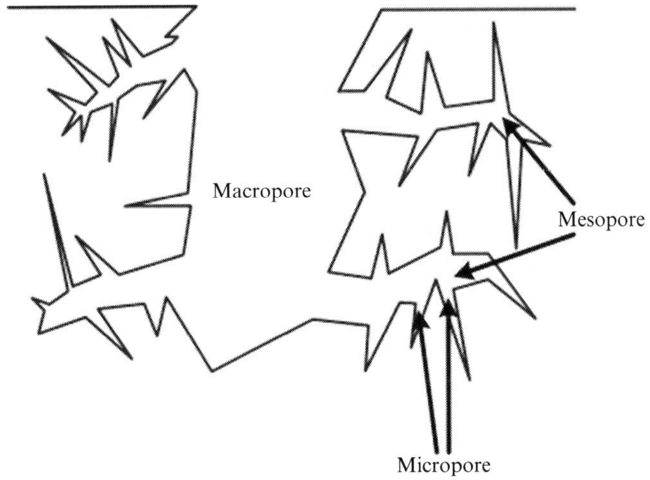

Figure 5.3 A detailed view of the double layer showing macro, meso, and microlevel pores. Adapted from Itagaki et al. (2007).

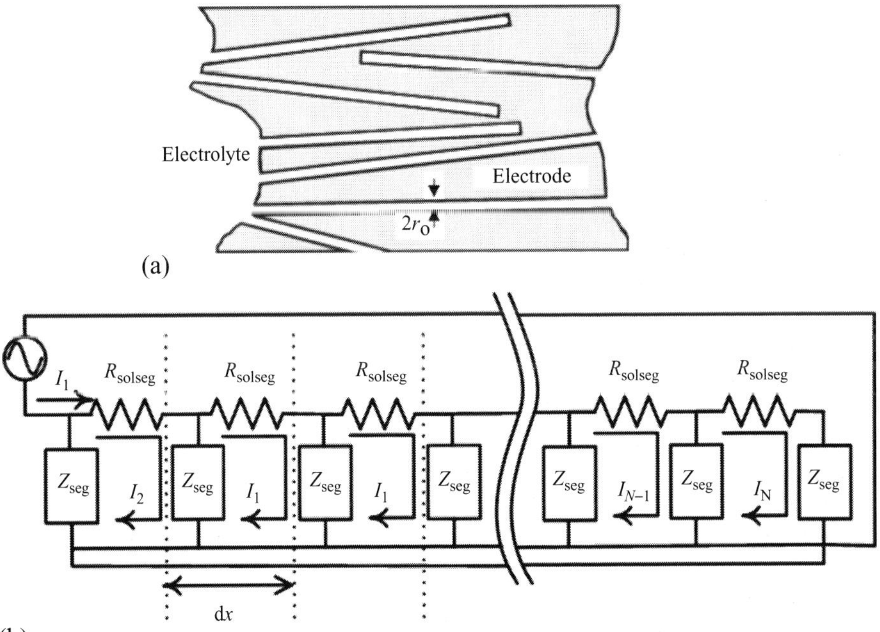

Figure 5.4 Simplification of porous structure providing a large specific surface area (SSA). (a) Simplified case of porous electrode by Levie (1963) with circular cylinders of identical diameter for developing transmission line model (TLM) and (b) transmission line model for porous electrode.
Adapted from Itagaki et al. (2007).

for the desired combination of energy and power characteristics, good electrical conductivity, and wettability by the electrolyte utilized in the device construction. The SSA and PSD are often considered to be two key factors controlling the energy and power density of EDLCs (Wei and Yushin, 2012). To improve the state-of-the-art devices, various porous carbon materials, such as ACs, carbide-derived carbons (CDCs), zeolite-templated carbons (ZTCs), carbon nanotubes (CNTs), carbon onions, carbon aerogels, and graphene have been studied extensively as electrodes in EDLCs (Wei and Yushin, 2012). Out of these, though CDCs and ZTCs offer tight pore size control, high SSA, high conductivity etc., these techniques are yet to get commercialized (Wei and Yushin, 2012).

5.3.1.1 Device construction

A practical DLC consists of two electrodes attached to current collectors, and separated by a separator soaked in electrolyte, as shown in Figure 5.5(a). By selecting a large SSA-type material, such as AC, capacitance per gram of material is increased. The inner surface of each electrode is not a smooth surface, but is rather padded with activated porous carbon, resulting in a surface area that is about 100,000 times as large as the surface area of an ordinary electrostatic capacitor (Bakhourn, 2009). Figure 5.5(b) depicts the actual construction of a cylindrical supercapacitor such as from Honda (Itagaki et al., 2007).

5.3.1.1.1 Electrode materials

Electrode material is the key for high capacitance and other important properties of electrochemical capacitors. High surface area, based on the porosity of the electrode, is the basic criteria to achieve, among other properties. Three different types of supercapacitors are possible based on electrode material, namely, (i) carbon/carbon, (ii) metal oxide, and (iii) conducting polymers. Carbon/carbon type devices with organic electrolyte are quite common.

Activated carbon While there are several electrode material options as already discussed, AC is the most commonly used type of porous carbons in supercapacitors. While there were recorded use of AC in Egypt as far back as 1550 BC, industrial production of ACs in USA was starred in 1913 (Wei and Yushin, 2012). At present, nearly all commercial EDLCs utilize high surface area AC due to their well-developed manufacturing technologies, easy production in large quantities, relatively low cost, and great cycle stability. Petroleum coke, pitch, and coal used to be the most common precursors for commercial AC productions, but the decreasing availability of fossil fuels, the growing global energy demand, and an increased awareness of the environmental impacts of fossil fuel combustion led to the AC productions from sustainable and renewable resources, such as nutshells, wood, starch, sucrose, cellulose, corn grain, banana fiber, coffee grounds, sugar cane bagasse, and others. The costs of some of the raw materials and the carbon yields of pyrolysis from these carbon sources are presented in Table 5.1 as per Wei and Yushin (2012). It is likely that renewable natural sources in Table 5.1 will likely dominate the future AC production.

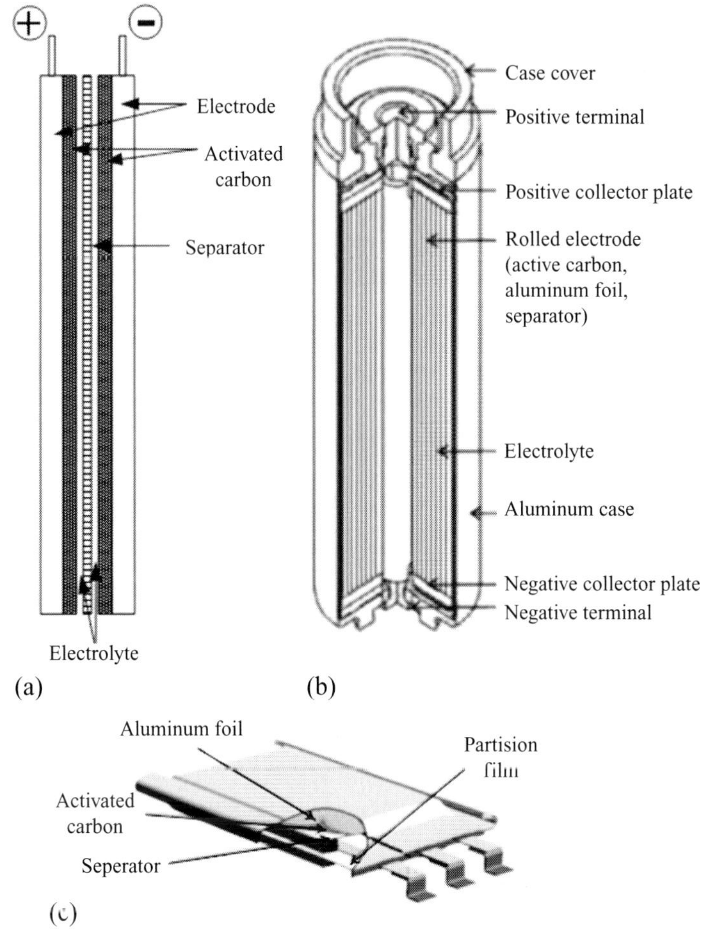

Figure 5.5 Practical construction of a supercapacitor. (a) Basic idea, (b) cross section of a practical device by Honda (Itagaki et al., 2007), and (c) cross section of a practical device by Murata (http://www.murata.eu/downloads/DMF-DMT_TechnicalGuide_MEED.pdf).

Table 5.1 Raw material costs and carbon yield from pyrolysis of natural products used for activated carbon (AC)

Raw material	US$ cost (kg)	Carbon yield (percentage by weight)
Petroleum coke	1.4	90
Charcoal	1.2	90
Lignite	0.75	50
Coconut shell	0.25	30
Wood	0.8	25
Potato starch	1.0	45
Sucrose	0.25	<45
Cellulose	0.65	<45
Corn grain	0.25	<45
Banana fiber	4	<45

Source: Wei and Yushin (2012).

There are different types of ACs, with different SSA values based on the processing of AC materials. SSA values could vary from about 1200 to 2300 m^2/g as with resistivity values varying from 0.4 to 3 Ω cm for range of ACs from the same manufacturer depending on the processes used for each case (Gamby et al., 2001). This performance variation depends on the PSD in each case and other properties of the material in general. Over the last quarter century, carbon has proved as a very useful and lower cost electrode material due to its (a) chemical stability, (b) easy processability, (c) well-established physical and chemical methods for activation, (d) operational temperature range, and (e) usability without any binding material (Frackowiak and Béguin, 2001). As per studies in mid 1990s, while petroleum coke could provide SSA in the range of 2600-3100 m^2/g, coconut shell-based AC provides SSA in the range of 950-2100 m^2/g (Morimoto et al., 1996). Up to the time this book was finalized, AC was the most commonly used commercial electrode material for EDLCs. Raw material and activation methods are two of the most important factors which may influence the final structure and cost of ACs. For example, coal variations such as hypercoal, which is an ashless version obtained by mild thermal extraction of coal (hypercoal process), has very much acceptable properties for AC. Zhao et al. (2014) discusses a process using KOH activation of hypercoal to achieve SSA of 2440/m^2 with specific capacitance (capacitance per gram of material) of around 44 F/g.

Most current research into electrode material places a heavy emphasis onto different forms of ACs, metal oxides, and conducting polymers.

Graphene Since carbon materials can meet the requirements of EDLC electrodes, various carbonaceous materials such as ACs (in different forms), aerogels, xerogels, nanotubes, nanofibers, mesocarbon microbeads, expanded graphite, and ordered mesoporous carbons have been studied extensively as electrode materials (Du et al., 2010). Although ACs always have high SSA values, due to PSD, compared to the size of ions in relation to pore sizes, all the available surface area is not effectively usable for EDLCs. This is simply because EDL is formed by the ions coming in direct contact with the electrode surface. If any micropores are too small to provide contact with ions, EDL is not formed to give high capacitance (Du et al., 2010).

Graphene is a two-dimensional carbon plane with one atomic thickness. Since the paper on the freestanding graphene was published by Geim and coworkers (Novoselov et al., 2004), the preparation, structure, and property of graphene have attracted great attention owing to its particularly useful properties such as mechanical hardness, electrical conductivity, and so on. These properties suggest wide applications, including field effect transistors, adsorbents, lithium-ion batteries, and supercapacitors. Du et al. (2010) discusses the possibility of mass producing grapheme nanosheets as electrode material for EDLCs.

Graphene-based EDLCs with ultra-high-performance EDLCs were developed around 2010, to demonstrate the possibility of efficient 120 Hz line frequency component filtering, which is not well achieved with most commercially available SCs (Miller et al., 2011). As discussed in Miller et al. (2011), due to time constants of values around 1 s in commercially available SCs, they are not too efficient in filtering twice the line frequency components in AC to DC converters. Figure 5.6(a) depicts the

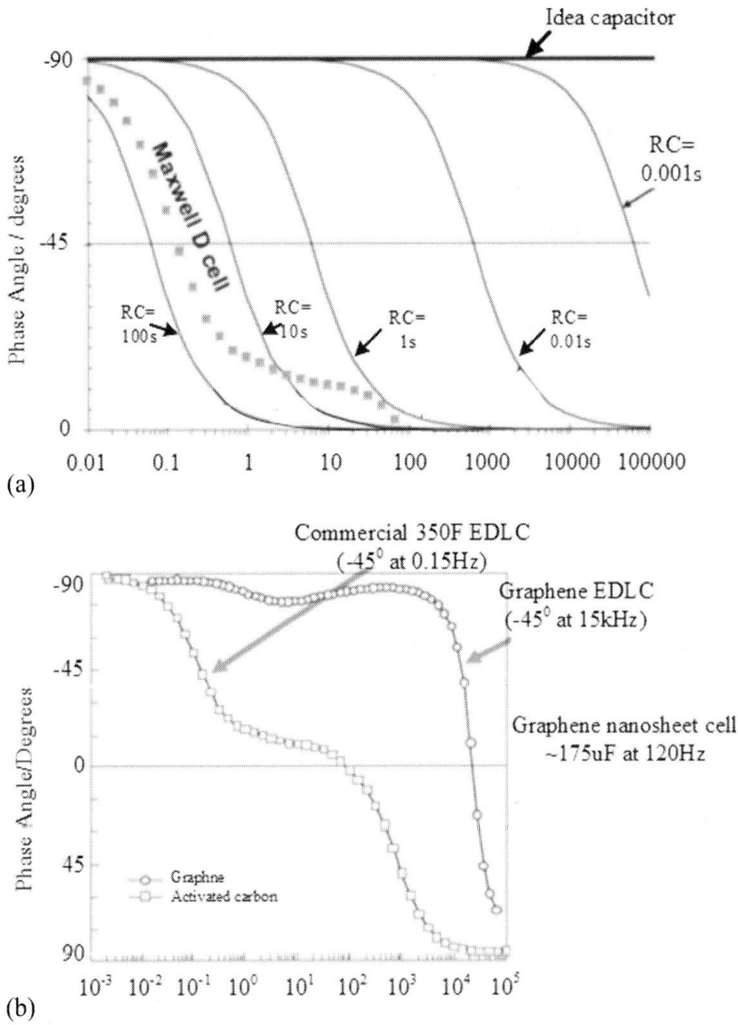

Figure 5.6 Phase angle plots of capacitors and AC line frequency filtering capability of graphene-based EDLCs. (a) Phase angle of a series *RC* circuit and (b) Comparison of phase angle versus frequency of commercial EDLC of 350 F and graphene-based EDLC. *Source:* Miller et al. (2011).

case of phase angle plot for various capacitors with various time constants including the case of ideal capacitor. Figure 5.6(b) depicts the phase angle performance of a graphene-based EDLC research (Miller et al., 2011) compared with a commercially available 350 F EDLC. While 45° phase angle for a commercial ELDC occurs around 0.15 Hz, this 175 F grapheme nanosheet capacitor provides an EDLC with 45° phase angle occurring at 15 kHz. More details on the subject are available in Miller et al. (2011), and basics related to this subject are in Appendix A.

Carbon nanotubes In general, ESR of an EDLC results mainly from the electrical contact resistance in electrodes, the resistance in the solution bulk phase, and the resistance to ion migration in carbon micropores. In order to reduce the ESR, CNTs could be used to reduce ESR since they are known to have high electrical conductivity and chemical stability. These two features render CNTs an ideal additive for electrodes to improve the conductivity. For this reason, there is an ongoing research to explore CNTs as suitable electrode-enhancing material in EDLCs (Huang et al., 2008). In this work, Huang et al. (2008) developed a technique to grow CNTs on a polyacrylonitrile (PAN)-based AC fiber, in an attempt to enhance the electronic conductivity between the constituting fiber threads. For EDLCs assembled with this carbon fiber, the CNT-grafting modification is shown to significantly enhance the conduction of both electrons and electrolyte ions during double-layer formation. In this experiment, PAN-based AC fiber was used in the form of a woven cloth with a thickness ranging from 0.4 to 0.6 mm, with an SSA of 1200 m^2/g and pore volume of 0.59 cm^3/g. By using nickel as a catalyst for the growth of CNTs on carbon cloth, CNT-grafted AC electrodes were achieved. Figure 5.7 depicts the comparison of this CNT-grafted fiber and the bare fiber based on Nyquist plots.

Recently, the miniaturized EDLCs are suggested for powering microelectromechanical systems, because of their dimensions and energy density. Liu et al. (2011) and Sung et al. (2006) provide details of microcapacitors suitable for MEMs and flexible microsupercapacitors, and work in Liu et al. (2011) is based on CNTs. Figure 5.8 shows the work described in Sung et al. (2006) by Sung et al. This all-polymeric cell can have any shape and it does show resistance to change of capacitance due to bending, etc. Zheng et al. (2014) indicates more recent works related to porous graphene and AC composites with high packing density and large surface area for SC electrode material.

Metal oxides Transition metal oxides such as oxides of Fe, Cu, Ni, Mn, Cu, and RuO_2 for electrode materials offer rich redox reactions such as in electrochemical cells providing high specific capacitance values for supercapacitors. In these situations, EDL effect is combined with a Faradaic process like in batteries and this additional effect is called *pseudocapacitance*, which will be discussed later. To date, it is well established that ruthenium oxide (RuO_2) is a prominent material exhibiting high specific capacitance and excellent reversibility. However, this metal oxide is expensive and toxic, and the search for environmentally friendly metal oxides is ongoing. Nowadays, transition metal oxides such as CuO are researched for supercapacitors (Dubal et al., 2013). CuO-based supercapacitors have recently attracted significant attention due to promising capacitance, environmentally friendliness, good electrochemical performance, and lower cost (Dubal et al., 2013). More details are available in Dubal et al. (2013).

Another new approach using TiO_2 nanodots on the walls of multiwalled carbon nanotubes (MWNTs) to develop a TiO_2/MWNT nanocomposite for electrodes is discussed in Sankapal et al. (2014). This seems to be a further development of many past efforts to fabricate different metal oxide/MWNT composites. MnO_2/MWNT, SnO_2/MWNT, NiO/MWNT, and V_2O_5/MWNT are other composites developed by

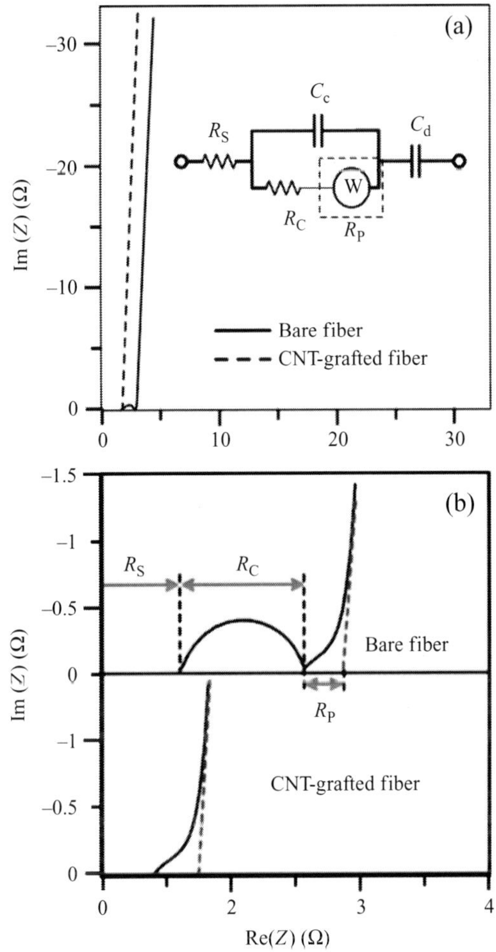

Figure 5.7 Nyquist plot-based comparison of CNT-grafted fiber and bare fiber from 10 mHz to 100 kHz. (a) Equivalent circuit showing resistive components R_s, ion migration resistance; R_c, electrical contact resistance; R_p, distributed pore resistance. (b) Magnified view of the high frequency region of Nyquist plot. *Source:* Huang et al. (2008)

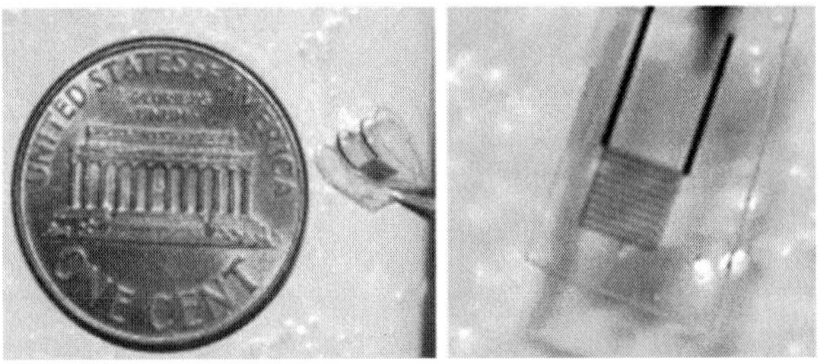

Figure 5.8 Flexible microsupercapacitor.
Source: Sung et al. (2006).

researchers worldwide (Zhao et al., 2014). In these efforts, pseudocapacitance and electrostatic capacitance effects are combined to get an overall capacitance of high value. New interest on TiO_2 is mainly due to its higher stability, high abundance, better safety, and environmental friendliness. In the binder-free approach of TiO_2/MWNT, capacitance yield is reported to be 329 F/g and the scalability for larger area for industrial fabrication looks feasible (Sankapal et al., 2014).

Conducting polymers Electronically conducting polymers (ECPs) are promising materials for the realization of high-performance supercapacitors. Conducting polymers such as polypyrrole, polyaniline, polythiophene, polyindole, etc. are typical prototype materials for supercapacitors based on the Faradaic process called pseudocapacitance. In particular, conducting polymers are especially suitable in view of their ease of fabrication and flexibility (Ramya et al., 2013). The ability to store charges in the case of conducting polymers arises from a doping process.

ECPs are characterized by high specific capacitances since their charging processes pertain to the whole polymer mass and not only to the surface, as in the case of double-layer ACs, and, by high conductivities in the charged states. Furthermore, their charge-discharge processes are generally fast. These features help to develop devices with low ESR and high specific energy and power. The most promising polymer supercapacitor configuration in terms of charge storage capacity and cell voltage able to outperform the double-layer carbon supercapacitors is the n/p type, i.e., the configuration in which an n-doped polymer and a p-doped one are used as negative and positive electrodes, respectively (Mastragostino et al., 2002). However, this is not an easy-to-realize configuration and the difficulties are mainly related to the n-doping process. The charge-discharge processes of polymer-based supercapacitors are Faradaic and those of the n/p devices take place in a cell potential range settled by the n- and p-doping-undoping process potentials of the polymer electrodes. Thiophene-based polymer materials are used in these and specific capacitance values of 95-220 F/g were achieved for p-doping and 80-165 F/g were achieved for n-doping as detailed in Mastragostino et al. (2002).

Composites of CNTs and conducting polymers can be prepared via chemical synthesis, electrochemical deposition on preformed CNT electrodes, or by electrochemical codeposition. These composites could combine the large pseudocapacitance of the conducting polymers with the fast charging/discharging double-layer capacitance and excellent mechanical properties of the CNTs. More details are available in the review paper by Peng et al. (2008).

5.3.1.1.2 Electrolytes

Different kinds of electrolytes have been used in EDLCs with AC electrodes: (i) organic (most commonly solution of tetraethyl-ammonium-tetrafluoroborate (TEATFB) salt either in anhydrous acetonitrile (AN) or propylene carbonate (PC) solvents) (ii) aqueous (solution of acids, bases, and salts), and (iii) ionic liquid (IL). Aqueous electrolytes are typically stable to 0.6-1.0 V in symmetric EDLCs, organic electrolytes to 2.2-2.9 V, and ILs to 2.6-4.0 V.

Organic electrolytes Organic electrolyte-based EDLCs offer life cycle in excess of 5000,000 and a moderately high operational voltage. They can operate in the temperature window from −30 to +50 °C (with an acetonitrile solvent) and are used in the majority of commercial devices with AC electrodes. However, they seem to suffer from high flammability and potential explosion risk because organic solvents exhibit high vapor pressure. Most practical devices today are based on organic electrolytes. Morimoto et al. (1996) provides some details on early coin-cell type capacitors with PC solutions for memory backup applications. These early coin-cell devices were achieving capacitance values up to 0.5 F with ESR values from 10 to 20 Ω. Based on same approach, with stacking layers of basic anode-cathode pairs, power capacitors with capacitance of 4700 F were also achieved. Acetonitrile (AN) is widely used as solvent for electrolytes in most commercial supercapacitors, with a dissociating salt such as tetraethylammonium tetrafluoroborate ($Et_4N^+BF_4$). The main advantage of this combination is low viscosity leading to high conductivity, but unfortunately it is too volatile and flammable. Organic electrolytes provide a greater potential window when compared to aqueous electrolytes, but low thermal stability, flammability, and toxicity limit their use in high-performance supercapacitors.

Aqueous electrolytes The advantages of using aqueous electrolytes include their very low cost and high ionic conductivity. Serious disadvantages, however, include their low maximum applied voltage and the corrosion of EDLC electrodes observed at higher temperatures and voltages (particularly for acid-based electrolytes, such as H_2SO_4 solutions), which limits their life time and life cycle of the EDLCs. These have the potential for bulk storage capability.

Ionic liquids ILs are room-temperature molten salts, which are entirely composed of cations and anions, the nature of which influence the chemical/electrochemical and physical properties. Particularly, the cation limits the negative potential window and the anion affects the positive potential window as well as the melting point, which in turn affects the working temperature range. ILs typically show a very low vapor pressure, high thermal stability, wide electrochemical window, and good conductivity at temperatures above 60 °C, which make them suitable for supercapacitors required for applications such as in EVs (Arbizzani et al., 2007). Due to problems with common solvents used in organic electrolytes such as AN (which releases toxic elements released when burned) or γ-butyrolactone (which is more viscous, heavier, and presenting less separating effect) room-temperature ILs are currently researched as alternatives for organic electrolytes (Abdallah et al., 2012).

The high voltage stability of ILs makes IL-based EDLCs of very high energy density. In addition, ILs are nonflammable, nonvolatile, and nontoxic, which are important attributes for many mobile applications, including their potential use in hybrid engines (Wei and Yushin, 2012). Serious shortcomings of ILs are their high (often prohibitively high) cost and commonly low ionic mobility at room temperature. More details are available in Wei and Yushin (2012) and the citations therein. Supercapacitors based on reduced graphene oxide electrodes and ILs as electrolyte are reported in Paulo et al. (2014) with electrochemical stability of 3 V with a specific capacitance of 71 F/g. In carbon-carbon (AC in both electrodes) electrodes with $PYR_{14}TFSI$ as an IL-specific capacitance value of 60 F/g was reported in 2007 with a maximum potential voltage of 4.5 V at 60 °C, without any deterioration of ESR up to 40,000 cycles (Balducci et al., 2007). More recent work combines graphene nanosheets (GNSs) with IL to achieve specific capacitances in the range of 114 F/g (Kim and Kim, 2014).

5.3.1.1.3 Summary of research directions: electrodes and electrolytes

Table 5.2 depicts a summary of research achievements as per citations in the chapter, comparing the specific capacitance, maximum voltage, and cycle life based on the work of worldwide electrochemists. It is important to note that specific capacitance depends on scan rate and information in the table may show highest achieved in the specific work.

While early research efforts were concentrated toward increasing the properties of carbon materials by surface modification (Fang et al., 2006; Prabaharan et al., 2006), voltage capability toward 4 V (Prabaharan and Michael, 2004) and the use of high SSA material like C-Cloth (Niu et al., 2006), on another front application of CNTs, aerogel, xerogel, and nanoporous carbon materials, etc. (Chang et al., 2013; Du and Pan, 2006; Fang and Binder, 2006; Nasibi et al., 2013) were investigated. After 2010, graphene, graphene oxide, and GNSs combined with metal oxides were investigated (Gund et al., 2013; Kim and Kim, 2014; Zhang et al., 2014) for enhanced performance and using pseudocapacitance effects.

Overall research into supercapacitors is presently divided into two main areas that are based primarily on their mode of energy storage, namely: (i) the redox supercapacitor (pseudocapacitor) and (ii) the electrochemical double-layer capacitor.

Manufacturing aspects of AC-based devices While the manufacture of supercapacitors aims at long life cycles of over a million charge-discharge cycles, cost of manufacturing is one of the important business criteria. While lot of new electrode materials and electrolytes are researched, still the industry is inclined to use cheap materials (like AC) for electrodes with organic electrolyte based on acetonitrile. A standard setup has emerged over the last decade: aluminum contact foils (coated on both sides with AC (Pandolfo and Hollenkamp, 2006)), an organic electrolyte, and a separator (e.g., paper). We here focus on tetraethyl ammonium tetrafluoroborate (TEA+ TFB−) in acetonitrile (AN), which has the advantage of high conductivity even at low temperatures. The electrodes are not only composed of AC but also contain a polymeric binder in order to allow facile coating of the aluminum foils, and a conductive additive, usually carbon black, to enhance electrical interconnections between AC particles. EDLCs can show a short "burn-in" phase where their electrical characteristics change considerably; however, this is very fast (hours to days). It is clear that the ultimate performance of carbon-based supercapacitors will be closely linked to the physical and chemical characteristics of the carbon electrodes. A review on AC properties and their role in EDLCs is available in Pandolfo and Hollenkamp (2006). Aging of EDLCs is discussed in Bittner et al. (2012). Impact of pore structure on series resistance is discussed in Celzard et al. (2002). It is found that microporous

Table 5.2 Summary of research achievements in SCs as per research work during 2004-2014

Electrodes	Electrolyte	Specific surface area (SSA)	Specific capacitance (F/g)	Maximum voltage	Number of test cycles performed measured	Reference source
Carbon-carbon	IL ($PYR_{14}TFSI$)	1428	90	3.5	40,000	Balducci et al. (2007)
SFG6-SFG6	LiPF6/EC + DMC		14.5	4.5	1000	Prabaharan and Michael (2004)
C-Cloth	H_2SO_4	2500	230-485	1.15		Niu et al. (2006)
MnO_2 NPs-Fe_3O_4 NPs	K2SO4 aq.		21.5	1.8	5000	Aiguo et al. (2009a)
GNS/IL	KOH		114		3000	Kim and Kim (2014)
AC	KOH		265	1	10,000	Xianzhong et al. (2012)
AC	$LiSO_4$		210	1.6	10,000	Xianzhong et al. (2012)

carbon helps develop high capacitance slow charge-discharge, while mesoporous carbon helps high rate devices in organic electrolytes (Wang et al., 2007).

In practical devices, separator material characteristics influence the high and low frequency series and parallel resistance, low and high frequency capacitance, time constant, energy, and power density. PSD, the thickness of the separator, and the SSA are important parameters for device performance, and research findings are available in Tõnurist et al. (2013) and Tõnurist et al. (2009).

5.4 Pseudocapacitance and pseudocapacitors

In electrochemical double-layer capacitance discussed so far, actual positive charges due to electron deficiency on one plate and negative charges due to excess electrons on the other plate are separated by a large area material to give very high capacitance with long life cycles. In comparison to batteries where redox reactions (or Faradaic processes) occur at electrode surfaces, these devices are similar to electrostatic capacitors, where a dielectric separator is used in between plates. As stated in Burke (2000) the capacitance, $\frac{dQ}{dV}$ in this case is relatively constant and independent of voltage.

In parallel with the development of DLCs, another type of large capacitors termed *pseudocapacitors* were developed by Conway et al. (1997) where Faradaic currents from reactions and the resultant charge transfer give rise to a *pseudocapacitance*. In general, this capacitance cannot be measured, but is utilizable as a capacitance and the value is voltage dependent. However, there are number of metal oxides-based pseudocapacitive materials with constant capacitance with changing voltage (Ruiz et al., 2013).

Pseudocapacitance arises in cases where Faradaic charge-transfer processes that lead to passage of extents of charge, q, that, for thermodynamic reasons, depend on potential, unlike ideal Nemstian processes, where a singular, (ideally) constant electrode potential, e.g., for a metal/metal-ion equilibrium, arises, independent of the extent of the reaction. Examples of processes which are associated with pseudocapacitance are as follows:

(i) two-dimensional depositions (so-called under-potential deposition) of adatom arrays on electrode surfaces, e.g., H or Cu on Pt, Pb on Au, Bi on Au, Bi on Ag, H on Rh or Pt;

(ii) redox processes in liquid or solid solutions where the electrode potential is a function of the log of the ratio of reductant converted to oxidant (or vice versa) in a redox system; and

(iii) where, in some cases, chemisorption of anions at electrode interfaces takes place with potential-dependent partial Faradaic transfer of an electronic charge related to the so-called electrosorption valency. In general, there can be coupling between non-Faradaic double-layer charging and Faradaic surface processes (Conway et al., 1997). Electrochemistry related to these devices can be found in Conway et al. (1997) and Conway (1990). In general, while EDLCs based on non-Faradaic processes can provide specific capacitances from 20 to 50 μF/cm^2 (which is effectively giving large capacitance devices due to the use of large SSA materials), pseudocapacitance-based devices with Faradaic processes can offer values in the range of 200-500 μF/cm^2 and for multistate overlapping processes around 2000 μF/cm^2 (Conway and Pell, 2003). These devices lead to the development of hybrid devices.

5.5 Hybridization of electrochemical capacitors and rechargeable batteries

Rechargeable batteries are typically characterized by low specific power and high specific energy, while the opposite applies for the electrochemical capacitors. Similarly, a myriad of different applications has different demands in terms of the power spectrum to be provided by the power source. It is therefore clear that a single system cannot fulfill the requirements of the different applications. High energy-high power devices being able to fulfill the requirements of more applications are highly desirable. In this respect, the goal of having high energy and high power in the same systems has been pursued in the last decades with different approaches. One of these is the hybridization of electrochemical capacitors with rechargeable batteries, with a view to combining the high specific power of the electrochemical capacitor with the high specific energy of the battery.

ECs are characterized by high specific power typically above 10 kW/kg but low specific energy values of less than 10 Wh/kg. Rechargeable batteries such as Li-ions today provide specific energies above 250 Wh/kg for industrial products (Cericola et al., 2011). However, they are characterized by low specific power such as 1 kW/kg. In an electrochemical cell, whole volume of the active material is used for charge storage and therefore the specific energy of a battery is much higher than EDLCs (Cericola et al., 2011). However, the rate of discharge in a battery is typically limited by the solid-state diffusion of Li in the bulk of the active materials, which is inherently slow and therefore specific power of batteries is significantly lower than the case of EDLCs.

Both the lithium-ion battery and the EDLCs have similar technical aspects. Both current practical systems utilize an aprotic organic electrolyte, and the charging processes of the respective electrodes occur in similar potential ranges. These are important considerations for the practical hybridization of the two systems.

The hybridization of electrochemical capacitors and rechargeable batteries can, in principle, be performed according to different approaches. Readily available batteries and electrochemical capacitors can be externally hardwire connected, in serial or in parallel.

By analogy, the same approaches can be proposed at the internal level within one device, leading to the internal serial or the internal parallel hybrid. Between these four possible approaches, only two were widely exploited in the available scientific literature (Cericola et al., 2011).

The hybridization of electrochemical capacitors and rechargeable batteries has been proposed by several authors in the last decades (Cericola and Kötz, 2012). Many different approaches utilizing many different materials have been discussed but there is no unambiguous definition of an electrochemical hybrid energy storage device. A simple sorting of possible electrochemical capacitor-rechargeable battery hybrid systems is proposed in Figure 5.9. The hardwired connection of a ready available electrochemical capacitor and a ready available rechargeable battery is defined as an "external hybrid." Such a connection can be arranged in a serial way where the

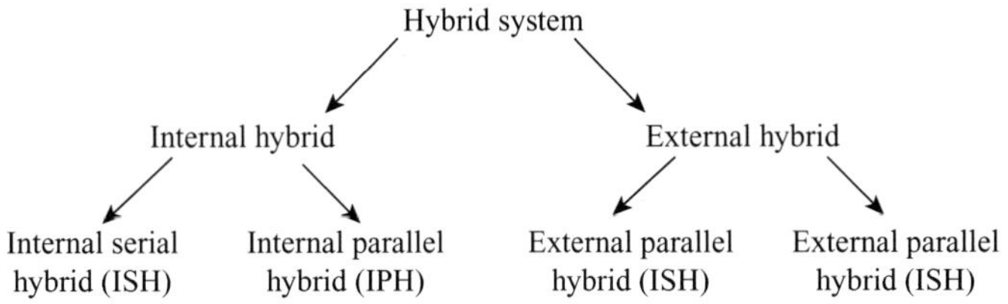

Figure 5.9 Hybridization approaches in EDLCs and rechargeable batteries (Cericola and Kötz, 2012).

electrochemical capacitor is electrically connected in series with the battery resulting in "external serial hybrid" device, or they can be connected in a parallel arrangement giving an "external parallel hybrid" device.

A similar way of hybridization may be imaged at the internal level. In this case, the "internal hybrid" devices are developed by the hybridization on the electrode level. Therefore, a device based on a battery electrode and an electrochemical capacitor electrode will be defined as an "internal serial hybrid." These kinds of systems were firstly named *asymmetric electrochemical capacitors* shortened in *asymmetric capacitor* and introduced in Razumov et al. (2001). The term *hybrid capacitor* was introduced and trademarked by Evans Capacitor Company to describe an electrolytic capacitor using a pseudocapacitive negative electrode (Razumov et al., 2001). The name "lithium-ion capacitor," sometimes abbreviated in "lithium capacitor" was introduced to describe a device using a graphite negative electrode and an AC-based positive electrode in combination with a Li+-containing electrolyte. Nowadays, however, these names and other are often loosely used in the scientific literature to describe various internal hybrids. Figure 5.10 depicts the serial and parallel possibilities for two devices. Benefit of the hybridization can be discussed by examining the two circuits in Figure 5.10. The two devices in the circuits idealize either an actual battery or electrochemical capacitor, or one of their electrodes. In Figure 5.10(a), devices 1 and 2 are connected in series and in Figure 5.10(b), the two devices are connected in parallel. Both devices can be described by a certain capacitance (C), resistance (R), and mass (m). The ratio r is defined as the ratio between the specific capacitance of device 2 over the specific capacitance of device 1, where $r = \frac{(C_2/m_2)}{(C_1/m_1)}$. In this example, device 1 has a specific capacitance always smaller than or equal to device 2. Both circuits for a given set of capacitance and resistance values can be simplified to an equivalent RC circuit where energy and power are derived as $E = \frac{1}{2}C_{\text{tot}}V^2$ and $P = \frac{V^2}{4R_{\text{tot}}}$, respectively. Where C_{tot} and R_{tot} are the total capacitance and the total resistance of the equivalent circuit, V is the voltage applied to the equivalent circuit. The so calculated energy E and power P are then normalized to the mass of the two devices in the corresponding circuits in Figure 5.10(a) or (b). Figure 5.10 depicts the variation of specific power and specific energy versus the mass ratio, m_r, where $m_r = \frac{m_1}{m_1 + m_2}$.

Cericola et al. (2011) describe a case of hybrid devices attempted using the electrode material used for Li-ion batteries such as LTO and LMO (discussed in Chapter 3), and Cericola and Kötz (2012) provide an excellent summary on

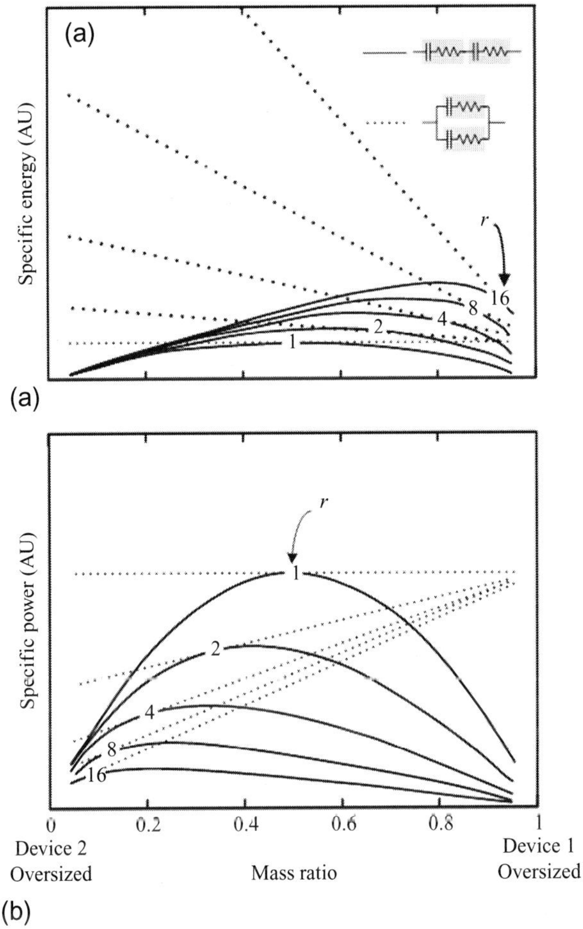

Figure 5.10 Specific energy and specific power of a hybrid device. (a) Specific energy versus mass ratio and (b) specific power versus mass ratio.
Source: Cericola and Kötz (2012).

fundamentals of hybrid devices together with a performance comparison of various electrode materials such as MnO_2, Ni based, Pb based, variety of metal oxides, and Li insertion materials, based on research papers published up to 2011. As per concluding remarks of Cericola and Kötz (2012), following provides a summary of a complex subject addressed by several studies:

- **(i)** External parallel approach is beneficial with respect to battery (alone) in systems where pulse-type charge/discharge cycles occur at high current.
- **(ii)** The use of additional electronic circuitry could enhance the performance of an external direct parallel system.
- **(iii)** Internal serial hybridization allows to achieve devices with specific energy greater than an EC capacitor, usually with twice the capacity. Battery electrode performs better than a complete battery with respect to life cycle and the power capability due to lower SOC achieved.
- **(iv)** The replacement of a capacitor electrode with a battery electrode typically allows the cell voltage to increase (this has allowed practical devices over 4 V into the commercial market). Electrolyte depletion is an issue.

(v) Internal parallel hybridization is an attractive approach to come up with devices with moderately high specific energy and power. Devices are highly tunable.

Sikha and Popov (2004), Pmar et al. (2010), and Catherino et al. (2006) provide a detailed discussion on external-type battery-capacitor hybrid systems and their applicability in cases like EVs, etc. Pasquier et al. (2004) and Michael and Prabaharan (2004) provide some details on early work related to Li electrodes combined with EDLCs for higher voltage capability.

5.6 Modeling and equivalent circuits

Discussions so far provide a background to the development of supercapacitors and the electrochemistry-based fundamentals applicable to EDLCs. Application of these devices requires simplified models suitable for design calculations, estimation of their limitations in a given application, and most importantly the calendar and cycle lifetimes. In Chapter 3, we have discussed the electrochemistry-based concepts leading to simplified models of rechargeable battery chemistries. Almost all those concepts apply to EDLCs also, except for the case that Faradaic reactions or redox reactions are not dominating in commonly used commercial devices. One exception is that pseudocapacitors where limited redox reactions are used to achieve high specific capacitance values. Zhang and Zhao (2009), Sharma and Bhatti (2010), and Sevilla and Mokaya (2014) provide an overview of current state of the art in supercapacitors, which indicates the domination of carbon-based materials in the commercial devices.

Conventional capacitors store little by way of energy, due to the limited charge storage areas and geometric constraints of the separation distance between the two charged plates. However, supercapacitors based on EDL mechanism can store significantly more because of the large interfacial area and the atomic range of charge separation distances. As schematically illustrated in Figure 5.11(a), the concept of the

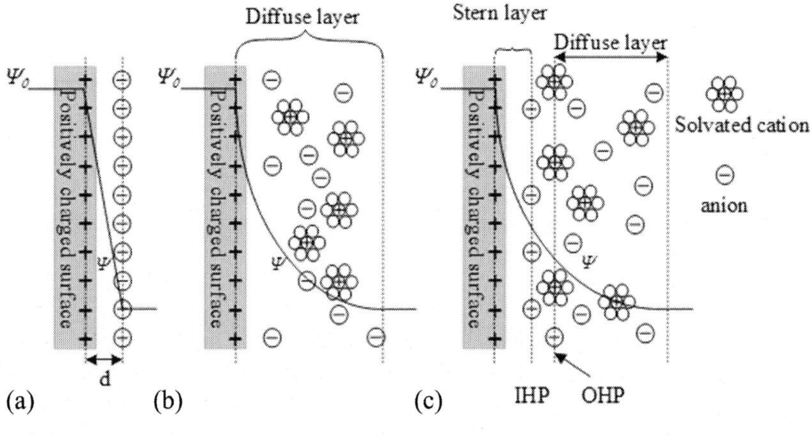

Figure 5.11 Physical models for the electric double layer at a positively charged surface. (a) The Helmholtz model, (b) the Gouy-Chapman model, and (c) the Stern model.

EDL was first described and modeled by von Helmholtz in the nineteenth century, when he investigated the distribution of opposite charges at the interface of colloidal particles (Sharma and Bhatti, 2010). As shown in Figure 5.11(a), Helmholtz double-layer model states that two electrodes of opposite charge form at the electrode/electrolyte interface and are separated by atomic distances. This double layer can be thought of as a molecular capacitor, where one plate is represented by the charges in the metal and the other plate by the ions at a minimum distance in the solution. The distance is of the order of only 0.5 nm, so the capacitance values are enormous. This simple model will be valid only for rather high concentration electrolyte solutions. In this simple model, the double layer is depleted from charges.

In more dilute solutions, the transition will not be so abrupt and the thickness of the double layer will increase, as shown in Figure 5.11(b). The thickness is related to the distance from the metal surface at which the ions can escape to the bulk by thermal motion. In that case, the counter-ion atmosphere will be more like the ionic atmosphere around an individual ion, and this is commonly referred to as the *diffuse electric layer*. In order to account for this situation, simple Helmholtz EDL model was further modified by Guoy and Chapman (Sharma and Bhatti, 2010). In the combined theory of Gouy and Chapman, the exchange of counter-ions between the double layer and the bulk solution due to thermal motion is taken into account. Both coulombic forces and thermal motion hence influence the equilibrium distribution of counter-ions in their model for the diffuse double layer as shown in Figure 5.11(b). However, the Gouy-Chapman model leads to an overestimation of EDL capacitance. Discussion on the assumptions for their theory is available in Grimnes and Martinsen (2008).

Later in 1923, Stern combined these two models to explicitly recognize the two regions of ion distribution—the inner region called the compact layer or the Stern layer and the diffuse layer or the Gouy-Chapman layer (Grimnes and Martinsen, 2008; Sevilla and Mokaya, 2014; Sharma and Bhatti, 2010). Figure 5.11(c) depicts the scenario. The inner Helmholtz plane and outer Helmholtz plane are used to distinguish the two types of adsorbed ions. The capacitance in the EDL, C_{dl} can be treated as a combination of capacitances from two regions, the Stern type of compact double-layer capacitance, C_H and the diffusion layer capacitance, $C_{diff.}$ Thus, a simple relationship can be expressed as

$$\frac{1}{C_{dl}} = \frac{1}{C_H} + \frac{1}{C_{diff}} \tag{5.1}$$

In a practical EDLC, electrode is a porous material with a high SSA, and the EDL behavior at the porous electrode is much more complex than that of an infinite planar electrode assumed in models of Figure 5.11. This behavior is much affected by a number of parameters such as mass transfer path, space constraints inside the pores, the ohmic resistances associated with the electrolyte, and the wetting behavior of the porous electrode. Figure 5.12 schematically depicts the behavior of a practical EDLC composed of the two current collectors and electrolyte absorbed into the porous electrode material and the separator. The capacitance estimation is generally assumed to follow that of a parallel-plate capacitor with value C given by,

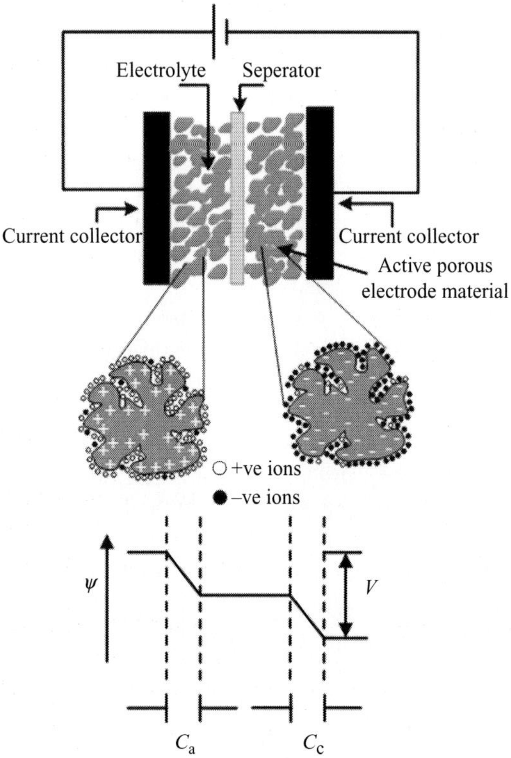

Figure 5.12 Schematic representation of an EDLC based on porous electrode materials and the two separate capacitors formed at each electrode.
Adapted from Zhang and Zhao (2009).

$$C = \frac{\varepsilon_r \varepsilon_0}{d} A \tag{5.2}$$

where, ε_r is the dielectric constant of the electrolyte, ε_0 is the permittivity of the vacuum, A is the surface area available to electrolyte ions, and d is the effective thickness of the EDL, which is also called the Debye length (Grimnes and Martinsen, 2008). Based on Eqn (5.2), there should be a linear relationship between specific capacitance and the SSA. However, experimental results discussed in Sharma and Bhatti (2010) indicate that this assumption is not accurate, and these new experimental findings cannot be fully interpreted by the EDL theory.

As shown in Figure 5.12, with two electrodes combined in an EDLC, we are achieving an overall capacitance of C_T based on the anode and cathode capacitances C_a and C_c, respectively, which gives rise to the relationship,

$$\frac{1}{C_T} = \frac{1}{C_a} + \frac{1}{C_c} \tag{5.3}$$

Performance of commercially available supercapacitors is mainly evaluated based on six important criteria: (i) power density substantially greater than batteries with acceptably high energy densities (usually >10 Wh/kg), (ii) very high life cycle (more than 100 times that of batteries), (iii) fast charge/discharge process within seconds, (iv) low self-discharge, (v) safe operation, and (vi) low cost. As depicted in the simplified *RC* equivalent circuit of Figure 5.12, with the equivalent circuits of the two electrodes shown separately, equivalent series resistance (R_s) and the terminal voltage determine the power density. Anode and cathode leakage resistances R_{Fa} and R_{Fc}, respectively, account for the self-discharge. To have the ability to fast charge transfer (charge/discharge), R_sC_T time constant of the device should be as small as possible. Since the capacitance of an SC is enormous, R_s value should be extremely small. However, the current commercial SCs have ESR values in the range of 0.1 mΩ to few mΩs.

5.6.1 Different types of equivalent circuits

Depending on the application, one can use different types of equivalent circuits with various complexities. Simplest version of ESR and the capacitance in series assumes a constant capacitance (independent of the voltage) and the ESR also to be constant. However, experimental behavior and measurements indicate that this is only a very first approximation, and this ignores the leakage. For this simplification, the equivalent circuit in Figure 5.12 ignores the values of R_{Fa} and R_{Fc} and uses ESR value and the total capacitance C_T as in Eqn (5.3).

It is important to understand that the simplified but independent electrode-based equivalent circuit is a major simplification of the overall theoretical model, which could be used to describe the porous electrodes and its behavior based on a theoretical model shown in Figure 5.13. Figure 5.13(a) shows the simplified arrangement of two current collectors, two electrodes, and the porous insulting membrane, and Figure 5.13 (b) shows the final distributed model of the complete system. This is related to various models described in Figure 5.11.

The different resistances that appear in Figure 5.13 depend on numerous parameters such as (a) electrode material resistivity, (b) pore sizes, (c) membrane porosity,

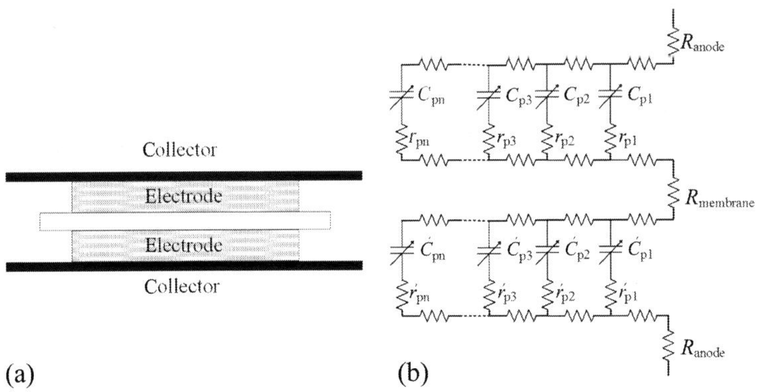

Figure 5.13 Supercapacitor packaging principle and detailed equivalent circuit. (a) Packaging principle and (b) theoretical model.
Source: Belhachemi et al.

and (d) packaging technology (impregnation and collector material resistivity). More detailed discussion can be found in Belhachemi et al.. As discussed in Figure 5.4(b), this is directly related to the TLM of the porous electrode. However, this is far too complex to handle when designing the circuits using SCs. Then we can simply classify this into different models such as two *RC* section or three *RC* section-based models for more accurate calculations. Literature indicates many different complex models based on including pore models and self-discharge-included models. Figure 5.14 indicates few popular and commonly discussed models as per discussions in Belhachemi et al., Yasser et al. (2008), and Lisheng and Crow (2008).

A practical observation on supercapacitors is that almost all important parameters such as ESR, capacitance, and the leakage are dependent on voltage, current, and temperature in general. Figure 5.15 depicts the dependence of ESR and capacitance on current and voltage.

Given the above situation, there are many publications which discuss different aspects of these dependencies as well as the differences in capacitor values based on the measurement method as well as the state of charge (SOC). Yang et al. (2010), Kurzweil et al. (2005), Spyker and Nelms (2000), and Nelms and Cahela (1999) provide a useful discussion on these aspects.

Useful practical models for design purposes are two or three branch series model and the voltage-dependent ladder model. In general, Figure 5.16 shows a general model that can be simplified into many different versions we have discussed, and how duration of charge and discharge affects the mode (Lajnef et al., 2007).

5.7 Testing of devices and characterization

Characterization methodologies applied to supercapacitors are not standardized and it is difficult to compare the laboratory test results with those of manufacturers. At present, there are not many international standards, except one set of IEC standards (IEC-40/1378/DIN IEC 62391-1, 2004; IEC-40/1379/DIN IEC 62391-2, 2004). Most supercapacitor parameters are dependent on parameters such as ambient temperature, current intensity, frequency, and voltage. Therefore, reliable and reproducible experimental results are required. This section describes some of the techniques that are used to estimate the parameters of supercapacitors. Some methods are mostly used by electrochemists in research laboratories, while some are used in industrial environments.

5.7.1 Charge-discharge method

This is one of the common methods used for capacitance and ESR measurement. In this process, a SC is discharged or charged at a constant current, and a solid-state load is used for this test. In this test, capacitor is simplified as a single *RC* time constant device, and voltage variation for constant current charge or discharge of the device is monitored. Figure 5.17(a) depicts this where the voltage across the supercapacitor

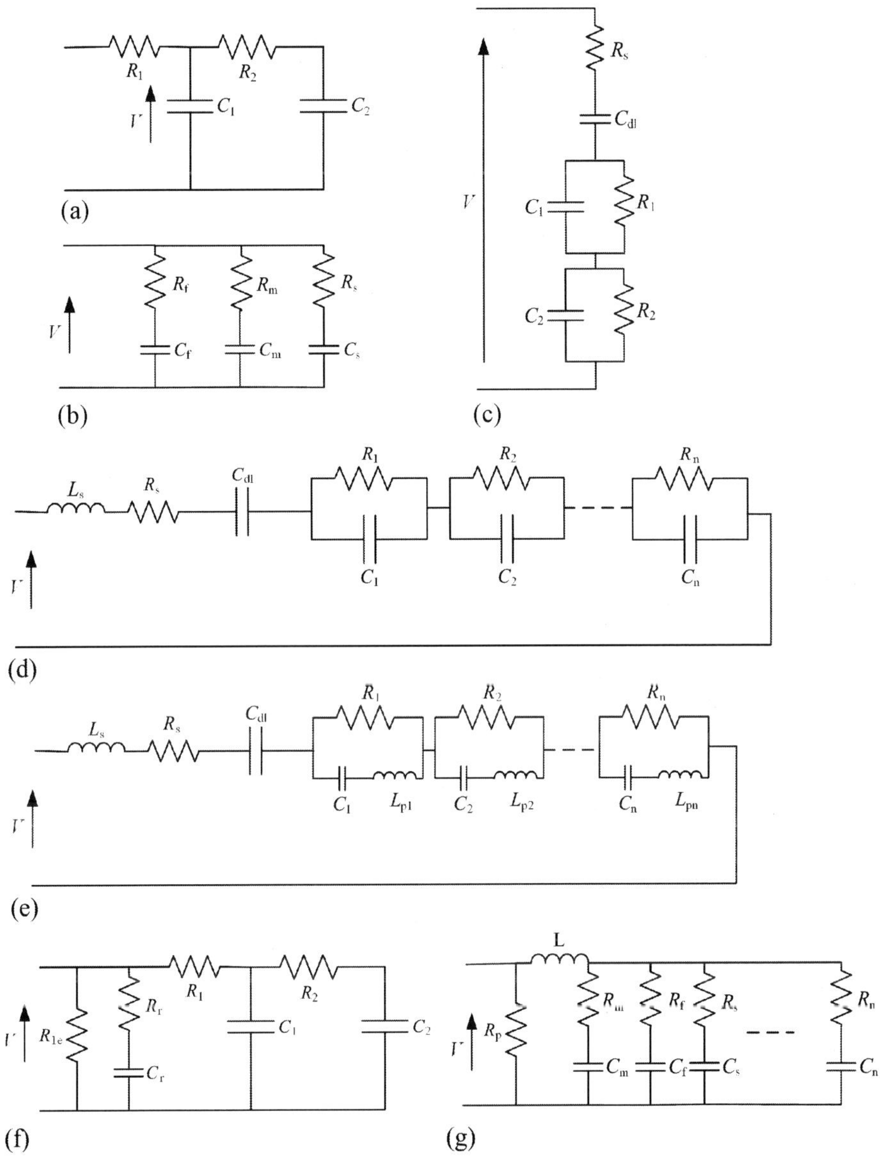

Figure 5.14 Different models for supercapacitors. (a) Two branch model, (b) three branch parallel model, (c) series-parallel model, (d) impedance pore model, (e) complex pore model with pseudo inductance, (f) self-discharge and diffusion controlled model, and (g) general *RC* network model.

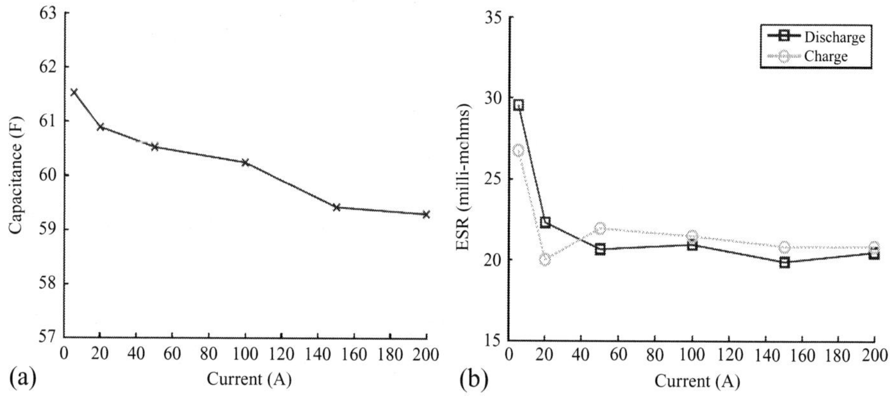

Figure 5.15 Capacitance and ESR variation of a commercial supercapacitor family. (a) Capacitance versus current and (b) ESR versus current.
Source: Yang et al. (2000).

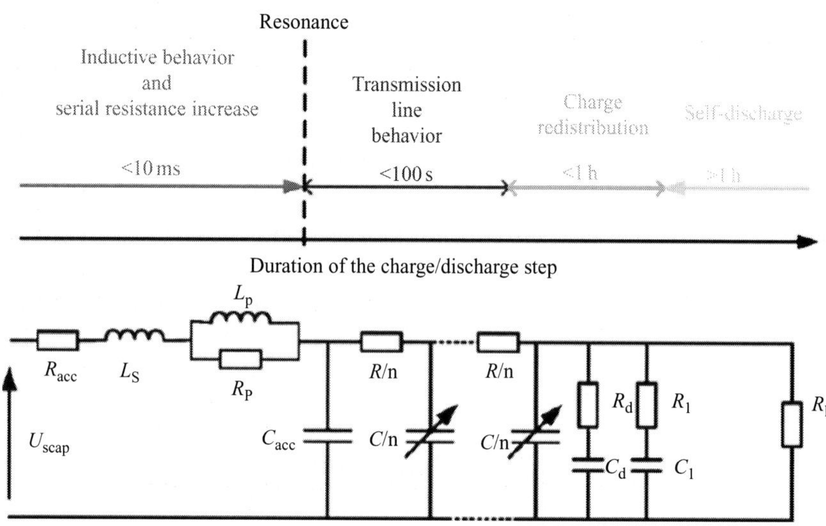

Figure 5.16 A generalized model for a supercapacitor and its operational durations and states.
Source: Lajnef et al. (2007).

is initially maintained at its rated voltage, V_R for 1 h and then the device is discharged at a constant current, I. The capacitance, C, determination, is estimated during constant current discharge between the voltage values from 0.8 V_R and 0.4 V_R according to the linear relationship $C = I \cdot \Delta t_c / \Delta V_c$ as shown in Figure 5.17(a). Dotted line in Figure 5.17 shows the linear discharge, while actual behavior is nonlinear as shown by the continuous trace.

Figure 5.17(b) shows the 200 A constant current charge and discharge curves for a 3000 F supercapacitor. With reference to the graphs in Figure 5.17(b), the ESR

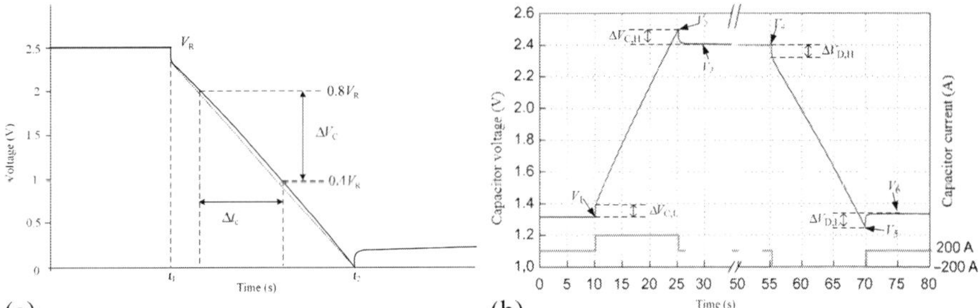

(a) (b)

Figure 5.17 Charge-discharge method. (a) Concept and (b) an actual measurement on a 3000 F capacitor.
Adapted from Lajnef et al. (2007).

measurement is obtained using the sudden change of voltage (ΔV) at the start of the charge $\Delta V_{C,L}$, at the end of the charge $\Delta V_{C,H}$, at the start of the discharge $\Delta V_{D,H}$, or at the end of the discharge $\Delta V_{D,L}$. For a constant charging or discharge current of I, the ESR value is given by $\Delta V/I$.

In literature, it is reported that these values obtained by charge-discharge tests are quite dependent on the current values used in the test. Figure 5.15 indicates this as adapted from Lajnef et al. (2007).

5.7.2 Constant power tests

Constant power tests are commonly used for the characterization of electric storage devices in general. In these tests, a solid-state load such as shown in Figure 5.18 could be used by setting its operation to constant power mode. For supercapacitors, the experimental procedure consists in discharging the element with constant power between two voltage limits of V_{nom} and $V_{nom}/2$ in order to measure the specific energy that this element can supply as function of discharge power. As detailed in Pell and Conway (1996), the corresponding results can be graphically reported in the

Figure 5.18 A solid-state load which can be used as constant power device.

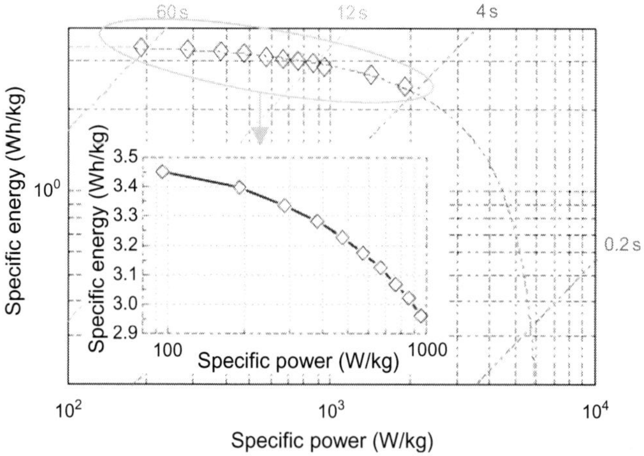

Figure 5.19 Ragone plot for 2600 F capacitor.

well-known Ragone diagram (Christen and Carlen, 2000) where the *x*-axis and the *y*-axis correspond, respectively, to the specific power (kW/kg) and to the specific energy (Wh/kg).

Figure 5.19 shows the Ragone plot obtained for a 2600 F/2.5 V cell and with a specific power that has been limited by the maximum current capability of the tester at 800 A. Then, the experimental points have been interpolated and extrapolated according to Eqn (5.4), where E is the energy, P is the power, R is the series resistance, and C is the rated capacitance (Christen and Carlen, 2000).

$$E_c(p) = \frac{C}{2}\left(RP \ \ln\left(\frac{RP}{U_0^2}\right) + U_0^2 - RP\right) \tag{5.4}$$

5.7.3 Impedance spectroscopy

Electrochemical impedance spectroscopy (EIS), summarized in Chapter 3, is often used in the characterization of electrochemical behavior of energy supercapacitor storage devices (Karden et al., 2002; Qu and Shi, 1998). To characterize a supercapacitor, the sweep in frequency must be done for various voltage levels and also for different temperatures. EIS allows the study of the influence of frequency on the electrode series resistance and on specific capacitance. A summary of test process is given below.

The supercapacitor is polarized with a DC voltage. A small voltage ripple, typically 10 mV, is superimposed on the DC component. The ripple frequency is swept between 1 mHz and 10 kHz. The measurement of the current amplitude and phase with respect to the injected voltage permits the determination of the real and imaginary part of the impedance as a function of the frequency (Buller et al., 2002). The measurements are usually performed in a controlled climatic chamber. Figure 5.20 shows the Nyquist plot (complex-plane representation) of measured impedance spectra of a 1400 F/2.5 V capacitor.

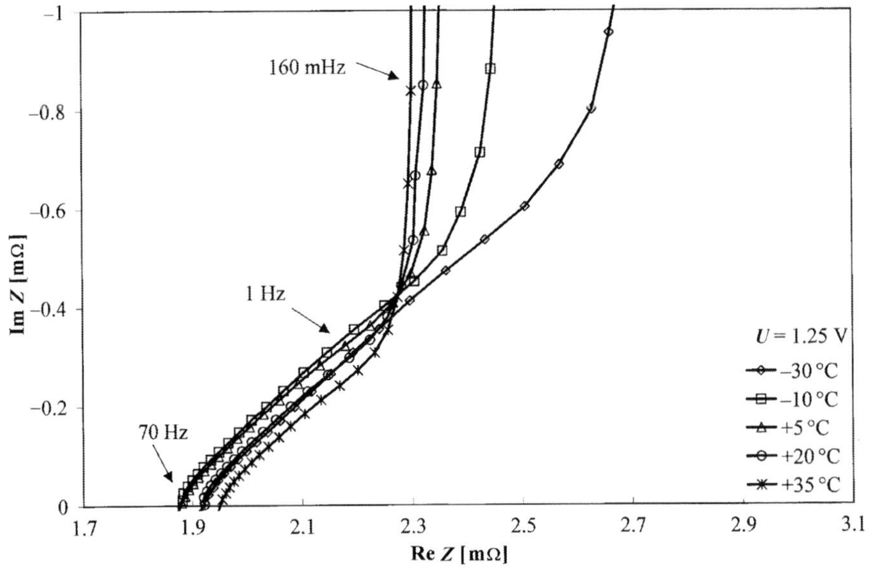

Figure 5.20 Complex-plane representation of measured impedance spectra of a 1400 F/2.5 V supercapacitor.
Reproduced with permission from IEEE (Buller et al., 2002).

5.7.4 Cyclic voltammetry

In this technique, the device under test is subjected to voltage rising at a fixed rate of sweep such as about 10 mV/s, and the corresponding device current is measured. The plot obtained by this test is quite frequently used in the process of device development stages. The graph obtained is called *voltammogram*.

This test is useful to validate the capacitive behavior, to investigate the symmetry between charge and discharge process, and to determine the potential limits of the device. It consists of linear voltage sweep between two limit values, V_1 and V_2. If the ultracapacitor is modeled with a constant capacitor C in series with an internal resistor R_s, the current is expressed by Lajnef et al. (2007).

$$I(t) = C\frac{\mathrm{d}v}{\mathrm{d}t}\left[1 - \exp\left(-\frac{t}{CR}\right)\right]$$

In this equation, there are two terms, the first is permanent $I(t) = C\mathrm{d}v/\mathrm{d}t$, the second is transient and disappears after a few time constants. When the capacitance depends on voltage and assumes very low serial resistance, the current is proportional to the capacitance by $I = C_{\mathrm{diff}}\frac{\mathrm{d}v}{\mathrm{d}t}$, given that the voltage drop due to the resistor can be neglected.

Figure 5.21 illustrates the voltammogram which is obtained by recording the current versus voltage with a sweep rate of 10 m/Vs. This figure shows that the current is constant between 0 and 0.3 V, i.e., the differential capacitance is constant within this

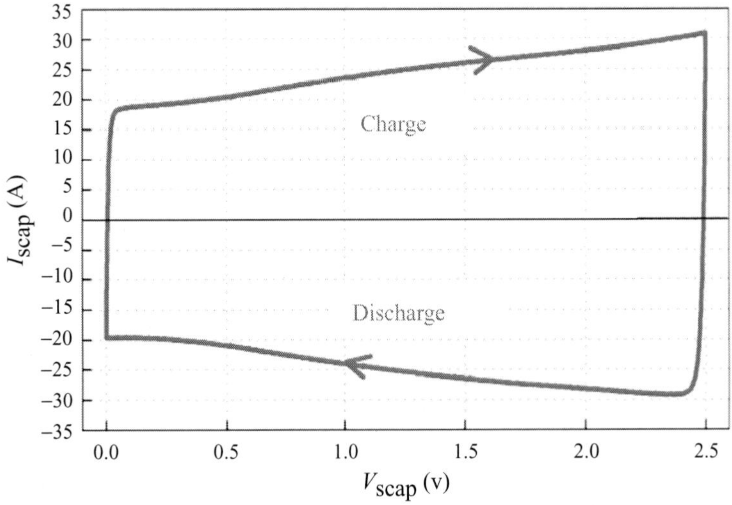

Figure 5.21 Voltammogram measured with a sweep rate of 10 mV/s on a 2600 F/2.5 V cell. *Source:* Lajnef et al. (2007).

voltage range. Then, the current increases as well as the differential capacitance. In the 1.25-2.5 V range, the linear fit of the differential capacitance, which has been determined from constant current tests, seems to be right. Also, the evolution shape of the current versus voltage seems to be symmetric between charge and discharge that corresponds to a reversible behavior, i.e., the differential capacitance in charge is the same as in discharge.

The Figure 5.22 shows the comparison of different parameter extraction techniques (Kurzweil et al., 2005). The molecules at the electrochemical interface between electrode and electrolyte show a nonlinear response to the electric field. This figure indicates that in a supercapacitor, capacitive and Faradaic charges could coexist. This charge, divided by the potential range from which it is transferred, constitutes a pseudocapacitance since it is not associated with a purely capacitive phenomenon. Transient techniques, which generate fast voltage changes, reflect the fast charge/discharge processes with "frozen" diffusion at the outer electrode surface. At slow voltage changes, the slow electrode reactions in the inner electrode surface (strongly depending on voltage) additionally contribute to the measured capacitance values. In Figure 5.23, SOC varies from full discharge to charge.

5.8 Modules and voltage balancing

Supercapacitor cells generally have very low operation voltage of between 2.5 and 3.0 V. However, most high power applications require considerably higher voltages. For example, a car battery pack situation may mean either 12 or 42 V, which means that we need anything from about 5 to 20 elements in series to meet the nominal

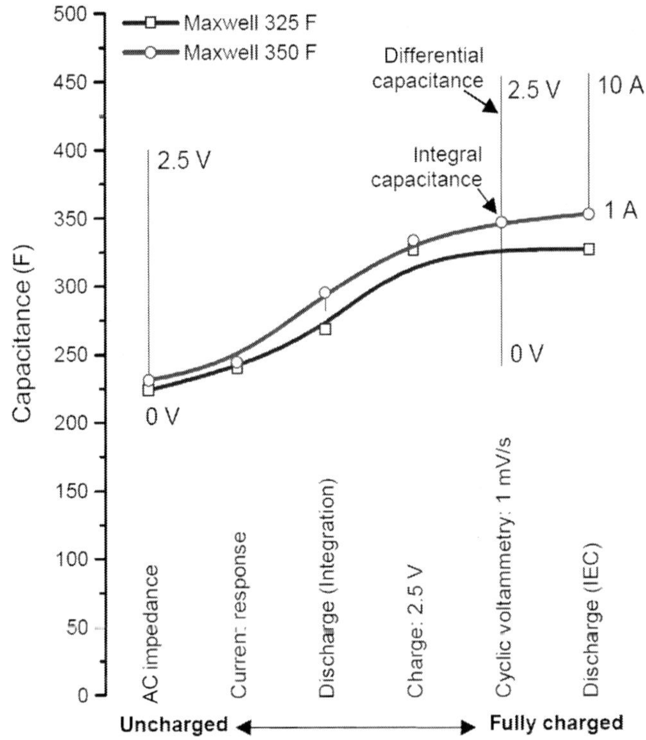

Figure 5.22 Capacitance of Maxwell capacitors (BCAP0350) determined by different measuring techniques, creating a different state-of-charge of the capacitor. Adapted from Kurzweil et al. (2005).

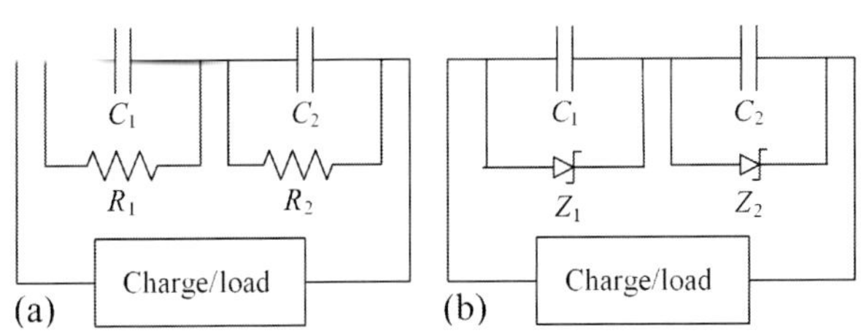

Figure 5.23 Passive resistance balancing. (a) Resistor based and (b) zener diode based (Sharma and Bhatti, 2010).

voltage value. In order to operate at higher application voltages super capacitors are connected in series and parallel configurations to form a module, to meet the DC voltage, energy capability and the total ESR value, which determines the maximum power capability of the bank.

Supercapacitors coming out of mass scale production lines are not exactly identical and hence the series-connected supercapacitor stacks develop unequal voltage distributions (Aiguo et al., 2009a,b; Chunhe et al., 2009; Gai Xiao et al., 2008; Rengui et al., 2008). Situation could become worse, due to device capacitance is also varying with different DC bias voltages. Voltage distribution in a series stack of super capacitors is initially a function of capacitance. After the stack has been held at voltage for a period of time, voltage distribution becomes a function of internal parallel resistance (leakage current). For example, consider a stack of 20 capacitors initially charged to 50 V. If the cells have identical capacitance, the voltage should divide evenly so that each capacitor charges up to 2.5 V. If the cells have any variation in capacitance, individual cell voltages will vary based on capacitance. The cells with greater capacitance will be charged to lower voltages, and the cells with smaller capacitance will be charged to higher voltages. This is because each cell conducts the same charging current, and the voltage is a function of current and capacitance.

Those undesired unequal capacitances could attribute to several factors as following ELNA and Ness Capacitor Co., Ltd.:

- the manufacturing tolerances of 5-10%
- the temperature gradient in the system, and
- the cell aging.

In this case, some supercapacitors will have an overvalue for their voltage, with an associated decrease of their lifetime. For the other supercapacitors, the maximum voltage will be under the voltage limit; the energy storage in supercapacitors will not be at its maximum level. A capacitor with a smaller capacitance needs a shorter charging time and, thus, reaches its maximum voltage faster than a larger capacitor. To avoid damaging the smaller capacitor, voltage equalization circuits have to be employed. Manufacturers of large devices supply SC modules similar to battery packs, with different types of cell-balancing circuits used inside the assembly.

5.8.1 Types of balancing

There are two common approaches to cell balancing in SC modules. They are (i) passive balancing and (ii) active balancing. Passive balancing circuits use passive components and diodes, etc. in general and they can be classified into two subgroups such as resistance balancing and zener diode balancing.

5.8.1.1 Passive balancing techniques

5.8.1.1.1 Resistance balancing

In order to prevent the adverse effects of unequalized charging of supercapacitors, individual supercapacitor needs to be maintained at an equalized charge level. This can be achieved by regulating the voltage of individual cells. A simple approach to equalize the supercapacitor stacks is to use a bypass resistor in parallel as shown in Figure 5.23. The amount of current drawn by the resistors is proportional to the cell voltage, which results in more current being diverted to the resistor as the cell voltage decreases. This will tend to reduce the voltage differences between the different cells within the stack since higher voltage cells will be further discharged by the resistors in parallel. One drawback of this approach is that the recovered energy is being converted into additional losses in the resistors. In addition, the amount of current drawn by the resistors is not regulated. As a result, the cell voltages are not fully regulated. The principle of equalization can be explained like this. As indicated in Figure 5.23, once the capacitors are fully charged, two resistors will share the input charger voltage by the resistor divider action combined with the leakage currents of the capacitors. This will help to distribute the total stack voltage evenly across the entire series of capacitors. The voltage equalization is closely achieved by connecting resistor in parallel across the capacitors, but great power dissipation is expected in the resistor.

5.8.1.1.2 Zener diode balancing

In order to better regulate the current drawn by the shunt elements, diode circuitry can be used. Use of zener diode is one such approach. Because of the breakdown voltage of the device, it is used as a voltage regulator when placed in parallel across a load to be regulated. Figure 5.23(b) indicates this concept where two zeners are connected across the two capacitors, C_1 and C_2. As soon as a capacitor C_1 reaches its maximum voltage, if the voltage is almost the breakdown voltage of the diode Z_1, it starts conducting and deviates the charging current. Similar situation works with C_2-Z_2 pair also. In this scheme, the amount of lost energy is minimized since the shunt circuitry is only active when the cell voltage exceeds the preset breakdown level. However, this approach also leads to increased losses and, furthermore, suffers from the temperature dependency of the Zener diodes.

Simulated energy losses for voltage balancing with resistors and zener diodes are compared in Figure 5.24(a) and (b). In both cases, four 1000 F and one 800 F devices were connected in series and charged to 12.5 V. Energy stored at the end of the charging process amounted to 15 kJ. In the case of the voltage-balancing resistors (each 0.1 Ω), 120 kJ was required to fully charge the EDLCs, resulting in an efficiency of 12.5%. When zener diodes were used, only 16.3 kJ was required, resulting in an efficiency of 92% (Barrade et al., 2000). Charging efficiency is therefore much improved by using zener diodes for voltage balancing instead of resistors.

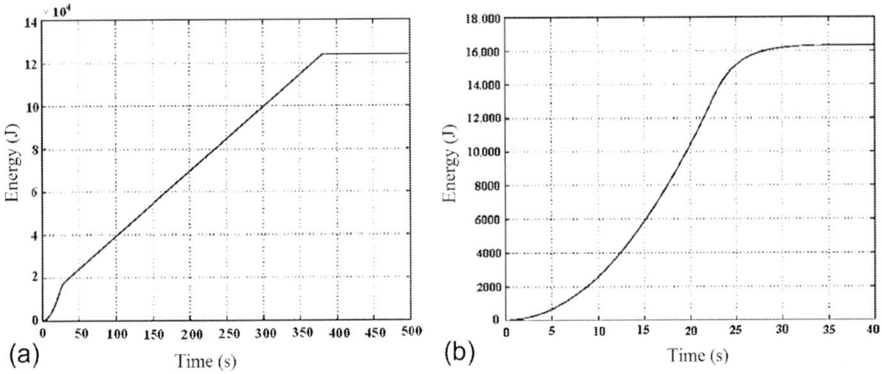

Figure 5.24 (a) Simulated energy loss while discharging an EDLC bank balanced with resistors (Sharma and Bhatti, 2010). (b) Simulated energy loss while discharging a bank balanced with zener diodes (Sharma and Bhatti, 2010).

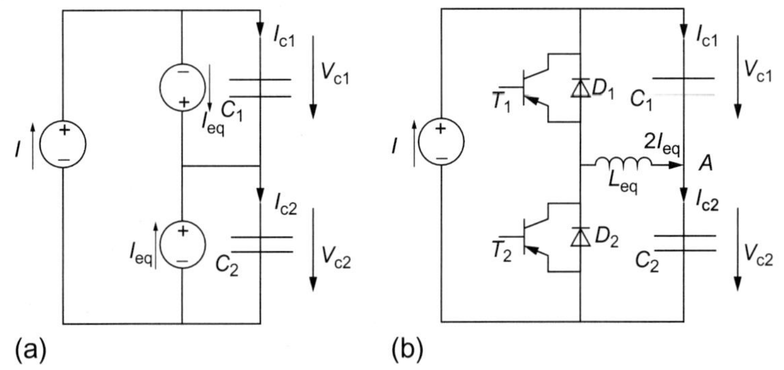

Figure 5.25 Active balancing. (a) Concept and (b) buck-boost topology for active voltage balancing (Sharma and Bhatti, 2010).

5.8.1.2 Active balancing

In passive balancing scheme, the recovery energy is dissipated into an energy dissipating resistive current shunts. As a result, the dissipated equalization schemes are best suited for low power applications or low current charge/discharge rates. This will tend to minimize the lost energy. In applications such as EVs, the current charge/discharge rates are pretty high (10-100%) and the charge/discharge times are quite short. As a result, the charge equalization currents could be of the same order of magnitude. Therefore, in order to minimize lost energy and optimize cell's performances, nondissipative equalization techniques (active balancing scheme) need to be used.

The general philosophy behind an active voltage-sharing device is shown in Figure 5.25(a) (Barrade et al., 2000). The main charging current I is augmented by the equalizing currents I_{eq}. The size and direction of the equalizing currents are controlled according to the local voltages present across the corresponding device. The equalizing current sources can be realised by the buck-boost topology as shown in Figure 5.25(b). If the voltage U_{c1} is detected as being significantly greater than voltage U_{c2}, transistor T_1 will be switched at a certain frequency to generate a positive equalizing current, $2I_{eq}$. If U_{c2} is greater than U_{c1}, then T_2 will be 80 switched to generate a negative equalizing current. This process continues until the voltages are balanced.

References

Abdallah, T., Lemordant, D., Claude-Montigny, B., 2012. Are room temperature ionic liquids able to improve the safety of supercapacitors organic electrolytes without degrading the performances? J. Power Sources 201, 353–359.

Aiguo, X., Shaojun, X., Xiaobao, L., 2009a. Dynamic voltage equalization for series-connected ultracapacitors in EV/HEV applications. IEEE Trans. Veh. Technol. 58 (8), 3981–3987.

Aiguo, X., Xiaobao, L., Shaojun, X., 2009b. Research on dynamic voltage equalization circuit for series connected ultracapacitors. In: IEEE International Conference on Industrial Technology, 2009 (ICIT '09).

Arbizzani, C., Beninati, S., Lazzari, M., Soavi, F., Mastragostino, M., 2007. Electrode materials for ionic liquid-based supercapacitors. J. Power Sources 174, 648–652.

Bakhourn, E.G., 2009. New mega-farad ultracapacitors. IEEE Trans. Ultrason. Ferroelectr. Freq. Control 56 (1), 14–21.

Balducci, A., Dugasa, R., Taberna, P.L., Simona, P., Plée, D., Mastragostino, M., Passerini, S., 2007. High temperature carbon-carbon supercapacitor using ionic liquid as electrolyte. J. Power Sources 165, 922–927.

Barrade, P., Pittet, S., Rufer, A., 2000. Energy storage system using a series connection of supercapacitors, with an active device for equalizing the voltages. In: International Power Electronics Conference, Tokyo, Japan.

Becker, H.I., July 23, 1957. Low voltage electrolytic capacitor. US Patent 2800616.

Belhachemi, F., Rael, S., Davat, B., 2000. A physical based model of power electric double layer capacitor. IEEE, 3069-3075.

Bittner, A.M., Zhu, M., Yang, Y., Waibel, H.F., Konuma, M., Starke, U., Weber, C.J., 2012. Aging of electrochemical capacitors. J. Power Sources 203, 262–273.

Buller, S., Karden, E., Kok, D., De Doncker, R.K., 2002. Modeling the dynamic behavior of supercapacitors using impedance spectroscopy. IEEE Trans. Ind. Appl. 38 (6), 1622–1626.

Burke, A., 2000. Ultracapacitors: why, how and where is the technology. J. Power Sources 91, 37–50.

Catherino, H.A., Burgel, J.F., Shi, P.L., Rusek, A., Xiulin, Z., 2006. Hybrid power supplies: a capacitor assisted battery. J. Power Sources 162, 965–970.

Celzard, A., Collas, F., Mareché, J.F., Furdin, G., Rey, I., 2002. Porous electrodes-based double layer capacitors: pore structure versus series resistance. J. Power Sources 108, 153–162.

Cericola, D., Kötz, R., 2012. Hybridization of rechargeable batteries and electrochemical capacitors: principles and limits. Electrochim. Acta 72, 1–17.

Cericola, D., Novák, P., Wokaum, A., Kötz, R., 2011. Hybridization of electrochemical capacitors and rechargeable batteries: an experimental analysis of the different possible approaches utilizing activated carbon, $Li_4Ti_5O_{12}$ and $LiMn_2O_4$. J. Power Sources 196, 10305–10313.

Chang, Y., Wu, C., Wu, P., 2013. Synthesis of large surface area carbon xerogels for electrochemical double layer capacitors. J. Power Sources 223, 147–154.

Christen, T., Carlen, M.W., 2000. Theory of Ragone plots. J. Power Sources 91, 210–216.

Chunhe, C., et al., 2009. Research of supercapacitor voltage equalization strategy on rubber-tyred gantry crane energy saving system. In: Asia-Pacific Power and Energy Engineering Conference, 2009 (APPEEC '09).

Conway, B.E., 1990. Transition from "supercapacitor to battery" behavior in electrochemical energy storage. In: Proceedings of Power Sources Symposium, pp. 319–327.

Conway, B.E., Pell, W.G., 2003. Double layer and pseudocapacitance types of electrochemical capacitors and their applications to the development of hybrid devices. J. Solid State Electrochem. 7, 637–644.

Conway, B.E., Birss, V., Wojtowicz, J., 1997. The role and utilization of pseudocapacitance for energy storage by supercapacitors. J. Power Sources 66, 1–14.

Du, C., Pan, N., 2006. Supercapacitors using carbon nanotubes films by electrophoretic deposition. J. Power Sources 160, 1487–1494.

Du, X., Song, H., Chen, X., 2010. Graphene nanosheets as electrode material for electric double layer capacitors. Electrochim. Acta 55, 4812–4819.

Dubal, D.P., Gund, G.S., Lokhande, C.D., Holze, R., 2013. Cuo cauliflowers for supercapacitor application: novel potentiodynamic deposition. Mater. Res. Bull. 48, 923–928.

ELNA. http://www.elnaerica.com/company.htm.

Endo, M., Takeda, T., Kim, Y.J., Koshiba, K., Ishii, K., 2001. High power electric double layer capacitor (EDLC's); from operating principle to pore size control in advanced activated carbons. Carbon Sci. Vol. 1. (3 & 4), 117–128.

Fang, B., Binder, L., 2006. A modified activated carbon aerogel for high-energy storage in electric double layer capacitors. J. Power Sources 163, 616–622.

Fang, B., Wei, Y.-Z., Kumagai, M., 2006. Modified carbon materials for high-rate EDLCs application. J. Power Sources 155, 487–491.

Frackowiak, E., Béguin, F., 2001. Carbon materials for the electrochemical storage of energy in capacitors. Carbon 39, 937–950.

Gai Xiao, D., et al., 2008. Analysis on equalization circuit topology and system architecture for series-connected ultra-capacitor. In: IEEE Vehicle Power and Propulsion Conference, 2008 (VPPC '08).

Gamby, J., Taberna, P.L., Simon, P., Fauvarque, J.F., Chesneau, M., 2001. Studies and characterizations of various activated carbons used for carbon/carbon supercapacitors. J. Power Sources 101, 109–116.

Grimnes, S., Martinsen, Ø.G., 2008. Bioimpedance and Bioelectricity Basics, second ed., Elsevier- Academic Press, 471 p.

Gund, G.S., Dubal, D.P., Patil, B.H., Shinde, S.S., Lokhande, C.D., 2013. Enhanced activity of chemically synthesized hybrid graphene oxide/Mn_3O_4 composite for high performance capacitors. Electrochim. Acta 92, 205–215.

Huang, C., Chuang, C., Ting, J., Teng, H., 2008. Significantly enhanced charge conduction in electric double layer capacitors using carbon nanotube-grafted activated carbon electrodes. J. Power Sources 183, 406–410.

IEC-40/1378/DIN IEC 62391-1, June 2004. Fixed electric double layer capacitors for use in electronic equipment. Part I: generic specification.

IEC-40/1379/DIN IEC 62391-2, June 2004. Fixed electric double layer capacitors for use in electronic equipment. Part II: Sectional specification: electric double layer capacitors for power application.

Itagaki, M., Suzuki, S., Shitanda, I., Watanabe, K., Nakazawa, H., 2007. Impedance analysis of electrical double layer capacitor with transmission line model. J. Power Sources 164, 415–424.

Kanthal Globar. Maxcap® Double Layer Capacitors-Product Information & Application Data, Document M-2004A-09/07.

Karden, E., Buller, S., De Doncker, R.K., 2002. A frequency domain approach to dynamical modeling of electrochemical power sources. Electrochim. Acta 47, 2347–2356.

Kim, J., Kim, S., 2014. Preparation and electrochemical property of ionic liquid-attached graphene nanosheets for an application of supercapacitor electrode. Electrochim. Acta 119, 11–15.

Kurzweil, P., Frenzel, B., Gallay, R., 2005. Capacitance characterization methods and ageing behaviour of supercapacitors. In: The 15th International Seminar on Double Layer Capacitors.

Lajnef, W., Vinassa, J.-M., Briat, O., Azzopardi, S., Woirgard, E., 2007. Characterization methods and modelling of ultracapacitors for use as peak power sources. J. Power Sources 168, 553–560.

Levie, R.D., 1963. On porous electrodes in electrolyte solutions. Electrochim. Acta 8, 751–780.

Lisheng, S., Crow, M.L., 2008. Comparison of ultracapacitor electric circuit models. IEEE, 1 (6), 20–24.

Liu, C., Tsai, D., Chung, W., Li, K., Lee, K., Huang, Y., 2011. Electrochemical microcapacitors of patterned electrodes loaded with manganese oxide and carbon nanotubes. J. Power Sources 196, 5761–5768.

Mastragostino, M., Arbizzani, C., Soavi, F., 2002. Conducting polymers as electrode materials in supercapacitors. Solid State Ionics 148, 493–498.
Michael, M.S., Prabaharan, S.R.S., 2004. High voltage electrochemical double layer capacitors using conductive carbon as additives. J. Power Sources 136, 250–256.
Miller, J.R., Outlaw, R.A., Holloway, B.C., 2011. Graphene electric double layer capacitor with ultra high performance. Electrochim. Acta 56, 10443–10449.
Morimoto, T., Hiratsuka, K., Sanada, Y., Kurihara, K., 1996. Electric double-layer capacitor using organic electrolyte. J. Power Sources 60, 239–247.
Nasibi, M., Golozar, M.A., Rashed, G., 2013. Nanoporous carbon black particles as an electrode material for electrochemical double layer capacitors. Mater. Lett. 91, 323–325.
Nelms, R.M., et al. 1999. A comparison of two equivalent circuits for double-layer capacitors. In: Fourteenth Annual Applied Power Electronics Conference and Exposition (APEC '99): Vol.2. p. 692–698.
Ness Capacitor Co., Ltd. http://www.nesscap.com/prod/ba3.htm.
Niu, J., Pell, W.G., Conway, B.E., 2006. Requirements for performance characterization of C double-layer supercapacitors: applications to a high specific-area C-cloth material. J. Power Sources 156, 725–740.
Novoselov, K.S., Geim, A.K., Morozov, S.V., Jiang, D., Zhang, Y., Dubonos, S.V., Grigorieva, I.V., Firsov, A.A., 2004. Electric field effect in atomically thin carbon films. Science 306, 666–669.
Pandolfo, A.G., Hollenkamp, A.F., 2006. Carbon properties and their role in supercapacitors. J. Power Sources 157, 11–27.
Pasquier, A.D., Plitz, I., Gural, J., Badway, F., Amatucci, G.G., 2004. Power-ion battery: bridging the gap between Li-ion and supercapacitor chemistries. J. Power Sources 136, 160–170.
Paulo, C.J., Rodrigo, T., Lavall, L., Silva, G., 2014. Supercapacitors based on modified graphene electrodes with poly(ionic liquid). J. Power Sources 256, 264–273.
Pell, W.G., Conway, B.E., 1996. Quantitative modeling of factors determining Ragone plots for batteries and electrochemical capacitors. J. Power Sources 63, 255–266.
Peng, C., Zhang, S., Jewell, D., Chen, G., 2008. Carbon nanotube and conducting polymer composites for supercapacitors. Proc. Natl. Acad. Sci. U.S.A. 18, 777–778.
Pmar, N., Mierlo, J.V., Verburugge, B., Bossche, P.V.D., 2010. Power and life enhancement of battery—electrical double layer capacitor for hybrid electric and charge-depleting plug-in vehicle applications. Electrochim. Acta 55, 7524–7531.
Prabaharan, S.R.S., Michael, M.S., 2004. High voltage electrochemical double layer capacitors using conductive carbons as additives. J. Power Sources 136, 250–256.
Prabaharan, S.R.S., Vimala, R., Zainal, Z., 2006. Nanostructured mesoporous carbon as electrodes for supercapacitors. J. Power Sources 161, 730–736.
Qu, D., Shi, H., 1998. Studies of activated carbons used in double-layer capacitors. J. Power Sources 74, 99–107.
Ramya, R., Sivasubramanian, R., Sangaranarayanan, M.V., 2013. Conducting polymers-based electrochemical supercapacitors—progress and prospectus. Electrochim. Acta 101, 109–129.
Razumov, S.N., Klementov, A., Litvinenko, S., Beliakov, A., 2001. US Patent No. 6222723.
Rengui, L., et al., 2008. A new topology of switched capacitor circuit for the balance system of ultra-capacitor stacks. In: IEEE Vehicle Power and Propulsion Conference, 2008 (VPPC '08).
Rightmire, R.A., November 29, 1966. Electrical energy storage apparatus. US Patent 3288641.
Ruiz, V., Roldán, S., Villar, I., Blanco, C., Santamariá, R., 2013. Voltage dependence of carbon based supercapacitors for pseudocapacitance quantification. Electrochim. Acta 95, 225–229.

Sankapal, B.R., Gajare, H.B., Dubal, D.P., Gore, R.B., Salunkhe, R.R., Ahn, H., 2014. Presenting highest supercapacitance for TiO_2/MWNTs nanocomposites: novel method. Chem. Eng. J. 247, 103–110.

Sevilla, M., Mokaya, R., 2014. Energy storage applications of activated carbons: supercapacitors and hydrogen storage. Energy Environ. Sci. 7, 1250–1280.

Sharma, P., Bhatti, T.S., 2010. A review on electrochemical double-layer capacitors. Energy Convers. Manag. 51, 2901–2912.

Sikha, G., Popov, B., 2004. Performance optimization of a battery-capacitor hybrid system. J. Power Sources 134, 130–138.

Spyker RL, Nelms RM., 2000, Classical equivalent circuit parameters for a double-layer capacitor. Aerospace and Electronic Systems, IEEE Transactions on. 36(3): p. 829-836.

Sung, J., Kim, S., Jeong, S., Kim, E., Lee, K., 2006. Flexible micro-supercapacitors. J. Power Sources 162, 1467–1470.

Tõnurist, K., Jänes, A., Thomberg, T., Kurig, H., Lust, E., 2009. Influence of mesoporous separator properties on the parameters of electrical double-layer capacitor single cells. J. Electrochem. Soc. 156, A334–A342.

Tõnurist, K., Thomberg, T., Jänes, A., Romann, T., Sammelselg, V., Lust, E., 2013. Influence of separator properties on electrochemical performance of electrical double-layer capacitors. J. Electroanal. Chem. 689, 8–20.

Wang, L., Morishita, T., Toyopda, M., Inagaki, M., 2007. Asymmetric electric double layer capacitors using carbon electrodes with different pore size distributions. Electrochim. Acta 53, 882–886.

Wei, L., Yushin, G., 2012. Nanostructured activated carbons from natural precursors for electrical double layer capacitors. Nano Energy 1, 552–565.

Xianzhong, S., Xiong, Z., Haitao, Z., Dacheng, Z., Yanwei, M., 2012. A comparative study of activated carbon-based symmetric supercapacitors in Li_2SO_4 and KOH aqueous electrolytes. J. Solid State Electrochem. 16, 2597–2603.

Yang, F., Lnguang, L., Yuping, Y., He, Y., November 2010. Characterization, analysis and modeling of an ultracpacitor. In: Proceedings of EVS25, pp. 1–12.

Yasser, D., Venet, P., Gualous, H., Rojat, G., 2008. Electrical, frequency and thermal measurement and modeling of supercapacitor performance. In: Proceedings 3rd ESSSCAP, 2008. [CD-ROM].

Zhang, L.L., Zhao, X.S., 2009. Carbon-based materials as supercapacitor electrodes. Chem. Soc. Rev. 38, 2520–2531.

Zhang, Q., Xu, C., Lu, B., 2014. Super-long life supercapacitors based on the construction of Nifoam/graphene/Co_3S_4 composite film hybrid electrodes. Electrochim. Acta 132, 180–185.

Zhao, X., Huang, S., Cao, J., Xi, S., Wei, X., Kamamoto, J., Takarada, T., 2014. KOH activation of hypercoal to develop activated carbons for electric double layer capacitors. J. Anal. Appl. Pyrolysis 105, 116–121.

Zheng, C., Zhou, X., Cao, H., Wang, G., Liu, Z., 2014. Synthesis of porous graphene/activated carbon composite with high packing density and large surface area for supercapacitor electrode material. J. Power Sources 258, 290–296.

Supercapacitor as a lossless dropper in DC-DC converters

6

Kosala Gunawardane

6.1 Introduction

Supercapacitors are typically used as energy storage devices (ESD) and such applications are well documented. However in state of the art devices, with the capacitance values increased by a factor of nearly one million (compared to large electrolytic capacitors), less significant fact is that the equivalent series resistance (ESR) values have dropped down to a range of 0.1 to about 100 mΩ. Given this important observation, when a supercapacitor of the value C is charged or discharged with a constant DC current of I_C we can expect an approximate voltage change across the terminals given by $\frac{I_C \Delta t}{C}$, where Δt is the time. When the capacitance value is very large, and the time duration is short we can expect only very small voltage change across the terminals of the capacitor. Combining this effect with the milliohms order ESR, we can expect the device to have a near constant voltage drop across a terminal, together with a very low resistive drop. In other words, if we place a supercapacitor in series with an electronic circuit (supplied by a voltage source value of maximum value of V_p), which typically consumes a finite current the capacitor will take very long time to block the path by developing a voltage difference across its terminals which is equal to the difference between V_p and $V_{\text{in}}^{\text{min}}$ where $V_{\text{in}}^{\text{min}}$ is the minimum input voltage required to operate the circuit as expected. Figure 6.1 simply depicts this scenario, where the supercapacitor can be acting as a lossless dropper in a practical circuit.

Next important situation here is to realize that the typical voltage drop across the ESR of the supercapacitor, could be much less than the typical DC voltage drop across a semiconductor power switch such as a metal oxide semiconductor field effect transistor (MOSFET), insulated gate bipolar transistor (IGBT) or a bipolar. Sate of the art power MOSFETs have typical DC on resistances ($R_{\text{DS(on)}}$) within the range of 1 to few 100 mΩs.

Combining these two cases of (i) voltage change across supercapacitor can be small due to charge/discharge currents over a short period and (ii) voltage drop across the ESR due to charge/discharge current could be much smaller than typical static DC losses across a power semiconductor switch has given rise to a unique DC-DC converter topology, developed by the supercapacitor applications team at Waikato (Kularatna and Fernando, 2009, 2011). This chapter provides a discussion on how a supercapacitor can be used as a lossless dropper in a linear DC-DC converter where end-to-end efficiency (ETEE) of a linear regulator based on low dropout regulators (LDO) can be improved by a significant multiplication factor.

The chapter commences with a brief overview on DC power management and DC-DC converter topologies, and directly proceeds to a designers view point on

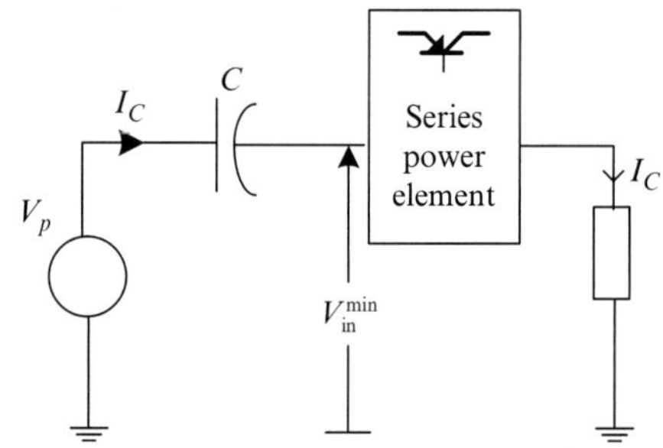

Figure 6.1 The basic concept: supercapacitor as a lossless dropper in a practical circuit.

the novel application of supercapacitor as a lossless dropper technique with some discussion on results achievable in typical DC-DC converters such as 12-to-5 V and 5-to-3.3 V cases.

6.2 DC-DC converters and DC power management

All electronic systems require a very stable DC source with constant DC value, minimum ripple and noise at the output, and a fast response to changes of load current without a dip or an overshoot at the output rail. Another important requirement in modern electronic systems is energy efficiency in the converters, and, longest possible battery run-time for battery powered systems. Energy efficiency is a very hot topic among product designers where ETEE from AC power source to final DC rails is to be kept as high as possible, while the battery run time is optimized by many advanced techniques combining different types DC-DC converters (Kularatna, 2011).

6.2.1 *Commonly used DC-DC converters*

In modern electronics, the following three basic approaches are available for DC-DC conversion:

1. Linear regulators
2. Switched-mode power supplies (SMPS)
3. Switched capacitors (charge pumps)

In linear regulators, a power semiconductor device such as a power MOSFET, IGBT, or a bipolar power transistor is used as a voltage dropper in the series path between the unregulated DC input and the regulated output. A control loop compares a sample of regulated output voltage with a constant voltage DC reference source, and feeds the control input of the power semiconductor switch to keep the output DC value at a constant value. Historically the linear regulators were the first development wave for DC-DC converters in the 1960s to 1970s, but these were very inefficient with typical efficiencies in the range of 30-60%. However, they had very low noise output with fast transient response to fast changes of load current.

Switched-mode converters use inductors and capacitors as ESDs to convert one DC voltage to another with a theoretical efficiency of 100%. There are several commonly used converter topologies such as buck converter, boost converter, and buck-boost converter and several related topologies. Kularatna (2011) provides a very good discussion on the subject and compares the techniques with linear regulators. All these topologies use power semiconductors as switches operating in the range of few tens of kHz to over 2 MHz.

Switched capacitor converters, or charge pumps use capacitor based switching techniques but they do not use bulky inductors or transformers which are typically used in switch-mode power supplies. These are low current capable converters, typically limited to few 100 mA only.

Table 6.1 compares the popular DC-DC converter techniques. More details can be found in Walt Kester (http://www.analog.com/static/imported-files/tutorials/ptmsect3.pdf), Jovalusky (2005), Linear and Switching Voltage Regulator Handbook on Semiconductor (2002), Simpson (2012), Kester et al. (1998), and Palumbo and Pappalardo (2010).

6.2.2 LDOs as a special case of linear regulators

Since the subject of this chapter is based on the use of LDOs, which is a special case of linear regulators let us start with a brief discussion on linear regulator designs. As shown in Figure 6.2, the major components of a typical linear regulator include the pass element (series power semiconductor which acts as voltage dropper element in the load current path), the precision reference, the feedback network, and the error amplifier. Chapters 1 and 2 of Kularatna (2011) give a detailed account on the design aspects of DC power supplies based on linear regulators.

In general when a linear regulator can be designed with a low DC voltage difference between the input and the output DC rails, it is called an LDO. Term low dropout comes from the fact that these single chip silicon devices, very commonly used in

Table 6.1 Comparison of popular DC-DC converter techniques

Feature	Linear regulators	Charge pump converters	Switched-mode power supplies
Design complexity	Low	Moderate	Moderate-to-high
Cost	Low	Moderate	Moderate
Noise	Lowest	Low	Low-to-moderate
Efficiency	Low-to-moderate	Moderate-to-high	High
Thermal management	Poor-to-moderate	Good	Best
Output current capability	Moderate	Low	High
Requirement of magnetic parts	No	No	Yes
Limitations	Cannot step up	V_{IN}/V_{OUT} ratio	Layout considerations

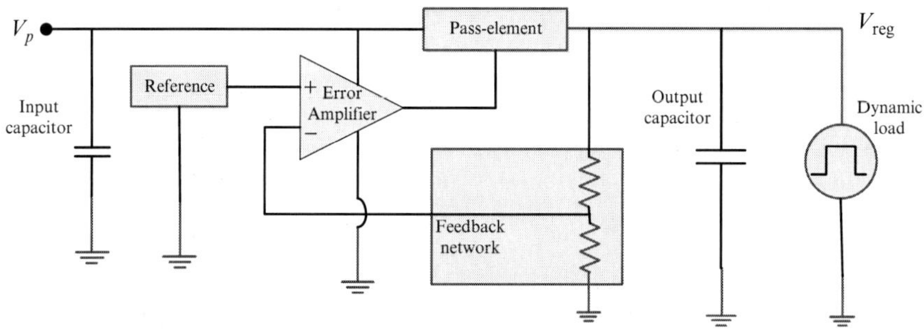

Figure 6.2 Linear regulator block diagram.

portable products such as phones, cameras, and laptops, usually combined with switch-mode types, can operate down to a very low voltage difference between the input and output, which is called the dropout voltage at which it stops working as a useful DC-DC converter.

Figure 6.3 illustrates the commercial utilization of LDO and switching regulator families with predictions for the future years as per market research reports from Databeans, indicating the total business potential in millions of dollars (lower trace-M), total unit sales in million units (middle trace-MU) and the average selling price (upper trace-ASP) of these components (Inouye et al., 2010).

Due to the difference in pass elements, in general, there are three basic types of linear regulator topologies:

1. standard (NPN Darlington) regulator,
2. quasi LDO regulator,
3. low dropout or LDO regulator.

In a linear regulator, the minimum possible difference between the input and the output voltages while retaining the regulatory function, is referred to as the dropout voltage. In the above topologies, the dropout voltage shows a remarkable difference across the three. A linear regulator, that operates with the highest dropout voltage, dissipates the most internal power so has the lowest efficiency.

Usually in linear regulator configurations, most of the energy is dissipated across the pass element. Typical linear regulator construction does not have any energy storing mechanisms, so the power that is not delivered to the load is dissipated as heat within the regulator.

The power which is dissipated within the regulator (P_D) can be described as follows (Falin, 2006; Lee, 1999; Marasco; Rincon-Mora, 2009; Texas Instruments (2009); Zendzian, 1999):

$$P_D = \left[V_p - V_{reg}\right] I_L + V_p I_q \tag{6.1}$$

where V_p, V_{reg}, I_L, and I_q are the input voltage, the regulated output voltage, the load current, and ground pin current (current drawn by the control circuits), respectively.

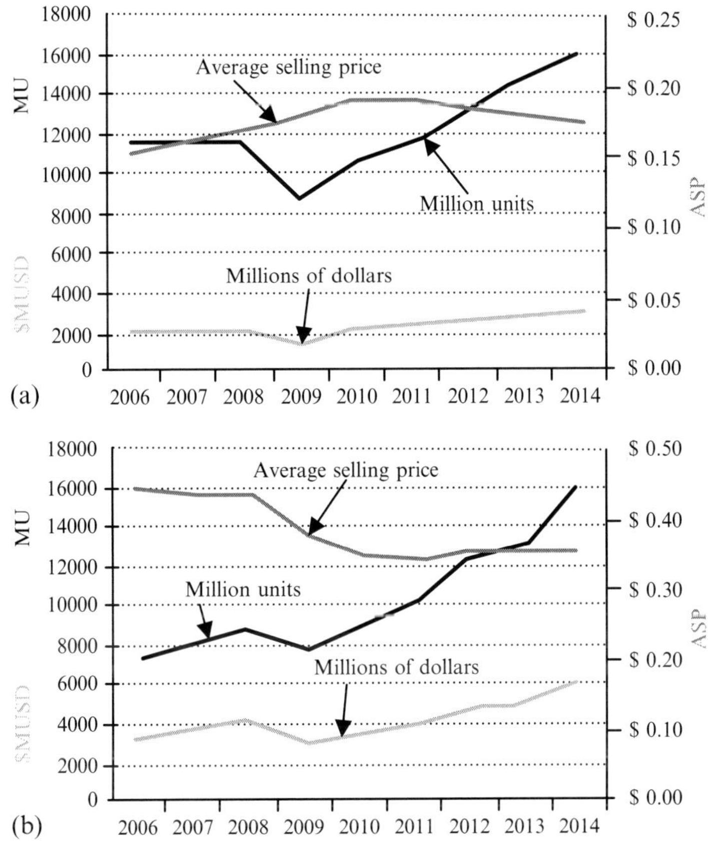

Figure 6.3 Worldwide market forecast: (a) switching regulators; (b) LDO regulators (Inouye et al., 2010).

The first part of this relationship is the dissipation of the pass device and the second part is the power consumption of the controller portion of the circuit. Linear regulators generate excessive heat if the voltage difference between input and output voltage is high. But, LDOs have extremely low drop-out voltages which enhances the efficiency.

The efficiency of a linear regulator can be approximated by,

$$\eta = \frac{V_{\text{reg}} I_{\text{L}}}{V_p I_p} \approx \frac{V_{\text{reg}} I_{\text{L}}}{V_p (I_{\text{L}} + I_q)} \approx \frac{V_{\text{reg}}}{V_p \left[1 + \frac{I_q}{I_{\text{L}}}\right]} \tag{6.2}$$

As shown in Eqn (6.2), the efficiency of an LDO regulator is also limited by the ground-pin current of the circuit. To have a high-efficiency LDO, the ground-pin current should be minimized (http://www.analog.com/library/analogdialogue/archives/43-08/ldo.html; Texas Instruments (2009)). Modern LDOs have reasonably low I_q,

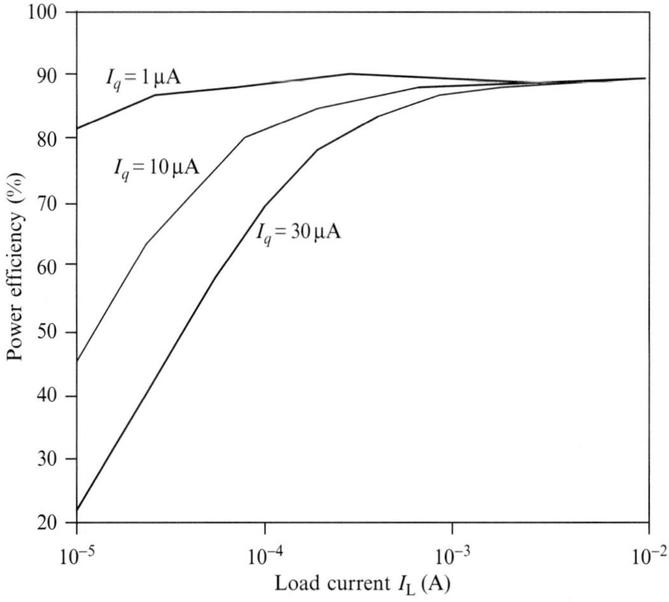

Figure 6.4 Effect of ground-pin current on efficiency of an LDO (Man et al., 2007).

and for simplicity, such a value can be neglected in efficiency calculations if I_q is very small compared to I_L (Man et al., 2007; Rincon-Mora, 1996).

Figure 6.4 shows the regulator efficiency as a function of load current (I_L) with different I_q under a particular output voltage (V_{reg}). It shows that the use of very low I_L improves the regulator efficiency at small I_q, which often happens when systems are operating in standby mode to extend battery life.

6.2.3 Losses associated with switch-mode converter-based DC power supplies

An important advantage of switching regulators is their relatively high power efficiency. Assuming the use of ideal energy storage elements such as capacitors and inductors with perfect lossless switches theoretically 100% efficiency can be achieved. But, in practical situations, inductors, capacitors and transistor switches do not show their ideal characteristics and thus deviate much from the theoretical maximum efficiency. Therefore, in practice the range of efficiencies could be in the range of 80-90% for the DC-DC converter stage (Walt Kester). If the AC-DC conversion stage is included as is required in a typical off-the-line power supply, ETEEs will be much lower. The following example illustrates this practical scenario.

There can be various sources of loss within a typical SMPS, depending on its power level, topology, and circuit complexity. The approximate loss contribution range of

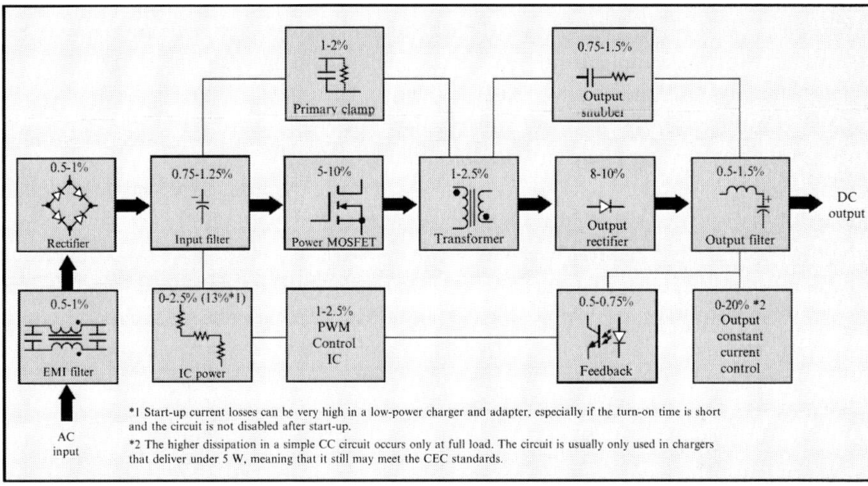

Figure 6.5 Block diagram of the power-consuming circuits within a SMPS (Jovalusky, 2005).

each is shown in Figure 6.5, as a percentage of the input power, for a typical supply that delivers 10 W of load power (Jovalusky, 2005; Kularatna, 2011).

In general, the major loss contributors in a switching power supply are:

1. static losses in diodes;
2. transistor losses such as $I^2R_{DS(on)}$ in MOSFETs or losses due to saturation voltage of BJTs;
3. dynamic losses in switching transistors and diodes occurring at high frequency due to the charging and discharging of different parasitic capacitances;
4. losses due to charge/discharge currents through ESR (in most cases AC ESR comes into effect);
5. magnetic components losses;
6. control circuits losses/PCB track losses.

Switched capacitor converters are capable of achieving efficiencies greater than 90%. However, these converters are not suitable to maintain high efficiency for a wide range of ratios of input to output voltages, unlike the case for switching and low dropout regulators. The load current capability is limited by the size of the capacitors and the current-carrying capacity of the switches. In practice, typical switched capacitor inverter chips are available with 10-500 mA load currents (Kester et al., 1998).

6.2.4 Power management approaches using combinations of switch modes and LDOs

Power supply designs for portable electronic products have become an exponentially growing segment due to the evolutional advancements in both switch-mode and linear voltage regulator designs. With sinking operating voltages and rising current demands, the designer must consider factors such as power wastage, long battery life, small size, and low cost. These design parameters change and rearrange the design priorities for power supplies and makes the choice between linear and switching mode power supplies ambiguous.

LDOs are frequently used in battery powered portable products such as point of load (POL) power supplies where adaptation of switch-mode power supplies becomes difficult, primarily due to noise performance, slew rate requirements, and the complex nature of their output impedance (Linear regulators, 2001; Travis, 1998). An example of a POL approach in a portable product with a 48 to 8 V intermediate bus converter is given in Figure 6.6 where all other voltages are derived from the 8 V bus. In this kind of situation, to minimize compound efficiency loss, which can be significant at times, the POL converters should offer efficiency close to or above 90% (Bull and Smith, 2003). The digital still camera (DSC) is shown in Figure 6.7, indicating the combined use of switch modes and LDOs is another example. Figure 6.8 depicts the part of power management solution offered for such a case from Texas Instruments TPS65010 IC. In TPS65010 IC, part of the circuitry allows configuring several switch-mode circuits, while the other part of the silicon is used to configure LDO circuits. More details are available in Day (2003) and (http://www.ti.com/product/tps65010). In some portable products where some parts of the load requires very limited currents, in the orders of 10 to few 100 mA only, and with DC power rails higher than the battery voltage or the bulk switching power supply's output rail voltage, switched capacitor converters are also used (Palumbo and Pappalardo, 2010).

In practical products switchers and LDOs are combined for many practical reasons such as,

1. noise level of switchers and the need for RFI/EMI filters,
2. LDO output quality,
3. fast transient response of LDOs,
4. simplicity and small PCB footprint consumed by LDOs,
5. inductors as bulky component in switchers.

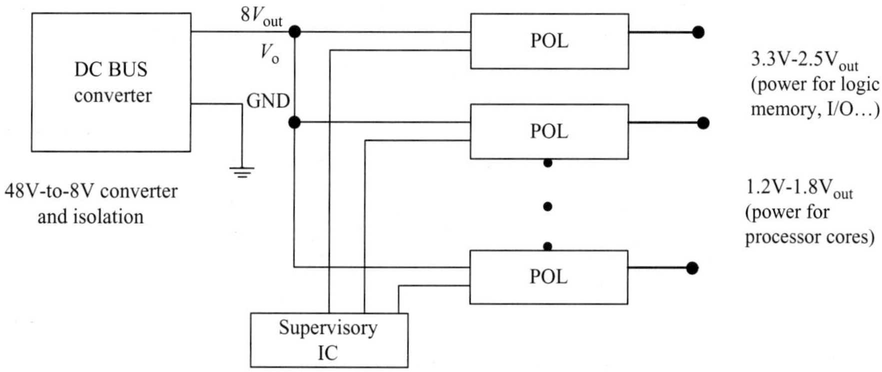

Figure 6.6 An 8 V intermediate bus architecture with 48-to-8 V bulk converter and POL stages (Bull and Smith, 2003).

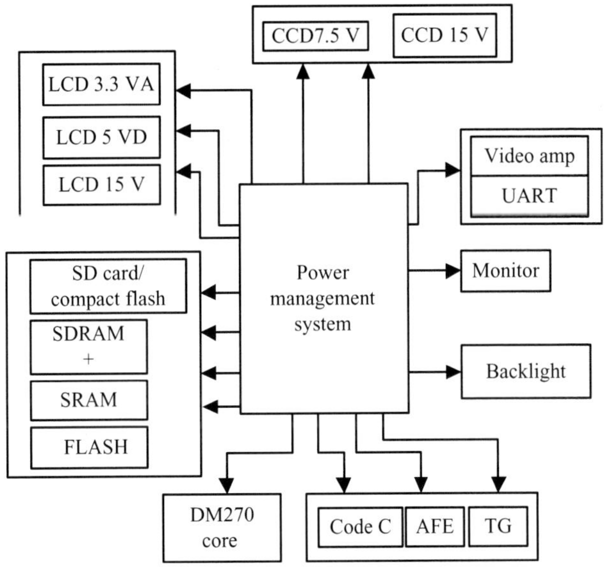

Figure 6.7 An example of a DSC architecture (Day, 2003).

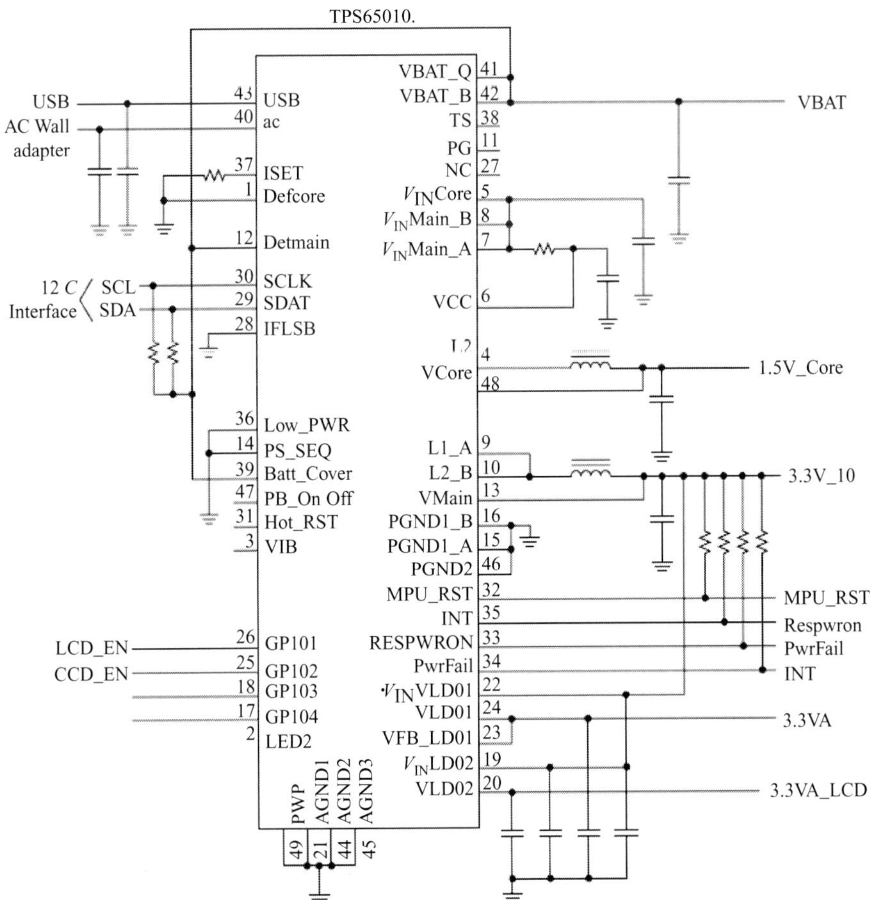

Figure 6.8 Power management solution, Texas Instruments TPS65010 IC (Day, 2003).

6.3 Supercapacitor assisted low dropout regulator (SCALDO) technique

6.3.1 Essential fundamentals

So far our discussion summarizes the fact that in practical electronic systems a combination of different DC-DC converter topologies are used to achieve the overall efficiency, noise minimization, transient response and the packaging issues. If the control circuit consumes only a negligible amount of power the best possible theoretical efficiency of a linear regulator is given by,

$$\eta = \frac{V_{\mathrm{reg}}}{V_p} \times 100\% \tag{6.3}$$

where V_{reg} and V_p are the regulated output and the input voltages, respectively.

Usually in a linear regulator configuration, most of the energy is dissipated across the pass-element or the series power semiconductor. Introducing a voltage dropper in series with the pass element will make the dissipation across the pass element lower.

Placing a resistor in series with the pass element as shown in Figure 6.9 would move some of the energy to the resistor, but this will not improve the overall ETEE.

For this purpose, the ideal voltage dropper element which can be used in the series path is an energy storing element, such as a capacitor or an inductor. If an inductor voltage is used as a voltage dropper, current through the inductor cannot be a constant. A capacitor is the matching candidate as a series dropper element, but in such a case if the capacitance is very small it will block the circuit quickly, stopping the current in a negligible period of time. As discussed in Section 6.1, an extremely large ideal capacitor can be logically used as a lossless dropper in a linear regulator in the series path. The fundamental concept of choosing a supercapacitor as the series dropper element is as follows.

The voltage change (dv) across a very large capacitance C such as a supercapacitor is very small when a finite charge (or discharge) current $i(t)$ flows during a finite time t

$$\mathrm{d}v = \int_0^t \frac{1}{C} i(t)\,\mathrm{d}t \tag{6.4}$$

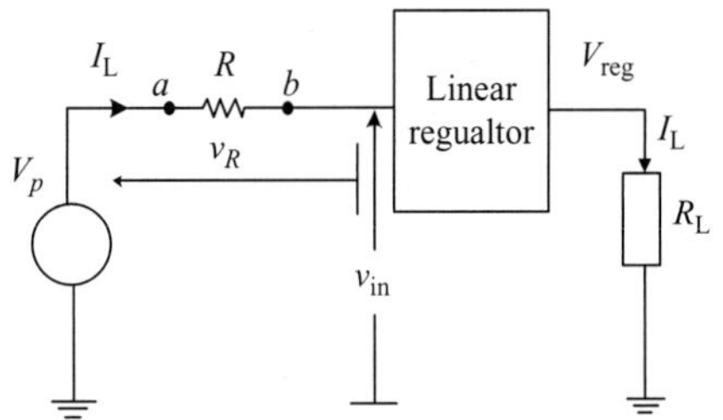

Figure 6.9 An arrangement of placing a resistor in series with the pass element to make the dissipation lower.

A large capacitor, such as a supercapacitor (with negligible ESR value) with a DC voltage can be used as a voltage dropper in the series path, then the effect of the voltage change (dv) due to charging the capacitor for a finite amount of time, can be neglected in terms of the operational principles applicable to that particular circuit. For example, when a large capacitor is connected in series to a circuit for a finite time, the capacitor will not act as a blocking element to that circuit due to charging during that time. Based on this simple principle, using a single supercapacitor or an array of supercapacitors in the series path to store and release the energy, the supercapacitor based efficiency improvement technique suitable for linear regulators was developed (Kularatna and Fernando, 2009, 2011).

6.3.2 Basic concept

Figure 6.10 illustrates the concept: a supercapacitor C_{SC} is placed in series with the input of the LDO. Since the series capacitor is large (of the order of farads), it will pass current as it charges for a reasonable period of time. Based on this simple principle, a single supercapacitor can cyclically store and release energy. As per Figure 6.10(a) in the first phase of the operation, the supercapacitor is charged from the unregulated power supply. When the value of LDO input voltage, V_{in} drops to the minimum allowable voltage V_{in}^{min} the operation is switched to the next phase. In this phase, disconnecting the unregulated power supply, the supercapacitor is connected in parallel with the input of the LDO as shown in Figure 6.10(b), to release the stored energy in the supercapacitor.

Assuming the initial voltage across the supercapacitor (C_{SC}) is $V_{SC}(0)$ after charging for a period t, the instantaneous voltage across the capacitor V_{SC} is given by,

$$V_{SC} = V_{SC}(0) + \frac{I_L t}{C_{SC}} \tag{6.5}$$

where I_L is the load current entering the capacitor from the unregulated power supply. (We can assume that the load acts as a constant-current sink, and that the bypass current from the LDO to ground can be neglected.)

The unregulated source voltage V_p, equals the sum of the supercapacitor voltage and the LDO input voltage,

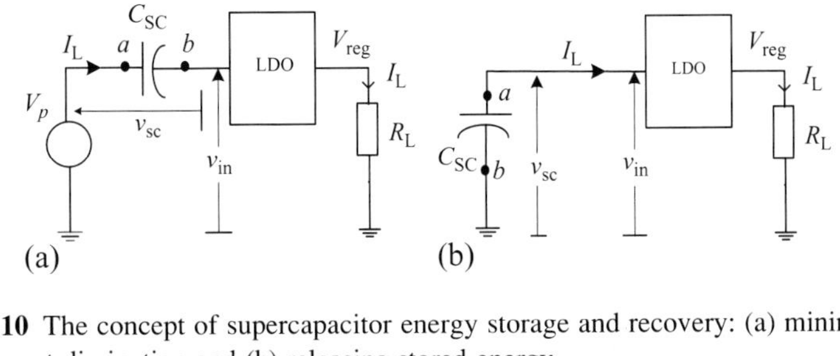

Figure 6.10 The concept of supercapacitor energy storage and recovery: (a) minimizing the series element dissipation and (b) releasing stored energy.

$$V_p = V_{SC}(t) + V_{in}(t) \tag{6.6}$$

The supercapacitor charges until V_{in} reaches V_{in}^{min} while the voltage across the capacitor reaches $V_p - V_{in}^{min}$ at the end of the charging time. In order to discharge this supercapacitor at the next stage down to V_{in}^{min}, the criterion $V_p - V_{in}^{min} > V_{in}^{min}$ must be satisfied. This creates the condition $V_p > 2V_{in}^{min}$.

At the end of the charging cycle, the stored energy can be released as shown in Figure 6.10(b). Discharging continues until the LDO input voltage drops back to V_{in}^{min} satisfying the following equation,

$$V_p - 2V_{in}^{min} = \frac{I_L \Delta t}{C_{SC}} \tag{6.7}$$

The circuit draws power from the unregulated input only during half the time of its operating period: during the supercapacitor charging phase, the supercapacitor draws current from the unregulated supply, whereas during the discharging, the supercapacitor delivers power to the LDO keeping the unregulated power supply disconnected from the system.

Considering the charge balance of the supercapacitor during the charge-discharge cycle,

$$I_L t_{ch} = I_L t_{dch} \tag{6.8}$$

where t_{ch} and t_{dch} are charge time and discharge time respectively.

This gives,

$$t_{ch} = t_{dch} \tag{6.9}$$

So that the average input current drawn from the unregulated power supply is,

$$I_{avg} = \frac{I_L t_{ch}}{t_{ch} + t_{dch}} = \frac{I_L}{2} \tag{6.10}$$

Assuming that the capacitors and switching elements are ideal and neglecting the power consumed by the control circuits, the approximate end-to end efficiency is given by,

$$\eta = \frac{P_{out}}{P_{in}} = \frac{V_{reg} I_L}{V_p \frac{I_L}{2}} = \frac{2V_{reg}}{V_p} \tag{6.11}$$

The discussion above applies to a practical DC-DC converter such as 12-to-5 V converter, where maximum ETEE possible is $\frac{5}{12}100\%$ which is approximately 42%. In a practical implementation, a supercapacitor can be used with a DC voltage between 6.7 (12-5.3 V) and 5.3 V for case with a commercial LDO chip with a minimum input voltage such as 5.3 V at the regulation. As we see from Eqn (6.11) maximum efficiency achievable is 84%. Practical values of 79-81% were achieved in preliminary proof of concept projects (Kularatna and Fernando, 2009; Kularatna et al., 2010). More discussion on this subject will be found in section 6.6. In general, the practical scenario is, the supercap keeps charging above its average DC voltage within the cycle, in the series mode, and later the supercap releases its charge until it reaches a minimum voltage required for the operation of the LDO, keeping the DC rail disconnected.

6.4 Generalized SCALDO concept

Simple case in the previous discussion requires only one SC and four low-speed switches. However, for general cases of step-down converters, we can extend the basic idea into general case applicable to almost all cases of step-down converters.

6.4.1 Series to parallel SC array

As shown in Figure 6.11 an array of n ideal series supercapacitors, C_{SC} each, is placed in series with the input of the LDO. This configuration is applicable to a SCALDO regulator, when it satisfies the criterion $V_p > 2V_{in}^{min}$. Basically, this is the extended version of the single supercapacitor, $n = 1$ case. The number of supercapacitors, n, can be determined using the following formula

$$n < \frac{V_p - V_{in}^{min}}{V_{in}^{min}} \tag{6.12}$$

As the equivalent capacitance of the series capacitor array is large (in the order of farads), it will pass current as it charges for a reasonable period of time. As per Figure 6.11(a) in the first phase of the operation, the series capacitor array is charged until the value of the LDO input voltage, V_{in} is dropped down to its minimum possible input voltage, V_{in}^{min}.

Assuming the initial voltage across a supercapacitor (C_{SC}) is $V_{SC}(0)$, after charging for a period t, the instantaneous voltage across the capacitor V_{SC} is given by,

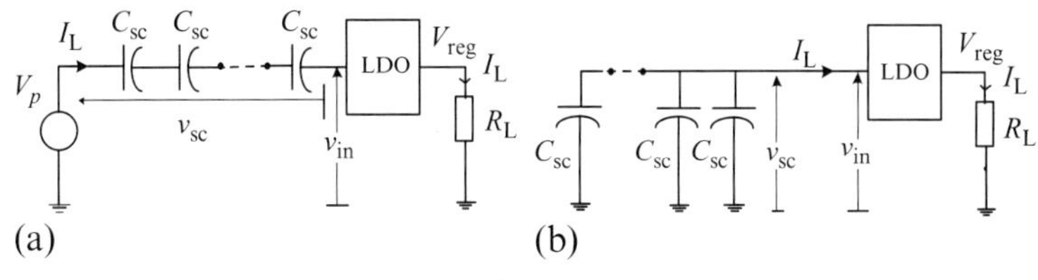

Figure 6.11 The concept of SCALDO regulator for $V_p > 2V_{in}^{min}$: (a) charging n capacitors series minimizing the series element dissipation and (b) discharging n capacitors parallel releasing stored energy.

$$V_{SC} = V_{SC}(0) + \frac{I_L t}{C_{SC}} \tag{6.13}$$

where I_L is the load current entering each capacitor from the power supply during the charging process assuming the ground pin current is negligible.

The voltage across a single capacitor at the end of the charging is,

$$V_{sc}^{max} = \frac{V_p - V_{in}^{min}}{n} \tag{6.14}$$

At the end of this phase, the array of supercapacitors is connected to the input of the LDO as an array of parallel capacitor as shown in Figure 6.11(b) to release the stored energy. This discharging process continues until the LDO input voltage drops back to V_{in}^{min}.

The circuit draws power from the unregulated supply during the supercapacitor charging phase (t_{ch}), whereas during the discharge phase (t_{dch}) the supercapacitor delivers power to the LDO keeping the unregulated power supply disconnected from the system.

Considering the charge balance of a single capacitor during charge-discharge cycles,

$$I_L t_{ch} = \frac{I_L}{n} t_{dch} \tag{6.15}$$

which gives,

$$t_{dch} = n t_{ch} \tag{6.16}$$

The average input current taken from the unregulated power supply is,

$$I_{avg} = \frac{I_L t_{ch} + 0 \times t_{dch}}{t_{ch} + t_{dch}} = \frac{I_L}{n+1} \tag{6.17}$$

Assuming that the capacitors and switching elements are ideal and neglecting the power consumed by control circuits, approximate ETEE for this configuration is given by,

$$\eta = \frac{P_{out}}{P_{in}} = \frac{I_L V_{reg}}{V_p I_{avg}} = \frac{I_L V_{reg}}{V_p \frac{I_L}{n+1}} = (n+1)\frac{V_{reg}}{V_p} \tag{6.18}$$

6.4.2 *Parallel to series SC array*

As shown in Figure 6.12(a) an array of n ideal parallel supercapacitors C_{SC} each, is placed in the input of the LDO. This is applicable to a SCALDO regulator

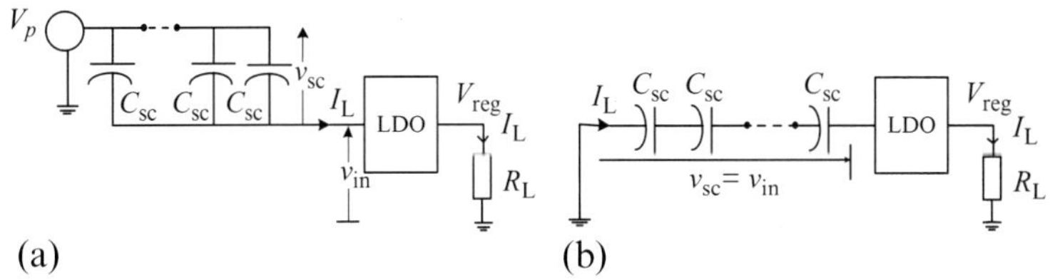

Figure 6.12 The concept of SCALDO regulator for $V_p < 2V_{in}^{min}$: (a) charging n parallel capacitor array minimizing the series element dissipation; (b) discharging n capacitors series releasing stored energy.

when $V_p < 2V_{in}^{min}$. The number of supercapacitors, n, can be determined using the following formula.

$$n > \frac{V_{in}^{min}}{V_p - V_{in}^{min}} \tag{6.19}$$

As the resultant of the parallel capacitor array is large (in the order of farads), it will pass current as it charges for a reasonable period of time. As per Figure 6.12(a) in the first phase of the operation, the parallel capacitor array is charged until the value of the LDO input voltage, V_{in} drops to V_{in}^{min}.

Assuming the initial voltage across a supercapacitor (C_{SC}) is $V_{SC}(0)$, after charging for a period t, the instantaneous voltage across each capacitor V_{SC} is given by,

$$V_{SC} = V_{SC}(0) + \frac{I_L t}{nC_{SC}} \tag{6.20}$$

where I_L/n is the portion of load current entering each capacitor from the power supply during the charging process assuming the ground pin current is negligible.

The voltage across a single capacitor at the end of charging is,

$$V_{sc}^{max} = V_p - V_{in}^{min} \tag{6.21}$$

As shown in Figure 6.12(b), in the next phase of operation, using an array of series capacitors, the stored energy is released until the LDO input voltage drops back to V_{in}^{min}.

Considering the charge balance of a single capacitor during charge discharge cycles,

$$\frac{I_L}{n} t_{ch} = I_L t_{dch} \tag{6.22}$$

which gives,

$$t_{dch} = \frac{t_{ch}}{n} \tag{6.23}$$

Table 6.2 The parameters values for the three different cases of SCALDO topologies

Parameter	$V_p > 2V_{in}^{min}$, basic config	$V_p > 2V_{in}^{min}$ gen. config	$V_p < 2V_{in}^{min}$ gen. config
n	1	$n < \frac{V_p - V_{in}^{min}}{V_{in}^{min}}$	$n > \frac{V_{in}^{min}}{V_p - V_{in}^{min}}$
k	4	$3n+1$	$3n+1$
V_{sc}^{max}	$V_p - V_{in}^{min} - I_L(2r + r_{sc})$	$\frac{1}{n}\left[V_p - V_{in}^{min} - I_L[(n+1)r + nr_{sc}]\right]$	$V_p - V_{in}^{min} - \frac{I_L}{n}(2r + r_{sc})$
t_{ch}	T_1	$\frac{T_n}{n}$	T_n
t_{dch}	T_1	T_n	$\frac{T_n}{n}$
η_r	2	$1+n$	$1+\frac{1}{n}$

n, number of supercapacitors; k, number of switches; V_{sc}^{max}, max voltage across the supercapacitor; t_{ch}, supercapacitor charge time, t_{dch}, supercapacitor discharge time; η_r, efficiency improvement factor.

Therefore, the average input current drawn from the unregulated power supply is,

$$I_{avg} = \frac{I_L t_{ch} + 0 \times t_{dch}}{t_{ch} + t_{dch}} = \frac{I_L}{1 + \frac{1}{n}} \tag{6.24}$$

Similar to $V_p > 2V_{in}^{min}$ case, under ideal conditions, the approximate ETEE is given by,

$$\eta = \frac{P_{out}}{P_{in}} = \frac{I_L V_{reg}}{V_p I_{avg}} = \frac{I_L V_{reg}}{V_p \frac{I_L}{1 + \frac{1}{n}}} = \left(1 + \frac{1}{n}\right)\frac{V_{reg}}{V_p} \tag{6.25}$$

Table 6.2 summarizes the characteristics of the basic configuration which satisfy $V_p > 2V_{in}^{min}$ criterion using a single supercapacitor and the two general configurations which satisfy $V_p > 2V_{in}^{min}$ and $V_p < 2V_{in}^{min}$ criteria using an array of supercapacitors. η_r the efficiency improvement factor is a benchmark indication of this technique, and for the two different general cases discussed in Section 6.4, these are $(1+n)$ and $(1+\frac{1}{n})$ for the two cases of $V_p > V_{in}^{min}$ and $V_p < V_{in}^{min}$ respectively.

6.5 Practical examples

The SCALDO technique can be easily configured for common DC-DC converters such as 12-to-5 V, 5-to-3.3 V, and 5-to-1.5 V and the technique provides efficiency improvement factors of 2, 1.33, and 3 respectively, compared with linear converters with the same input-output combinations (Kularatna and Fernando, 2009; Kularatna et al., 2010, 2011a,b,c; Kankanamge and Kularatna, 2012). For example, in a 5-to-

Table 6.3 **Some useful parameters and their values for a selected converter requirements**

Configuration	End-to-end regulation requirement	LDO min. input voltage	Capacitor requirements				Approximate efficiency at different stages of ope ration in discharging			Efficiency improvement factor		Max theoretical efficiency of a standard regulator
			Number of caps (n)	Voltage across a single cap in discharging (V)			@ Max voltage	@ Mid voltage	@ Min voltage			
				Max.	Mid	Min						
$V_p < 2V_{in,(\min)}$	5-3.3 V	3.5	3	1.50	1.40	1.20	73%	78%	92%	1+1/n	1.33	66%
$V_p > 2V_{in,(\min)}$	12-5 V	5.3	1	6.50	6.00	5.30	77%	83%	93%	2	2	42%
$V_p > 2V_{in,(\min)}$	5-1.5 V	1.6	2	1.70	1.65	1.60	88%	91%	94%	1+n	3	30%

Parameters: number of capacitors; capacitor voltages and approximate efficiencies of SCALDO regulators at transfer times.

1.5 V SCALDO regulator, using thin profile supercapacitors in the range of fractional farads to a few farads, this translates to an approximate ETEE of nearly 90%. Table 6.3 indicates the efficiencies at each transfer stage, assuming that LDO is not abruptly disconnected. This table gives a clear overview of how this novel topology can assist with increasing the ETEE of a linear regulator, with a view to getting the best out of linear-regulator-based DC-DC converters. Practical implementations of these topologies will be discussed in the next chapter.

6.6 SCALDO implementation examples

6.6.1 12-to-5 V SCALDO regulator

Figure 6.13 shows the schematic implementation of the 12-to-5 V SCALDO regulator which satisfies the criteria, $V_p > 2V_{in}^{min}$, in the form of its basic configuration, $n = 1$. In designing this prototype, an LDO of the type MCP1827 from Microchip Technology Inc., which has the specifications shown in Table 6.4, was used as the main low dropout regulator. For higher-efficiency LDO operation, an approximate LDO input voltage of 0.3 V above the regulated output voltage (5 V) was chosen, so that V_{in}^{min} was maintained at 5.3 V. As the power supply input voltage was around 12 V and the regulated output is 5 V, three supercapacitors of 4 F/2.5 V of the type PC-series from Maxwell Technologies, were connected in series to maintain the required voltage level of the series dropper element. This resulted in 1.33 F/7.5 V resultant capacitor as the series capacitor.

Solid state relays of the type PVN012 from International Rectifier were used as S_1, S_2, S_3, and S_4 switches. Switches S_1 and S_3 control supercapacitor charging, while S_2

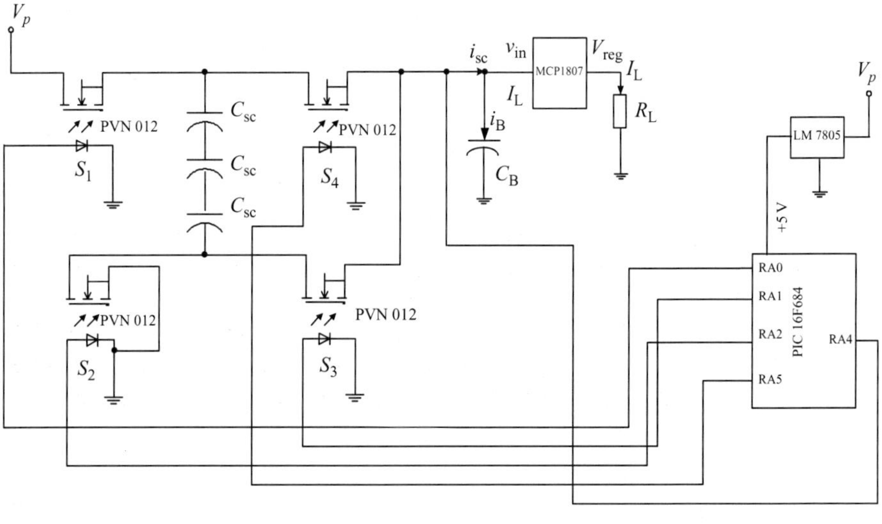

Figure 6.13 The schematic of the 12-to-5 V SCALDO regulator designed using MCP1827 LDO, 4 F/2.5 V Maxwell Supercapacitors, PVN012 Photovoltaic switches, PIC16F684 microcontroller and LM7805 regulator.

Table 6.4 Important specifications of the LDOs used for the SCALDO prototypes

Specification	MCP1827
Maximum output current	1.5 A
Input voltage range	2.3-6.0 V
Low dropout voltage	330 mV at 1.5 A load
Adjustable output voltage range	0.8-5.0 V
Transient response	Fast response to load transients
Stability	Stable with 1.0 μF ceramic output capacitor

and S_4 control the discharge phase. Solid state relays were used in this proof-of-concept prototype to avoid complications arising from the body diode of common power MOSFETS. A PIC microcontroller of the type PIC16F684 was used as the controller to drive the photovoltaic switches. An independent 5 V regulator (LM7805) was used to power the microcontroller. The S_1,S_2, S_3, and S_4 switches were controlled through the RA0, RA5, RA1, and RA2 port pins of the microcontroller respectively. The LDO input voltage was monitored through the RA4 port pin. The voltage level at this pin is converted to digital format using 10-bit analog to digital converter of the PIC which operates at 4 MHz oscillation frequency. The result is then compared with reference point (digital value of the minimum LDO working voltage, V_{in}^{min}) and the switching between the supercapacitor charge and discharge phases was made accordingly. The microcontroller firmware was developed to drive the switches based on algorithm shown in the flow chart in Figure 6.14.

A switch-over period, t_{sw} was designed to accommodate any transition delays of the PVN012 switches. This creates a short time duration just before the circuit is transferred from the supercapacitor charge mode to the discharge mode. Time duration is set by the PIC micro-controller according to the switch delay. The time duration is determined according to the data sheet of the switch and typically is about 300 ms. Figure 6.15 shows the delay characteristics of the PVN012 switch.

To keep the linear regulator powered during the switch transitions, a buffer capacitor (C_B) with sufficient capacitance was connected in between the linear regulator input and the ground terminal. In t_{sw} time duration, the buffer capacitor is expected to discharge into the LDO input. The value of the buffer capacitor (C_B) can be determined as shown in the Eqn (6.26) using the maximum load current capability of the LDO (I_L^{max}), switching time delays of a PVN012 switch (t_{sw}) and the minimum possible input voltage at the LDO input (V_{in}^{min}). 1000 μF electrolytic capacitor was chosen as the C_B for the 12-to-5 V SCALDO design.

$$C_B = \frac{I_L^{max} t_{sw}}{V_{in}^{min}} \tag{6.26}$$

Figure 6.16 shows a prototype of 12-to-5 V SCALDO regulator.

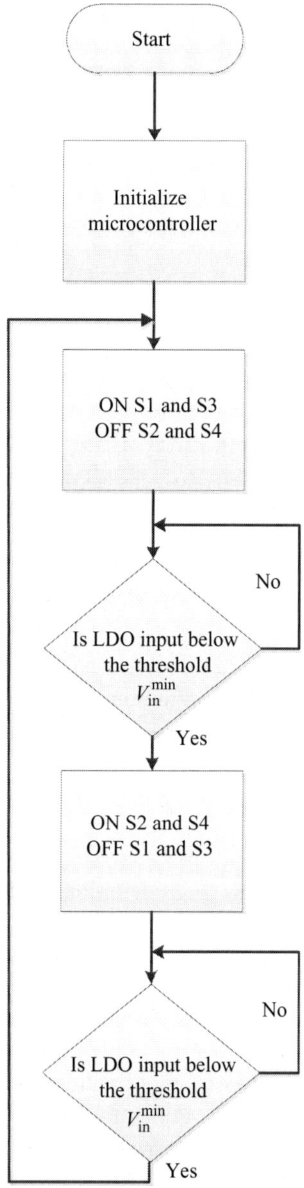

Figure 6.14 The flowchart of the PIC algorithm.

6.6.1.1 Load regulation

Figure 6.17(a)-(d) indicate the oscilloscope waveforms during the circuit operation for a load current of 200, 400, 600, and 800 mA respectively. At these stages the input power supply voltage was maintained at 12 V. As the load current increases, the

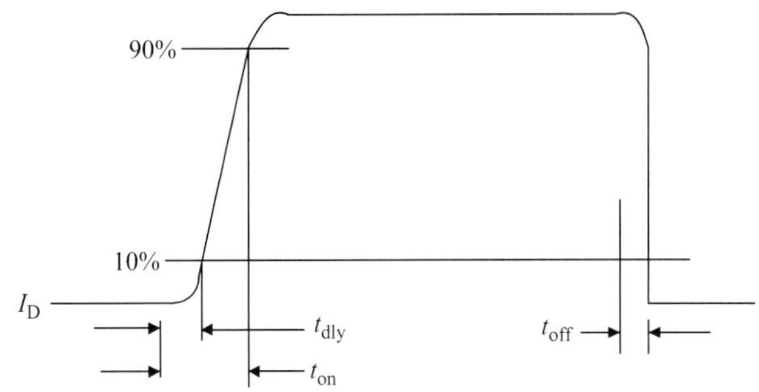

Figure 6.15 Delay characteristics of PVN012 switches.

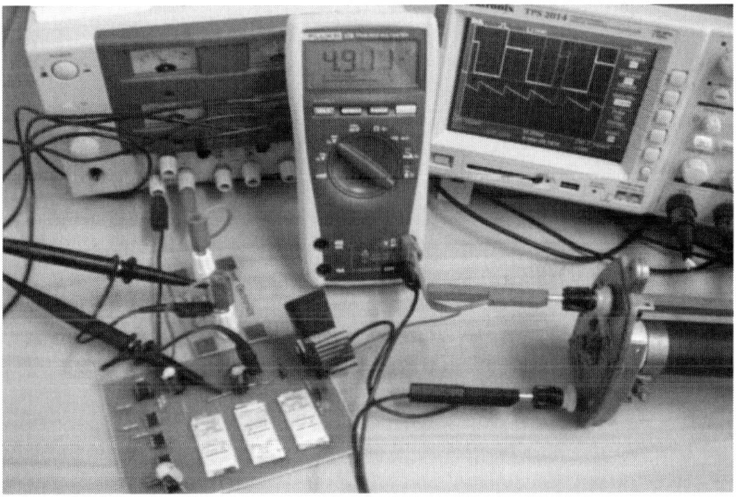

Figure 6.16 The prototype design of the 12-to-5 V SCALDO regulator.

switch operating frequency also increases. When the current flowing through the supercapcitor increases, the time taken to reach the threshold voltage level decreases. Therefore both the charging and discharging cycles of the supercapacitor take a shorter time period. In Figure 6.17 traces (1)-(4) represent the input voltage of the LDO (V_{in}), the voltage across one of the three supercapacitors (V_{SC}), the LDO output voltage (V_{reg}), and the input current drawn from the unregulated power supply (I_{in}), respectively. The trace (3), V_{SC}, clearly shows the supercapacitor charge-discharge cycles during the circuit operation. The trace (1) shows how the LDO input voltage changes during the charge-discharge cycles. The waveform of trace (4), I_{in} shows the current drawn from the unregulated power supply during the charge-discharge cycles. This clearly shows that the circuit draws a considerable current only in the charging phase of the supercapcitor. In the discharging phase only a little current is drawn for

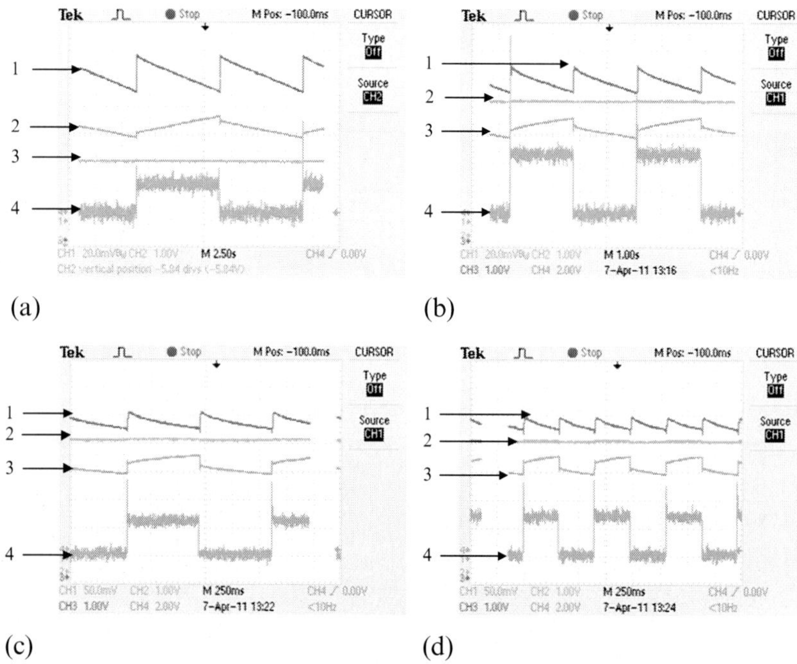

1 Input voltage of the LDO

2 Voltage across one of the three supercapacitors

3 LDO output voltage

4 Input current drawn from the unregulated power supply

Figure 6.17 Oscilloscope waveforms of 12-to-5 V SCALDO regulator, for a load current of (a) 200 mA, (b) 500 mA, (c) 800 mA, and (d) 600 mA at an input voltage of 12 V.

the control circuits to function. This I_{in} was measured by connecting a series resistor of 0.1 Ω in series with the unregulated power supply and the input terminal of the circuit.

In the SCALDO prototypes, the control circuit current was not set at a minimum value due to the time constraints and the limited budget. Therefore, in the calculations the control circuit current is not considered.

In the following calculations, assume that V_p, I_{in}, V_{reg}, I_L, and I_c are the input voltage of the unregulated power supply, the current drawn from the unregulated power supply, the output voltage, the output current and the control circuit current of the SCALDO regulator, respectively. Neglecting the control circuit current, the average current drawn from the unregulated power supply during a supercapacitor charge-discharge cycle is,

$$I_{avg} = \frac{[I_{in} - I_c]}{2} \tag{6.27}$$

The overall ETEE of SCALDO cycle can be estimated as,

$$\eta = \frac{V_{\text{reg}} I_{\text{L}}}{V_p I_{\text{avg}}} 100\% \tag{6.28}$$

The ETEE results of the 12-to-5 V SCALDO regulator which were estimated using Eqns (6.27) and (6.28) based on the experimental results. The circulation frequency increases from 0.05 to 4 Hz with increase of the load current to 100-800 mA respectively. Figure 6.18 shows the load regulation plot based on the experimental results (Gunawardane, 2014) and load regulation is about 0.04%/mA.

6.6.1.2 Line regulation

When the input voltage increases, the operating frequency of the switches decreases. The reason is, when the difference between the unregulated input voltage and the threshold voltage level at the input of the LDO increases, the amount of energy is stored in the supercapacitor also increases. Therefore, both the charge and the discharge cycles of the supercapacitor take longer time periods and this results in a lower operating frequency of the switches. Figure 6.19(a)-(d) indicate the oscilloscope waveforms during the circuit operation, for an input voltage of 13.5, 13, 12, and 11.5 V respectively. The load current is maintained at 250 mA at these stages. In Figure 6.19 traces (1)-(4) represent, the input voltage of the LDO (V_{in}), the voltage across one of the three supercapacitors (V_{SC}), the LDO output voltage (V_{reg}) and the input current of the system (I_{in}) respectively. The trace (3), V_{SC}, clearly shows the circuit operation during the supercapacitor charge and discharge cycles. The trace (1) shows how the LDO input voltage changes during the charge and discharge cycles.

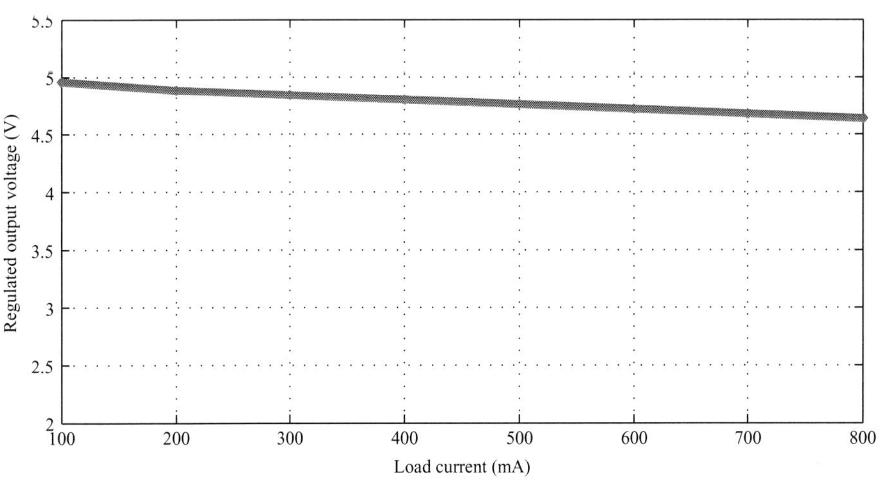

Figure 6.18 The load regulation plot of the 12-to-5 V SCALDO regulator at an input voltage of 12 V.

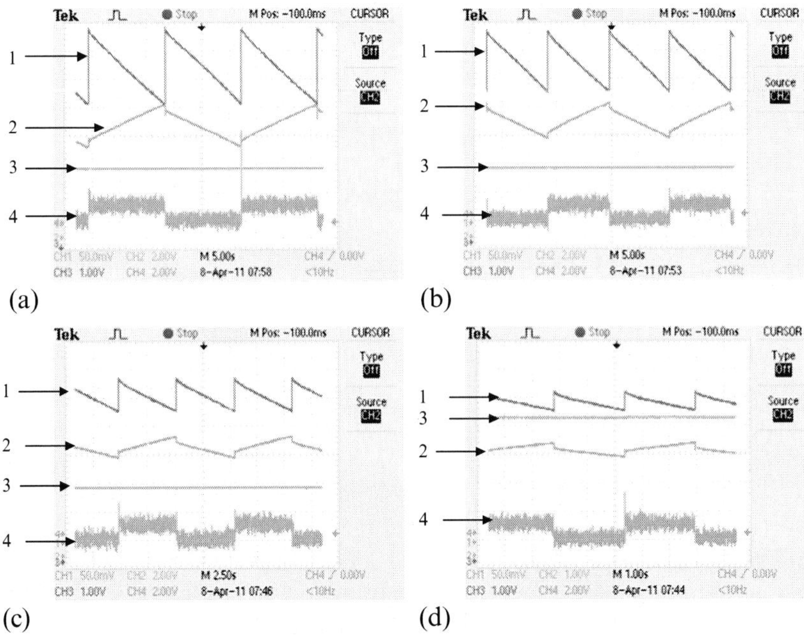

1 Input voltage of the LDO

2 Voltage across one of the three supercapacitors

3 LDO output voltage

4 Input current drawn from the unregulated power supply

Figure 6.19 Oscilloscope waveforms of the 12-to-5 V SCALDO regulator for an input voltage of (a) 13.5 V, (b) 13 V, (c) 12 V, and (d) 11.5 V at a load current of 250 mA.

The I_{in} trace clearly shows that the circuit draws a considerable current only in the charge phase of the supercapacitor and in the discharge phase only a little current is drawn for the control circuits to operate. Figure 6.20 depicts the corresponding line regulation plot. As with load regulation calculations, the control circuit current was neglected in the calculations.

6.6.1.3 End-to-end efficiency

According to the load regulation experimental results (Gunawardane, 2014), the ETEEs were calculated. The corresponding efficiency plotted against load current is shown in Figure 6.21. Using the line regulation experimental data (Gunawardane, 2014), efficiency plotted against the input voltage was drawn as in Figure 6.22. This figure shows a comparison of theoretical performance of SCALDO technique (upper trace), practical performance of SCALDO technique (middle trace) and performance of standard linear regulator (lower trace) for a 12-to-5 V regulator. The prototype achieves overall ETEEs in the range of 70-80%, compared to the

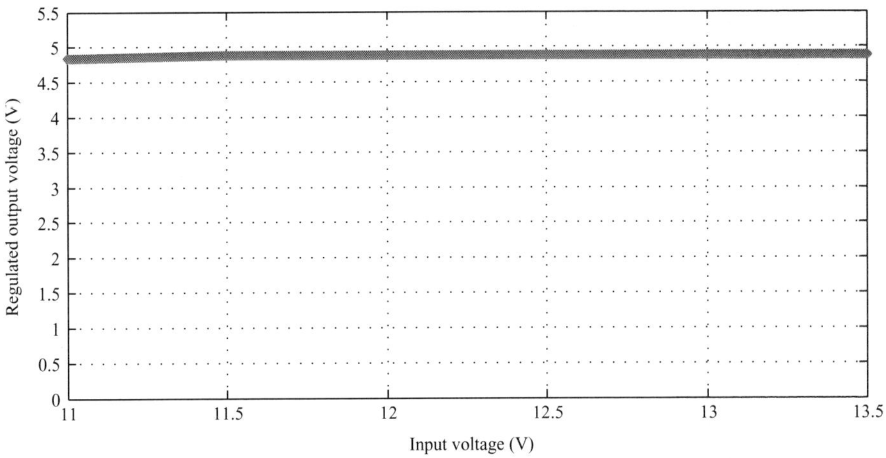

Figure 6.20 The line regulation experimental results of 12-to-5 V SCALDO regulator at a load current of 250 mA.

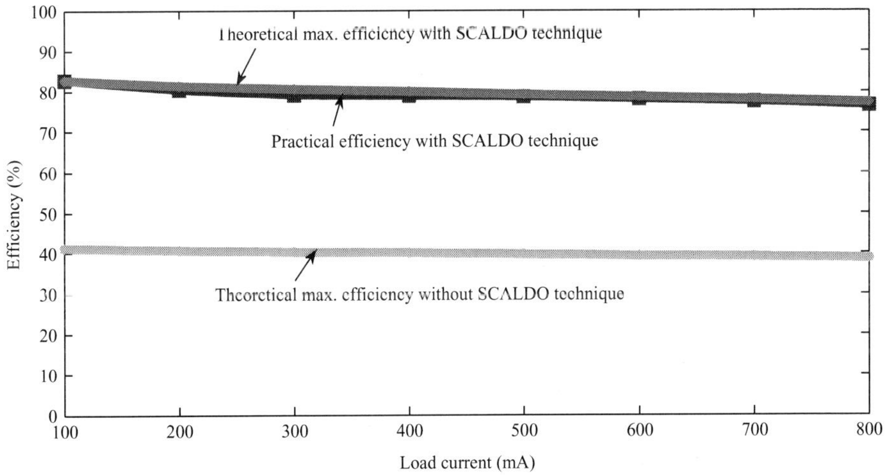

Figure 6.21 Efficiency vs Load current of the 12-to-5 V SCALDO regulator at an input voltage of 12 V.

maximum theoretical efficiency of 42% for a linear regulator used for the same input-output combination. The middle trace of Figure 6.22 shows the improved efficiency in the new SCALDO implementation. The graphs indicate a 3-5% difference between the theoretical and measured power efficiencies for the lossless-dropper supercapacitor technique. This discrepancy is due to fact that the theoretical curve ignores the losses in switch resistances and the capacitor ESR. It is useful to compare this case of only 3-5% loss contribution due to ESR and switch resistances, with that of commercial switched-mode configurations where the losses are typically 10-30% due to

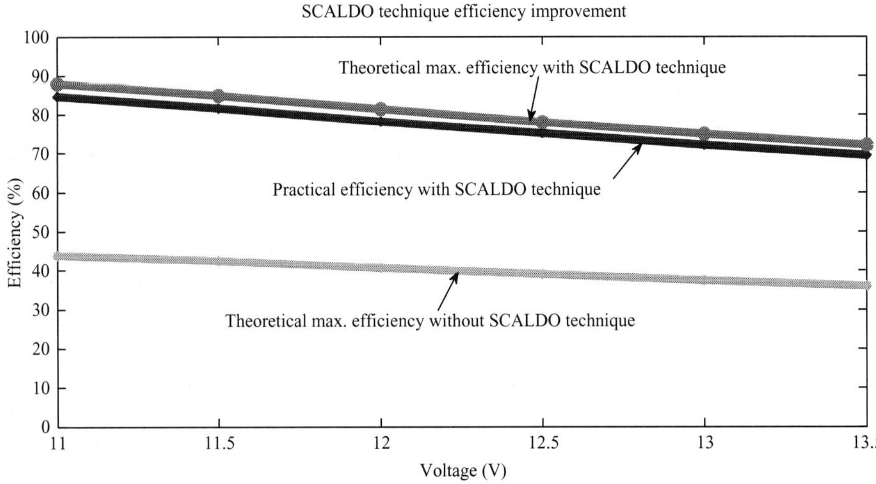

Figure 6.22 Efficiency comparison for a 12-to-5 V regulator: Theoretical results with SCALDO technique (upper trace), practical results with SCALDO technique (middle trace) and basic 12-5V linear regulator theoretical performance (lower trace) at a load current of 250 mA.

static and dynamic losses in diodes, high-frequency switching transistors, and other parasitic effects. This comparison demonstrates the clear advantage of the new supercapacitor-based lossless-dropper technique of the SCALDO regulator.

6.6.2 Load transient response

6.6.2.1 Output current slew rate of DC power supplies

A well-regulated power supply is critical for proper operation and reliability of modern high-speed microprocessors. They require fast delivery of large currents in a smaller time, tight supply-voltage tolerance, and intelligent voltage programming. This voltage regulation must be as tight as ±5% and this tolerance is the resultant of DC set point accuracy, accuracy over temperature and input voltage variations, and transient response. If the voltage on the processor is too high or low or even a processor voltage is temporarily too high or too low can cause problems with the end operation of the processor. In achieving such rigorous voltage requirements the total accuracy of the power supply including the accuracy of the output voltage in a steady state condition and the transient response are vital (Glaser, 2011). The latest CPU voltage-regulator specifications from Intel and AMD call for load-current slew rates of 50-200 A/μs and peak currents of 60 A to more than 120 A (Gentchev, 2000).

There is an increasing demand for point-of-load (POL) converters that maintain good regulation in the presence of fast varying load transients (di/dt). A POL converters that is capable of handling transients up to 300 A/μs is recently been introduced. This level of performance presents new challenges to engineers during the converter design, and also during the test and verification stages of product development. The combination of high-frequency converter operation and fast load transients demands stringent design practices and a thorough understanding of every element of the design and test setup (Callanan, 2004).

In modern microprocessors based systems, the operation at low voltages, GHz clock frequencies, high power, and the ability to survive fast current transients requires a new generation of voltage regulator modules (VRMs). Transient response is the most important VRM test and the most difficult to perform. A VRM must maintain regulation under high transition rate loading, so it is crucial to generate a fast load pulse or single transition to verify the voltage transient levels and transient response time under specified loading. As with traditional power supplies, different loading patterns can cause significant differences in transient response time and voltage level. Therefore, an electronic load that simulates fast load changes is necessary to verify the transient response of a VRM. In order to verify the transient response, measurements of the rise and fall times upon a step change in the load is necessary. Generally, this type of test requires a load that is able to produce a rise and fall time approximately five times faster than the power supply (Lee, 2001).

6.6.2.2 Load transient response of 12-to-5 V SCALDO regulator

The load transients of the SCALDO prototypes were measured using TEXIO PXL-151A DC electronic load.

Once the switch load current was set at the TEXIO DC load, it was connected to the SCALDO prototype and corresponding changes in the output voltage were observed from the Tektronix digital oscilloscope. For better observation of the variations, AC coupling mode of the oscilloscope was enabled. For a given slew rate in the rising and falling edges of the load, corresponding output voltage variations were recorded in terms of transient voltage ripple and transient time. The experimental set-up used for the transient measurement is shown in Figure 6.23. Typical current step of the TEXIO PXL-151A DC electronic load is shown in Figure 6.24.

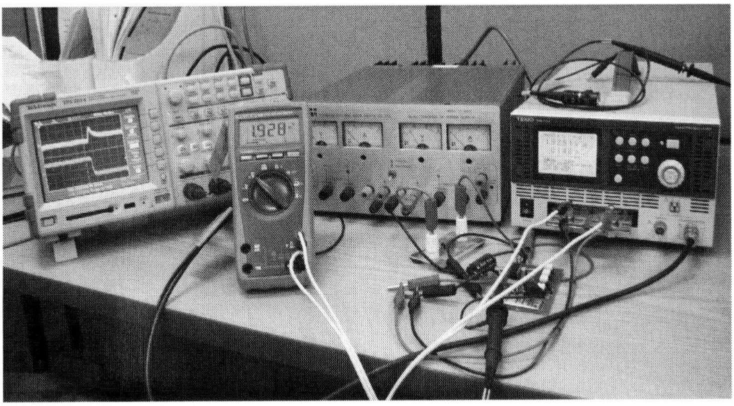

Figure 6.23 Transient measurement set up of a SCALDO regulator using TEXIO DC electronic load.

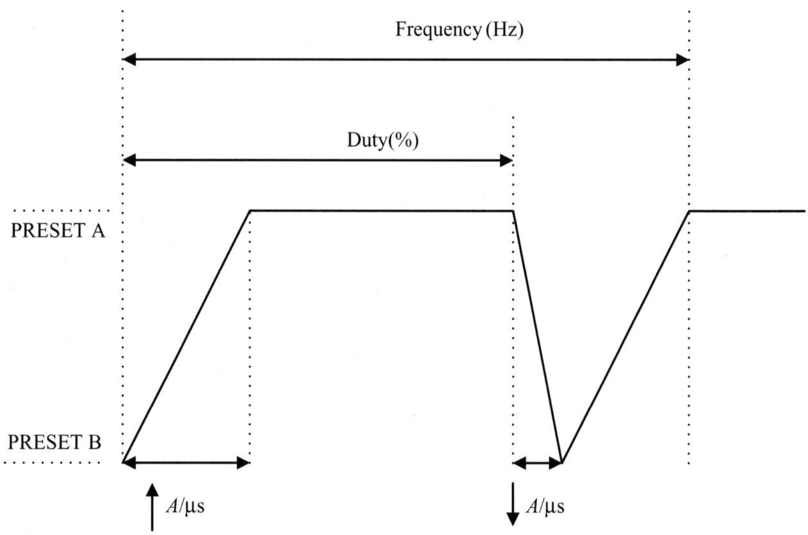

Figure 6.24 Typical current step of the TEXIO PXL-151A DC electronic load.

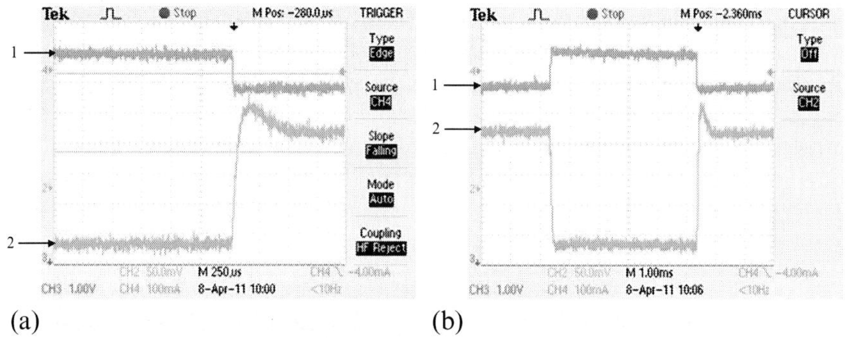

Figure 6.25 Transients measurements of 12-to-5 V SCALDO: (a) falling edge; (b) falling/rising edge.

Figure 6.25 shows the transient voltage changes for 12-to-5 V SCALDO regulator. where the trace (1) shows the current step and the trace (2) shows the corresponding transient voltage change.

6.6.3 5.5-to-3.3 V SCALDO regulator

Figure 6.26 block diagram shows the 5.5-to-3.3 V SCALDO regulator which is an implementation of $V_p < 2V_{in}^{min}$ criterion, $n = 3$ case. Figure 6.27 shows the schematic

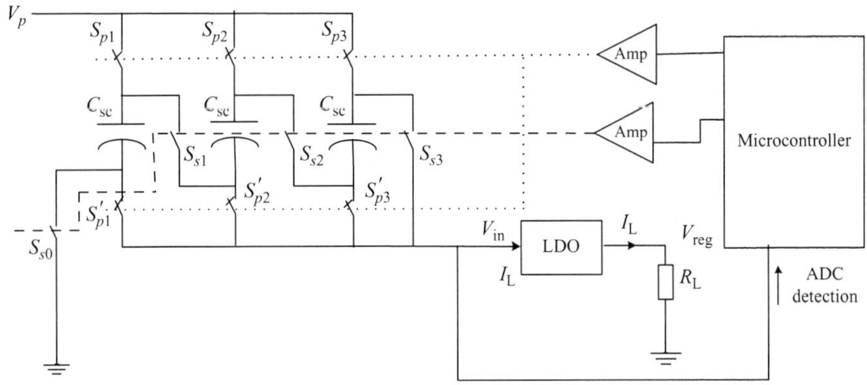

Figure 6.26 Block diagram of the 5.5-to-3.3 V SCALDO regulator.

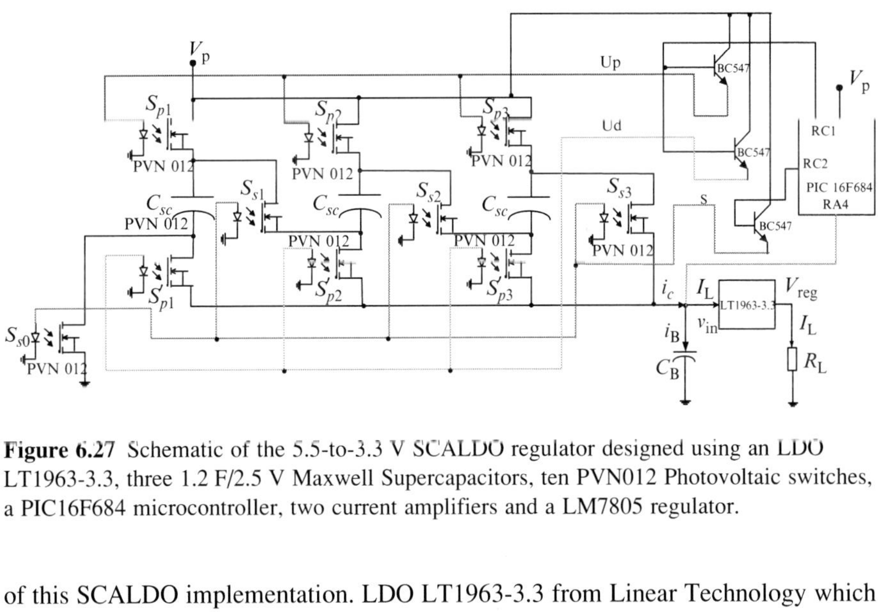

Figure 6.27 Schematic of the 5.5-to-3.3 V SCALDO regulator designed using an LDO LT1963-3.3, three 1.2 F/2.5 V Maxwell Supercapacitors, ten PVN012 Photovoltaic switches, a PIC16F684 microcontroller, two current amplifiers and a LM7805 regulator.

of this SCALDO implementation. LDO LT1963-3.3 from Linear Technology which has the following specifications was used as the main low dropout regulator in this prototype (Figure 6.28):

- output current of 1.5 A,
- fixed output voltages at 3.3 V,
- dropout voltage of 340 mV,
- quiescent current of 1 mA,
- optimized for fast transient response.

In order to maintain a better LDO operation, the LDO input voltage was kept at least 0.3 V above the required output voltage, so that, V_{in}^{min} was maintained at 3.6 V.

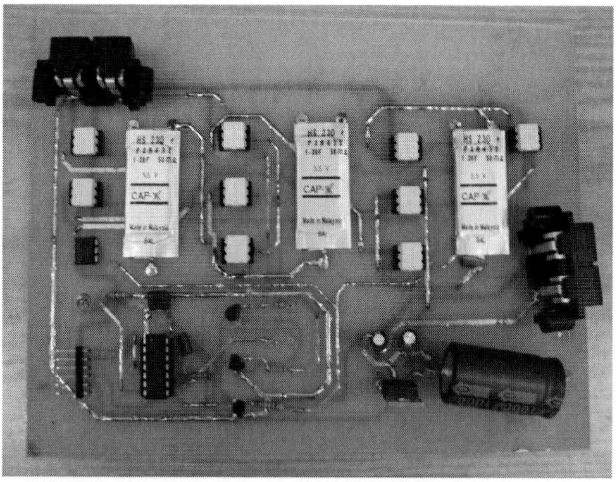

Figure 6.28 Prototype of the 5.5-to-3.3 V SCALDO regulator.

Originally, it was planned to design a 5.5-to-3.3 V regulator but due to switch resistance losses and supercapacitor ESR losses, the input voltage was maintained at 5.5 V to overcome the issues of series resistances which drops the $V_{\text{in}}^{\text{min}}$ to unacceptable level. Three supercapacitors of 1.2 F/2.5 V of the type Cap-XX, were used as the group of identical supercapacitors (C_{SC}). A PIC microcontroller of the type PIC16F684 was used as the controller to drive the photovoltaic switches. The LDO input voltage was monitored through the RA4 port pin of the PIC micro-controller. The switch set (6 nos) $S_{p1}, S_{p2}, S_{p3}, S'_{p1}, S'_{p2}, S'_{p3}$ and the switch set (4 nos) $S_{s0}, S_{s1}, S_{s2}, S_{s3}$ were controlled through the RA1 and RA2 port pins, respectively. In order to control several switches using a single control signal, two current amplifiers were used for the parallel and series set of switches respectively. The three supercapacitors were kept in parallel using the switch set $S_{p1}, S_{p2}, S_{p3}, S'_{p1}, S'_{p2}, S'_{p3}$. This capacitor group was connected in series with the LDO input to keep the supercapacitors charging from the unregulated power supply, while maintaining the requirement for the LDO operation. During the charge process, when 3.6 V for the LDO was reached, the controller switched on switch group $S_{s0}, S_{s1}, S_{s2}, S_{s3}$ and switched off the other group of switches to keep the three supercapacitors in series to release the stored energy. The stored energy was released until the 3.6 V condition for the LDO was met and then switched back to the charging configuration again. To keep the linear regulator powered during this transition, a buffer capacitor (C_{B}) with sufficient capacity was connected in between the linear regulator input and the ground terminal. Ten PVN012 photovoltaic relays were used as switches to keep the set of supercapacitors in charge and discharge configurations. The micro-controller firmware developed for this technique was based on the same algorithm as the $V_p > 2V_{\text{in}}^{\text{min}}$ configuration in the flow chart of Figure 6.14. Additional details including the micro-controller firmware is available in Gunawardane 2014).

6.6.3.1 Load regulation

Figure 6.29(a)-(d) indicate the oscilloscope waveforms during the circuit operation for a load current of 200, 600, 1000, and 1300 mA, respectively. At these stages input voltage was maintained at 5.5 V. As the load current increased, both the charge and the discharge cycles of the supercapacitor took shorter time period to operate. In Figure 6.29 traces (1)-(4) represent voltage across one of the three supercapacitors (V_{SC}), LDO output voltage (V_{reg}), the input current of the system (I_{in}), and input voltage of the LDO (V_{in}) respectively. Figure 6.30 shows the load regulation graph.

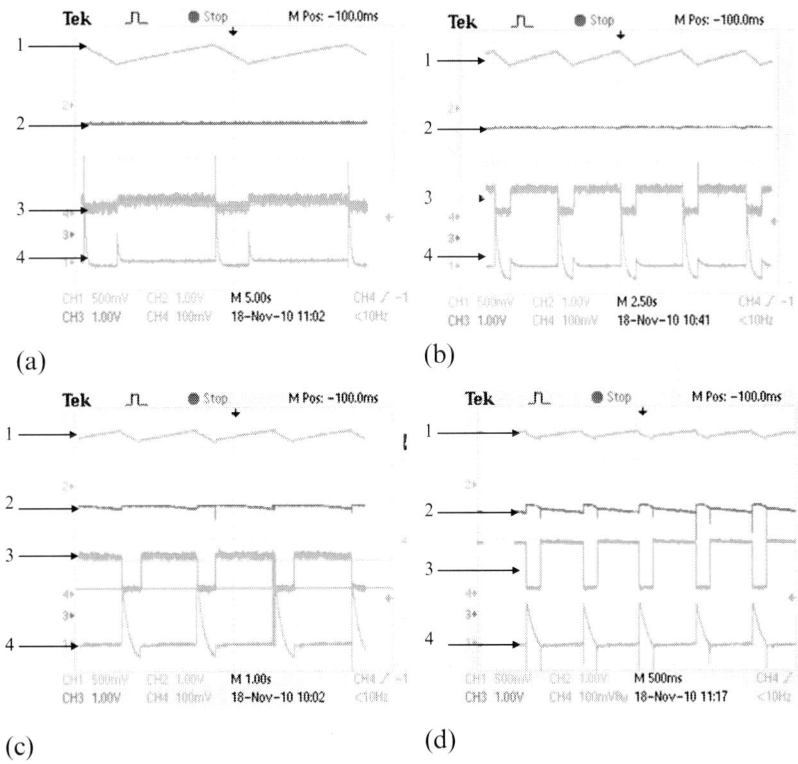

1 Voltage across the supercapacitor

2 LDO output voltage

3 Input current drawn from the unregulated power supply

4 Input voltage of the LDO

Figure 6.29 Oscilloscope waveforms of 5.5-to-3.3 V SCALDO Regulator for a load current of (a) 200 mA, (b) 600 mA, (c) 1000 mA, and (d) 1300 mA at input voltage of 5.5 V.

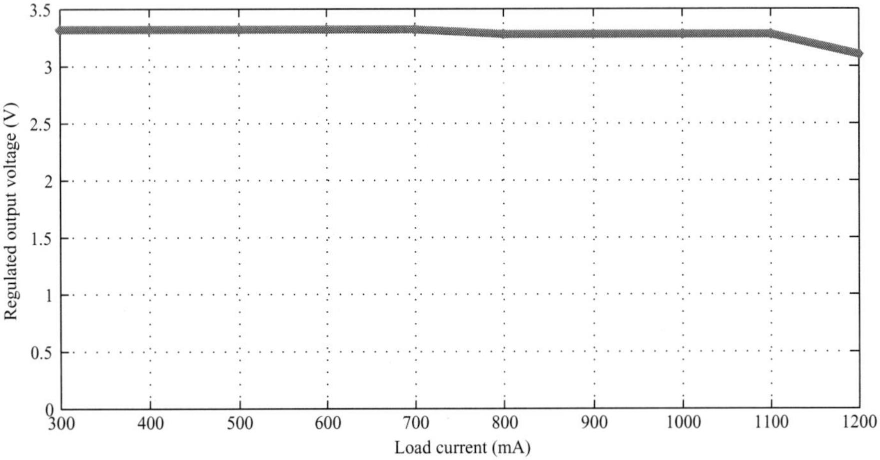

Figure 6.30 Load Regulation results of 5.5-to-3.3 V SCALDO regulator at input voltage of 5.5 V.

6.6.3.2 *Line regulation*

For 5.5-to-3.3 V SCALDO regulator Figure 6.31 depicts the corresponding line regulation plot based on the experimental results. The load current was fixed at 300 mA during these measurements.

6.6.3.3 *End-to-end efficiency*

Using the line regulation experimental data (Gunawardane, 2014), the ETEE plotted against the input voltage is shown in Figure 6.32. This figure shows a comparison of the theoretical performance of SCALDO technique (upper trace), the practical performance of SCALDO technique (middle trace) and the performance of a standard linear

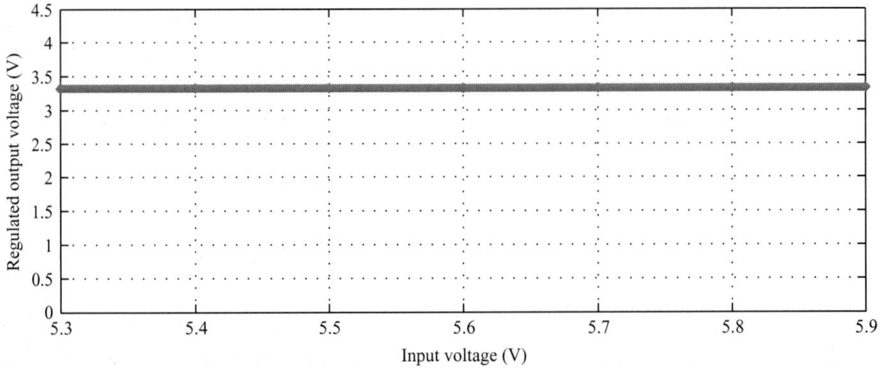

Figure 6.31 Line Regulation results of 5.5-to-3.3 V SCALDO regulator at a load current of 300 mA.

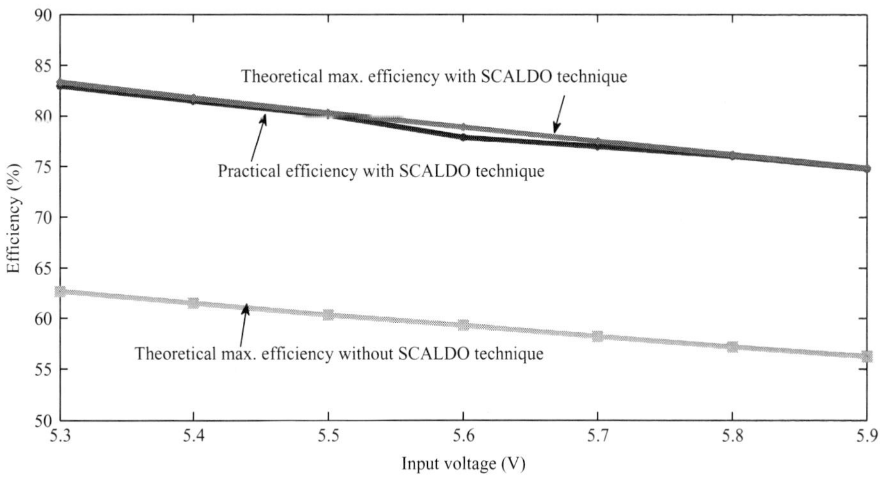

Figure 6.32 Efficiency comparison for a 5.5-to-3.3 V SCALDO regulator: theoretical results with SCALDO technique (upper trace), practical results with SCALDO technique (middle trace) and basic 5.5-to-3.3 V linear regulator theoretical performance (lower trace).

regulator (lower trace) for a 5.5-to-3.3 V regulator. The prototype achieves overall ETEEs in the range of 75-85%, compared to the maximum theoretical efficiency of 60% for a linear regulator used for the same inputoutput combination. The middle trace of Figure 6.32 shows the improved efficiency in the new SCALDO implementation. The graphs indicate a 1-2% difference between the theoretical and measured power efficiencies for the SCALDO technique. This discrepancy is due to fact that the theoretical curve ignores the losses in switch resistances and the capacitor ESR.

6.6.3.4 Transient response

Figure 6.33 shows the oscilloscope figures of the transient measurements for 5.5-to-3.3 V SCALDO regulator. For a given slew rate, in the rising and falling edges of the load, output voltage variations were recorded in terms of transient voltage ripple and transient time (Gunawardane, 2014). Transient response is a complicated situation based on the output capacitor and the behavior of the LDO itself. In this SCLADO analysis, further detailed analysis on finer details were not carried out due to the time constraints of the project.

6.7 Wider applications of SCALDO technique

SCALDO technique is a unique way to design high efficiency linear regulators without RFI/EMI issues. With commercial supercaps with capacitance values varying from 0.2 to over 500 F where size of the SC is nearly similar to electrolytic capacitors, and their ESR is in the range of less than 100 mΩ (for smaller sizes)to about 1 mΩ,

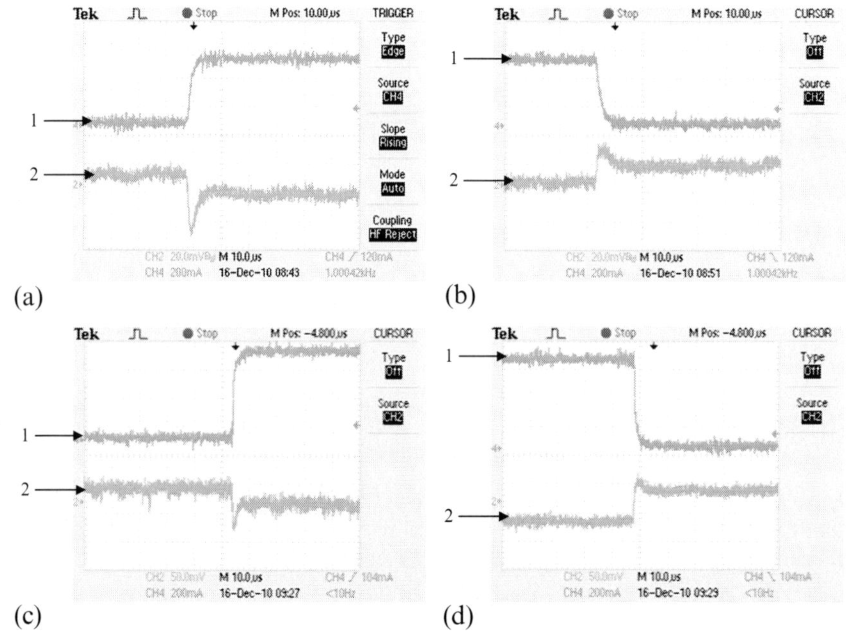

1 Transient load current

2 Change of output voltage

Figure 6.33 Transients of 5.5-to-3.3 V regulator using SCALDO technique: (a) 1A/μs 50-600 mA rising edge; (b) 1 A/μs 50-600 mA falling edge; (c) 20 A/μs rising edge 50-800 mA; and (d) 20 A/μs falling edge 50-800 mA.

designer can use the technique to come up with different types of SCALDO implementations. Possible applications are:

1. high current high efficiency linear regulators,
2. desk top computer power supplies,
3. linear high efficiency DC power supplies with DC UPS capability,
4. VRMs.

Since modern supercapacitors can charge and discharge at very high currents such as 10 A to few 100 A in general, if suitable power switches can be implemented the technique can be useful in linear regulators with output current ratings up to few 100 A.

Technique can be extended to desktop computer power supplies, with the target of ETEE exceeding the current capability for desktop computers. This is another major possible application of the SCALDO technique.

Since SCs are available up to about 5000 F with extra low ESR (less than 0.5 mΩ) the SCALDO technique can be simply extended to buffer a few cycle loss of AC mains in an AC mains derived DC supply, by over-sizing the supercaps. However this require additional circuit design extensions to take care of the pre-charging needs

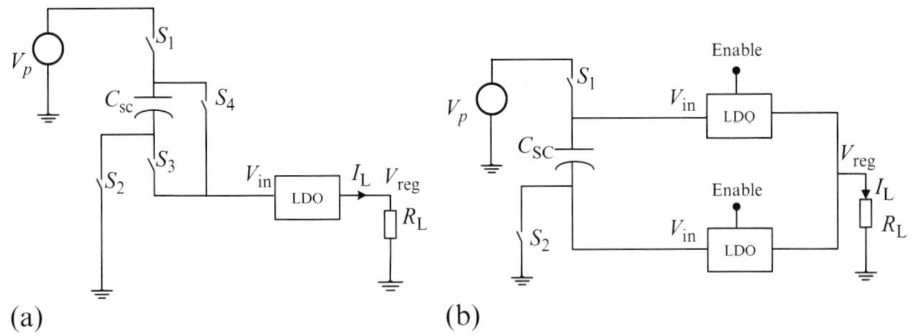

Figure 6.34 (a) Basic SCALDO configuration with single LDO and four switches; (b) modified RS-SCALDO (reduced switches) configuration with two identical LDOs and two switches.

of the supercaps used. Application of the technique to develop fully linear VRMs is possible by modifying the base technique to minimize the number of power switches. All these techniques are currently under development at the University of Waikato (Kularatna and Wickramasinghe, 2013).

6.7.1 RS-SCALDO technique

For high current capable LDO circuits, efforts should be taken to reduce the number of switches. One such improvement, applicable to the present SCALDO design has been identified. Using two identical LDOs, the number of switches can be reduced from $3n + 1$ to $2n$. This reduced-switch SCALDO (RS-SCALDO) topology has been suggested as a solution for modern VRM requirements.

More theoretical and experimental details on this new modification can be found in (Kularatna and Wickramasinghe, 2013). As shown in Figure 6.34, for the $V_p > 2V_{in}^{min}$ basic SCALDO design, using two identical LDOs, the number of switches can be reduced to 2 instead of the 4 switches used in the present design.

6.8 Comparison between SCALDO regulators and charge pumps

Supercapacitor-assisted low dropout regulators (SCALDO) were proposed as an alternative design approach of DC-DC converters, where a very low frequency supercapacitor circulation technique is combined with a commercial low dropout regulator IC to achieve significantly high ETEE. In this technique the supercapacitor is used as a lossless voltage dropper, and the energy reuse occurs at very low frequencies such as from a few to a few 100 Hz, eliminating any RFI/EMI issues. The efficiency advantage of this approach is very much closer to the efficiencies of practical switching regulators and also eliminates the utilization of bulky inductors.

It is important to note that the SCALDO technique does not use supercapacitors for any DC-DC conversion purpose, as in the case of switched-capacitor type DC-DC converters (charge pumps) such as voltage doublers, multipliers and inverters. Charge pumps utilize microfarad-order capacitors to achieve a voltage conversion using switching frequencies in the order of hundreds of kHz. The SCALDO technique utilizes the supercapacitor as a voltage dropper only, and the very small ESR in modern supercapacitors allows the capacitor losses in the series path to be minimized, well approximating a lossless voltage dropper. This SCALDO technique can be extended to provide high output currents in the order of 5-50 A at DC rail voltages suitable for modern processors. Given the low leakage currents in new families of supercapacitors, usually within a range of 5 to about 50 μA (Mars, 2012; http://www.cap-xx.com/products/products.htm), compared to their high charge /discharge current capabilities in the range of 10 to several 100 A, leakage current related losses are negligible. Also the supercapacitor is the best candidate for this new application, as the ESR values (in the range of fractional milliohm to less than 100 mΩ) are less than or comparable $R_{DS(on)}$ values of low voltage MOSFETs. Given that the new technique works in the range of fractional hertz to few 100 Hz order frequencies, and with the new supercapacitor families having very low leakage currents, dynamic losses will be significantly lower than a charge pump. Compared to linear or switching regulators, the output voltages of charge pump circuits are usually not regulated and are designed to operate at a fixed switching frequency of few 100 kHz with a rated output load current and voltage. Because of practical limitations on the size of capacitors and switches, applications of charge pumps are limited mainly to low and medium power levels of several tens of milliwatts.

Table 6.5 provides capabilities and limitations of SCALDO technique comparing the practical performance with typical switched-capacitor converters.

Table 6.5 A comparison summary of SCALDO technique and switched capacitor converters

SCALDO technique	Switched-capacitor techniques
A modified version of a linear regulator with an enormously large capacitor in the series path as a lossless voltage dropper. The capacitor and the switches do not convert the voltage	Basically a high frequency switching technique for voltage conversion
Always a step-down configuration	Practically used to step-up or invert a DC voltage
Enormously large capacitors (supercapacitors) are used	Capacitors used are in the range of a few nanofarads to a few tens of microfarads
Operation frequency is variable and depends on the load current. Switching operation is based on the case of a maximum/minimum voltage detected across the input of the LDO	Circuit design starts with a fixed oscillator supplying the switching frequency

Continued

Table 6.5 Continued

SCALDO technique	Switched-capacitor techniques
Very low frequency (10 to few 100 Hz) switching used for energy storage and reuse in capacitors	Switching frequency is in the range of 10 to several 100 kHz
Load always sees the precise output of a linear/low drop out regulator	Load regulation is not precise and requires a another voltage regulator (linear/low drop out type) for precise output voltage
Capacitor never comes parallel to the input unregulated supply	In one part of the cycle capacitor comes in parallel to the unregulated supply
Technique is applicable to very large load currents [high current capable LDO is required]	Technique is suitable only for very low load currents
Negligible dynamic losses in switches	Significantly high dynamic losses in switches
Theoretically an efficiency multiplication factor is defined for a given configuration	Theoretically a voltage conversion factor applies to a given configuration

Given the above simple summary, it is very clear to see that the SCALDO technique is quite different to the operation of charge pumps, for three primary reasons:

1. SCALDO technique uses a capacitor as a lossless voltage dropper, together with a linear low dropout regulator for precise output regulation.
2. Extremely low switching frequency is variable, and it depends on the load current.
3. If an ultra LDO can be developed for the required output current, there is no limit to precise output regulation at high load currents. (This is due to the availability of single cell supercapacitors (commercially available) which are in the range of fractional farads to a few thousand farads.)

References

Bull, C., Smith, C., 2003. Integrated Building Blocks for Dual-output Buck Converter. Power Electronic Technology.

Callanan, S., 2004. Testing high di/dt converters. Artesyn Technologies. Available: www.powerelectronics.com.

Day, M., 2003. Integration saves time and board space Converter. Power Management Special Supplement. Power Electronics Technology. pp. 64–67. Available: www.powerelectronics.com.

Falin, J., 2006. A 3-A, 1.2-V_{OUT} linear regulator with 80% efficiency and P_{LOST} <1w Converter. Analog Appl. J, 10–13. Available: www.ti.com/aaj.

Gentchev, A., 2000. Designing high-current, vrm-compliant CPU power supplies. Analog Devices. Available: www.edn.com.

Glaser, C., 2011. Optimal transient response for processor-based systems. Texas Instruments. Available: www.powerelectronics.com.

Gunawardane, K., 2014. Analysis on supercapacitor assisted low dropout regulators. The University of Waikato, Ph.D. dissertation.

Inouye, S., Robles-Bruce, M., Scherer, M., 2010. Power Management – General Purpose Analog Service. Databeans Incorporated. Tech. Rep. Available: www.databcans.net.

Jovalusky, S., 2005. New Energy Stand ards Banish Linear Supplies. Power Electronics Technology, Power Integrations, San Jose, CA. Available: www.powerelectronics.com.

Kankanamge, K., Kularatna, N., 2012. Implementation aspects of a new linear regulator topology based on low frequency supercapacitor circulation. IEEE Applied Power Electronics Conference, USA.

Kester, W., Erisman, B., Thandi, G., 1998. Switched capacitor voltage converters, Analog Electronics Technical Report.

Kularatna, N., 2011. DC Power Supplies Power Management and Surge Protection for Power Electronic Systems. CRC Press.

Kularatna, N., Fernando, J., 2009. A supercapacitor technique for efficiency improvement in linear regulators. IEEE 35th Annual Conference of Industrial Electronics, pp. 132–135.

Kularatna, N., Fernando, J., 2011. High current voltage regulator. US Patent 9707 430 B2, March 15.

Kularatna, N., Wickramasinghe, J., 2013. Supercapacitor assisted low dropout regulators (scaldo) with reduced switches: A new approach to high efficiency vrm designs. IEEE International Symposium on Industrial Electronics (ISIE), pp. 1–6.

Kularatna, N., Fernando, J., Kankanamge, K., Tilakaratna, L., 2010. Very low frequency supercapacitor techniques to improve the end-to-end efficiency of DC-DC converters based on commercial off the shelf LDOs. Proceeding of IEEE Industrial Electronics Conference, pp. 721–726.

Kularatna, N., Kankanamge, K., Fernando, J., 2011a. Supercaps improve LDO efficiency. Part 1: Low noise linear supplies. Power Electronics Technology Magazine, April.

Kularatna, N., Kankanamge, K., Fernando, J., 2011b. Supercapacitor enhance LDO efficiency. Part 2: Implementation. Power Electronics Technology, May.

Kularatna, N., Fernando, J., Kankanamge, K., Zhang, X., 2011c. A low frequency supercapacitor circulation technique to improve the efficiency of linear regulators based on LDO ICs. IEEE Applied Power Electronics Conference, Texas, USA, pp. 1161–1165.

Linear and Switching Voltage Regulator Handbook on Semiconductor. February, 2002, Available: http://onsemi.com.

Linear regulators in portable applications. Maxim Integrated Products. Available: http://www.maxim-ic.com/an751.

Lee, B.S., 1999. Understanding the Terms and Def initions of LDO Voltage Regulators. Texas Instruments.

Lee, J., 2001. High slew rate electronic load checks new generation voltage regulator modules. Power Electronics. Chroma ATE Inc., Irvine, CA.

Man, T.Y., Mok, P.K.T., Chan, M., 2007. High slew-rate push-pull output amplifier for low-quiescent current low-dropout regulators with transient-response improvement. IEEE Trans. Circuits Syst. 54 (9), 755–759.

Marasco, K., How to successfully apply low dropout regulators, Analog Devices. [Online], Available: http://www.analog.com/library/analogdialogue/archives/43-08/ldo.html; www.analog.com.

Mars, P., 2012. Coupling a supercapacitor with a small energy harvesting source. EDN Magazine[Online]. Available: http://www.edn.com/design/components-and-packaging/4374932/1/ Coupling-a-supercapacitor-with-a-small-energy-harvesting-source-.

Palumbo, G., Pappalardo, D., 2010. Charge pump circuits: An overview on design strategies and topologies. IEEE Circuits Syst. Mag., 31–45.

Rincon-Mora, G.A., 1996. Current Efficient, Low Voltage, Low Drop-out Regulators. Georgia Institute of Technology, Ph.D. dissertation.

Rincon-Mora, G.A., 2009. Analog IC Design with Low-Dropout Regulators. McGraw-Hill.

Simpson, C. (2012). Linear and Switching Voltage Regulator Fundamentals, National Semiconductor Application Note [Online], Available: http://www.ti.com/lit/an/snva558/snva558.pdf

Texas Instruments, 2009. Technical review of low dropout regulator operation and performance, Tech Report, August 2009.

Travis, B., 1998. Linear vs. switching supplies: weighing all the options. EDN, January 1, 40.

Walt Kester, B.E. Switching regulators, [Online], Available: http://www.analog.com/static/imported-files/tutorials/ptmsect3.pdf.

Zendzian, D., 1999. A high performance linear regulator for low dropout applications. Unitrode Corporation. [Online], Available: http://www.ti.com.cn/cn/lit/an/slua072/slua072.pdf.

Supercapacitors for surge absorption

7

7.1 Introduction

Supercapacitors can be used in applications where they are not typically used as energy storage devices or in hybridized form with batteries. In Chapter 6, we used it as a theoretically lossless voltage dropper in a DC-DC converter, and in this chapter supercapacitor's ability to absorb high-voltage (HV) transient surges is discussed with a unique application.

If we consider a supercapacitor as an ideal device, the device will have an energy storage capacity of $\frac{1}{2}CV_C^2$, where C is the device capacitance and V_C is the rated DC voltage of the device. Now if this capacitor is used in a simple circuit where an ideal DC voltage source of value V_S and a resistor of value R are used, the overall RC circuit will have a time constant of $\tau = RC$. In Figure 7.1(a), this arrangement is shown and the graph in Figure 7.1(b) indicates the exponential rise of the voltage across the capacitor, for the cases of $V_S = V_C$ and $V_S = mV_C$.

For the circuit shown in Figure 7.1, capacitor voltage is given by

$$v(t) = V_S\left(1 - e^{-\frac{t}{RC}}\right) \tag{7.1}$$

Equation (7.1) shows that the capacitor can reach the value V_S after an approximate time of $T_{final} = 5RC$. However, if the rated voltage of the capacitor $V_C < V_S$, the capacitor may be destroyed due to overcharging, and in most cases we can expect a blast! However, if we can control the duration of the voltage source connected to the capacitor using a switch to disconnect it when the capacitor reaches its rated voltage, no destruction will occur. Also if the voltage source acts as a step voltage source of duration T_{step} after which time the voltage abruptly drops to zero value, capacitor will not get destroyed. This scenario is depicted in Figure 7.2.

Based on the general validity of Eqn (7.1), if we can control the duration of occurrence of a HV source to a "short-enough" period, final voltage across the terminals of the capacitor will be kept within the limit of V_C. This discussion indicates us that a very large value capacitor, such as a supercapacitor in a circuit loop of finite series resistance, can be used to safely absorb energy from a HV (transient) source with a short-duration occurrence.

7.2 Lightning and inductive energy dumps in electric circuits and typical surge absorber techniques

In practical AC or DC power supplies, short-duration HV transient surges can get superimposed on the energy supply due to acts of gods such as lightning, or due to switching off of large inductive loads such as high-power motor drives. Most

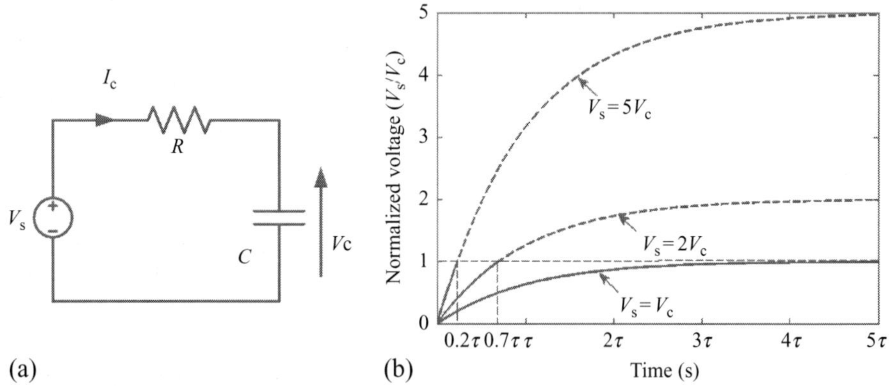

Figure 7.1 A supercapacitor of capacitance value C charged from a simple DC voltage source. (a) Simple circuit and (b) graph of capacitor voltage versus time.

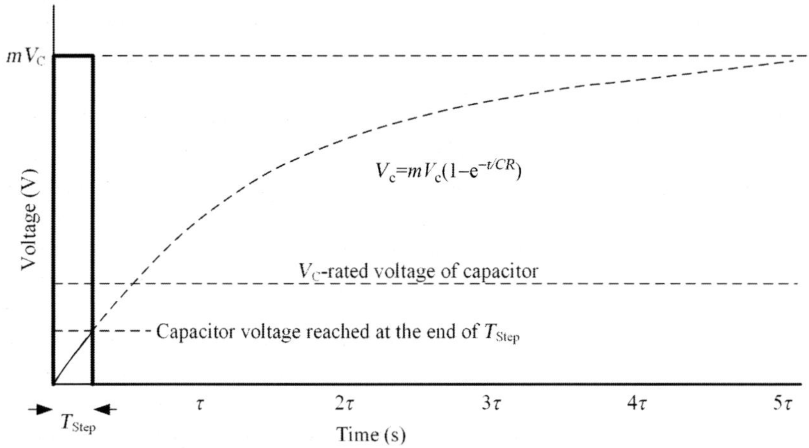

Figure 7.2 A supercapacitor connected to a step voltage source of duration T_{step} and a voltage value of mV_C.

electronic circuits designed with smaller feature size transistors such as the metal oxide semiconductor field effect transistors in high-performance processors and memories can be easily destroyed due to these energy dumps. Superimposed voltages such as lightning-induced transients on incoming mains (or data and telecommunication circuits) can come in the form of differential and common mode surges, which becomes lethal to smaller feature size transistors in modern integrated circuits. Figure 7.3 shows a typical circuit used for protecting electronic systems against surges and the two modes of surges (i) common mode and (ii) differential mode.

In a typical surge protector circuit, nonlinear devices (NLDs) such as metal oxide varistors (MOVs) and bidirectional break-over diodes (BBDs) are coupled with

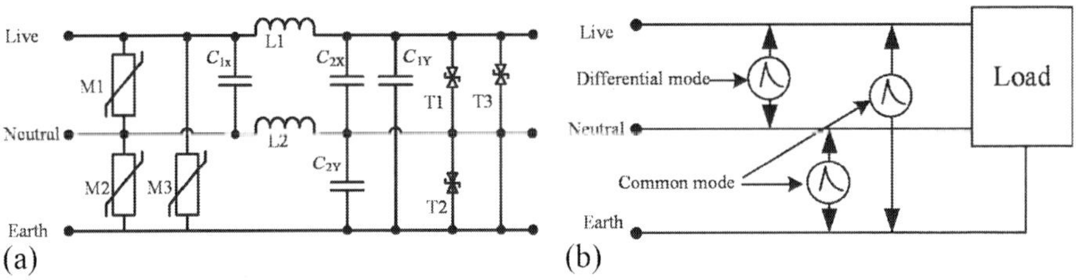

Figure 7.3 Typical surge protection circuit and modes of surges. (a) Surge protector with MOVs and BBDs coupled with filter stages and (b) common and differential mode surges.

LC-type filter stages as shown in Figure 7.3(a). Depending on the level of protection required, different sizes of NLDs are used in varied versions of this general configuration. When a HV surge such as a lightning gets induced on the wire pairs (differential form on the live-neutral pair, common mode on neutral-earth or live-earth pairs), if the peak voltage of the induced voltage exceeds the firing voltage of the corresponding MOV, it fires and starts conducting a high instantaneous current. A maximum voltage termed as *clamping voltage* develops across the terminals of the MOV and the MOV starts absorbing the surge energy based on the voltage-current product over the period of the surge. This mitigates the danger of surge voltage propagating toward the critical load. In fact, the inductor capacitor filters on the path filter another part of the surge energy due to impedance $L\omega$ of the series inductor and the $\frac{1}{C\omega}$ impedance value of the parallel capacitor absorbing some more energy of the HV transient. Any remaining HV transient is absorbed by the firing of BBDs at the critical load end, provided the peak of the remaining surge is above the firing voltage of the BBD.

Overall, such a surge protector that absorbs the energy of the HV transient is designed on the basis of transient energy absorption capability of the NLDs used. Table 7.1 summarizes the properties of a typical MOV from the component manufacturer Littelfuse, USA. In general, these NLDs are characterized by their transient energy absorption rating given for a short duration such as few milliseconds, which is typically related to transients lasting less than about 100–200 microseconds. During this time, the NLD heats up due to the absorbed transient energy, while maintaining the clamping voltage across the device. However, if repeated high-energy transients keep firing the NLD, device's transient energy absorption rating (usually called the *Joule rating*) will be exceeded and device ends up in a failure. In most general cases of MOVs, every time a transient is absorbed by the device, it deteriorates gradually, and after a limited period of time, its surge absorption capability could be totally lost.

Table 7.2 shows the characteristics of a BBD such as the 1.5KE440CA series from Littelfuse, where continuous power dissipation rating of back-to-back diode pair is in the order of 6.5 W, but its transient power rating, specified for a limited duration of 1 ms, is 1500 W. What this means is that surges can fire the junctions and absorb the transient energy and the absorbed energy dissipates as heat within a milliseconds-order time duration, provided surge is just a short duration of several 100 ms only.

Table 7.1 Summary of a typical MOV characteristics-device type TMOV14RP320 from littelfuse

Device specification	Value	Units	Remarks
Continuous AC voltage rating	320	V	
Continuous DC voltage rating	420	V	
Transient energy capability	136	J	For a period of 2 ms
Peak forward surge current	200	A	Single pulse (of 8/20 μs current wave)
	4500	A	Two pulses (of 8/20 μs current wave)
Continuous RMS current	100	mA	
Agency approvals	UL1449, IEC61051-1, IEC61051-2, IEC60950-1 (annex Q)		

Source: Littelfuse (2014a).

Table 7.2 Summary of a typical BBD characteristics-device type 1.5KE440CA from littelfuse

Device specification	Value	Units	Remarks
Steady state power dissipation	6.5	W	
Peak power dissipation rating	1500	W	Tested using a 10/1000 μs test waveform as per data sheet (Littelfuse, 2014b)
Breakdown voltage	418–462	V	With a test current of 1 mA
Maximum clamping voltage	602	V	
Maximum peak pulse current	2.5	A	

Source: Littelfuse (2014b).

If a longer duration surge hits the device, failure may result due to excessive energy dissipated over the safe duration.

Lightning or inductive energy dumps appearing on circuits are completely statistical. Standards related to surge protection such as IEEE C62.4X series (IEEE recommended practice, 2002a,b) define test waveforms with specific rise times, width, and peak amplitude with different value sets as suitable for different classes of protection. In general, IEEE C62 series standards define three different categories (or classes) of protection based on the distance from the power entry point. Figure 7.4 shows these details. Table 7.3 shows test data usable for different test processes as specified by the standards. For more details on surge protection devices, test processes, and standards, Kularatna (2012) is suggested.

It is important to realize that the international standards define the test waveforms suitable for the categories A, B, and C based on double-exponential shapes or decaying sine waves as shown in Figure 7.5. Usually an open-circuit voltage waveform, short-circuit current waveform, and a ring wave are used as per Figure 7.5(a)-(c).

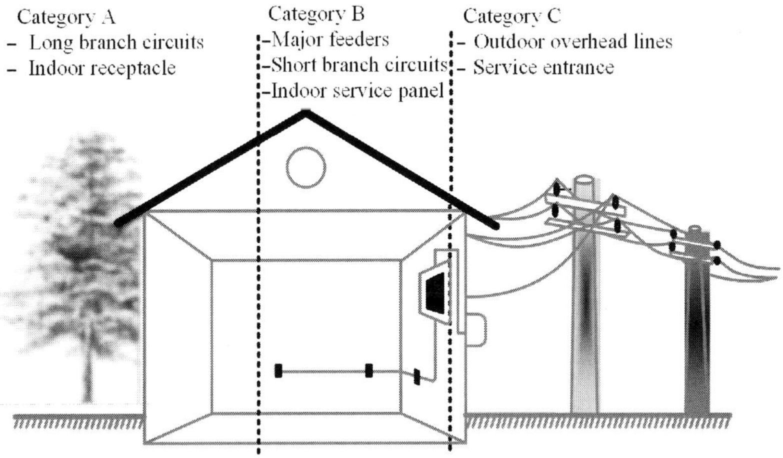

Figure 7.4 IEEE surge categories: classes A, B, and C based on the distance from surge entry point.

Table 7.3 **IEEE C62.41 location categories, frequency of occurrences, and surge waveform details**

IEEE location category	Level of exposure	Ring-type test waveform details		
		2 kV, 70 A, 0.63 J	**4 kV, 130 A, 2.34 J**	**6 kV, 200 A, 5.4 J**
A1	Low	0	0	0
A2	Medium	50	5	1
A3	High	1000	300	90
		Combination-type test waveform details		
		2 kV, 1 kA, 9 J	**4 kV, 2 kA, 36 J**	**6 kV, 3 kA, 81 J**
B1	Low	0	0	0
B2	Medium	50	5	1
B3	High	1000	300	90

7.3 Supercapacitor as a surge absorption device: summarized results of a preliminary investigation

At the University of Waikato, research team led by the author conducted some preliminary investigations on the ability of modern supercapacitors to absorb HV transient energy such as the lightning voltages superimposed on the power lines. In this process, the team used a lightning surge simulator (Models LSS 6230/6130) by Noiseken, Japan to apply repeated surge waveforms as per C62.4X series of standards

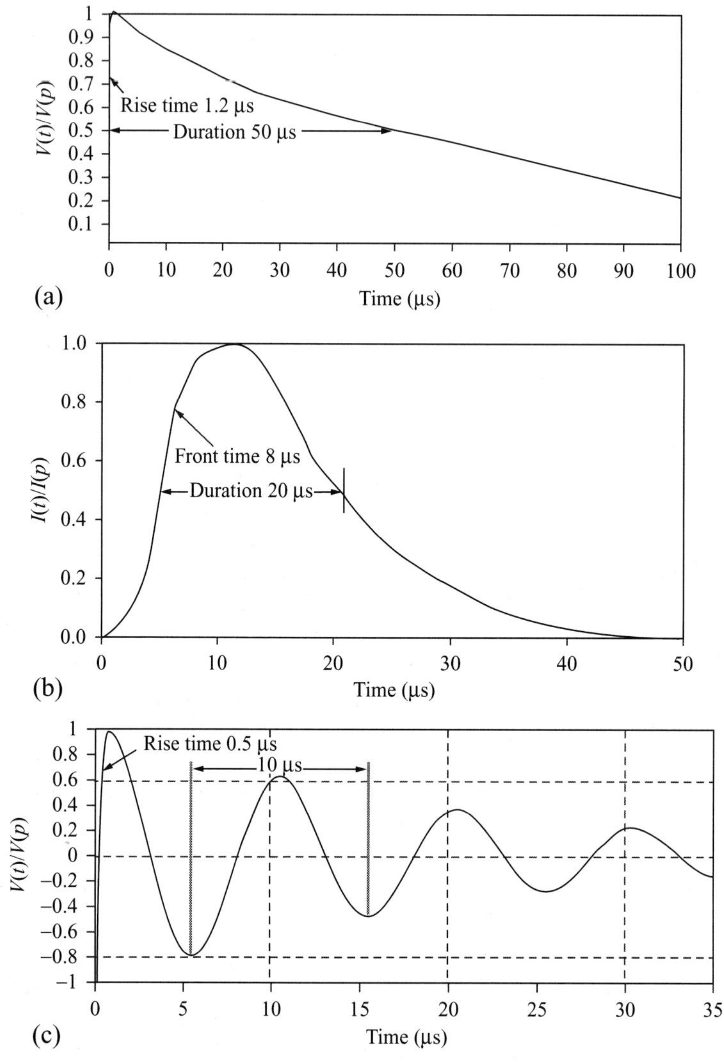

Figure 7.5 Few examples of wave shapes defined in surge test standards. (a) Open-circuit voltage waveform, (b) short-circuit current waveform, and (c) ring waveform.

to commercial supercapacitors from three different suppliers. As shown in Figure 7.4, a Category B lightning surge simulator (LSS) can generate repeated waveforms of well-defined double-exponential types, based on the principle of some finite energy stored in an internal HV power supply and a capacitor, passed through a waveform-shaping circuit inside the LSS. Such LSS are able to regenerate the waveforms repeatedly with either 10 or 20 s intervals up to peak voltages of 6.6 kV.

In an early investigation conducted by the author's team where several supercapacitors were subjected to a single-shot HV surge as well as multiple surges of identical

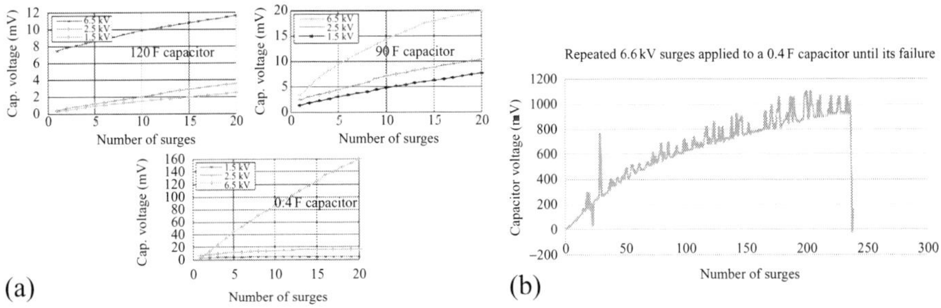

Figure 7.6 Terminal voltage development versus number of surges at different peak values of a combined waveform as per IEC61400-4-5. (a) Testing of three capacitors, where vertical axis units are in millivolts and (b) repeatedly applying surge to a small supercapacitor (0.4 F) until its failure.

shape from a LSS indicated that these devices do not get destroyed by transient surges with 100 μs-order time durations. The waveforms used were as prescribed in standards such as IEEE C62-41 and IEC 61400-4-5. Figure 7.6 indicates how three different supercapacitors developed their terminal voltage as we repeatedly surged them with identical HV transient waveforms with peak values from 1.5 to 6.5 kV. Devices used were (i) 120 F, 2.3 V device from Maxwell Technologies, (ii) 90 F 2.7 V device from NessCap, Korea, and (iii) 0.4 F 5.5 V device from Cap-XX, Australia. As shown in Figure 7.6(a), all what we could measure as voltage across the terminals was in the order of millivolts only.

Given this observation where the devices under test (DUTs) do not develop adequate DC terminal voltage even after 20 pulses that would exceed the rated DC voltage, it was decided to develop an automatic test interface for repeated application of standard surge waveforms to the DUT in order to determine their surge endurance. The purpose of this was to eliminate the tedious and time-consuming nature of repeated tests, allowing a LSS to recharge within about 20 s before each pulse was applied. In fact, the results shown in this preliminary test closely relates to the fact that an LSS output is a limited duration positive voltage source, which lasts less than 100 μs compared to the time constant of a commercial SC such as Cap-XX 0.4 F device with about 100 mΩ equivalent series resistance (ESR), giving a time constant of 40 ms. If we approximate the case of LSS output to be a constant 6.6 kV voltage, which is energized only for a period of less than 100 μs, we can expect only millivolts order voltages developed across the terminals of the device. This also indicates that the device still retains its capacitive behavior and not adversely affected by the transient HV at the terminals. However, as shown in the case of repeated surge application into small supercapacitors such as 0.4 F device failed as shown in Figure 7.6(b).

Given the preliminary results showing that a supercapacitor is not destroyed by the repeated application of HV transients as specified in C62.4X standards, etc., and the gradual voltage rise across terminals after each hit is in the order of millivolts, we were encouraged to develop a more detailed test system and test process to confirm this observation supported by theory (Kularatna et al., 2011a). Based on the simple formula for a capacitor's energy storage capability of $\frac{1}{2}CV^2$, a supercapacitor of 10 F with a terminal DC voltage rating of 2 V gives an energy storage capability of 20 J. A typical computer power supply's smoothing capacitor within the rectifier stage of value 220 μF with a voltage rating of 400 V will have a maximum energy storage capacity of 17.6 J. What this means is that a small supercapacitor could absorb almost the same energy, which is capable of destroying a smoothing capacitor in an off-the-line DC power supply. Author's long working experience in an equatorial island with two monsoon seasons indicates that lightning induced on incoming mains destroys electronic components in rectifier stages and switch-mode power converter components, irrespective of the presence of (limited) surge protection capability on the desktop computer power supplies, which is sometimes called the silver box. James et al. (2010) provides a summary of laboratory surge tests conducted on computer silver boxes with peak voltages up to 6.6 kV.

Above summary indicated us the confidence to see if a supercapacitor can be used to absorb part of the surge energy during a transient traveling through the incoming mains or the telecom/data circuits. From various models developed for supercapacitors, two common and useful models depicted in Figure 7.7 based on the publications (Kularatna et al., 2011a; Spyker et al. (2000)) depict the equivalent circuit representing the combination of the LSS, connection cables, and the supercapacitor under test (IEEE recommended practice, 2002a).

Table 7.3 indicates a summary of test waveforms usable to test various surge protectors used at different IEEE location categories, with different statistical levels of exposure such as low, medium, and high. It also indicates that a laboratory instrument like a LSS with a combination-type waveform or a ring waveform (as in Figure 7.5(a) and (c), respectively) with peak voltage capability up to about 6 kV and energy ratings up to about 90 J is used for testing surge protector devices.

The voltage and current values shown in Table 7.3 are the peak voltages applicable to Figure 7.5(a) and the peak current values applicable to Figure 7.5(b), respectively. The current values shown in Table 7.3 also refer to the short-circuit current capabilities of test systems such as an LSS used to test industrial surge protectors. Figure 7.8

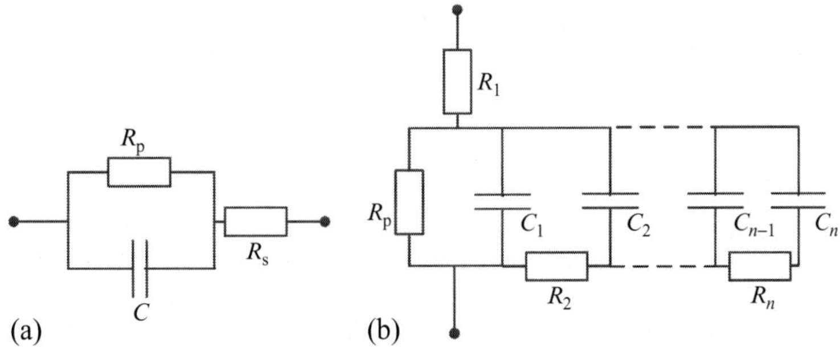

Figure 7.7 Supercapacitor equivalent circuits usable in developing a surge protection circuit topology. (a) Classical equivalent circuit and (b) ladder circuit.
Adapted from Spyker et al. (2000).

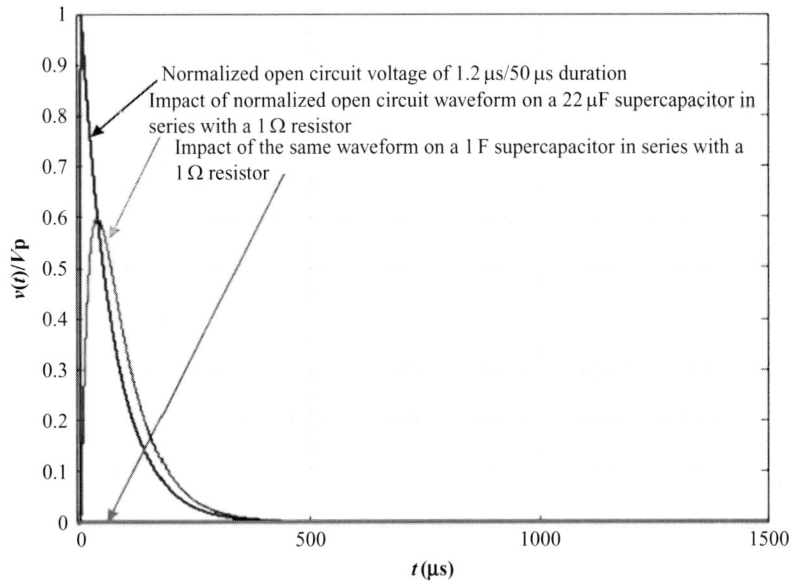

Figure 7.8 MATLAB simulation results—applying an IEC61400-4-5 combined surge waveform to 220 μF capacitor and 1 F supercapacitor in series with 1 Ω.

shows the case of an open-circuit voltage waveform shown in Figure 7.5(a) applied to a SC with a simplified model shown in Figure 7.7(a). A surge waveform as shown in Figure 7.3(a) can be represented mathematically as

$$v_{SG}(t) = \frac{\alpha\beta}{\alpha+\beta}\left(e^{-\alpha t} - e^{-\beta t}\right) \tag{7.2}$$

where α represents the inverted value of the time constant related to the decaying part of the exponential waveform and β is the same for the rising part of the exponential waveform. By differentiation and estimating the maxima of the time-varying surge, we can get the peak value of the surge waveform Vp as,

$$V_p = \frac{\alpha\beta}{\alpha+\beta}\left[\left(\frac{\alpha}{\beta}\right)^{\frac{\alpha}{\beta-\alpha}} - \left(\frac{\alpha}{\beta}\right)^{\frac{\beta}{\beta-\alpha}}\right] \tag{7.3}$$

From Eqns (7.2) and (7.3), we can arrive at a normalized open-circuit voltage as similar to Figure 7.5(a). By suitable mathematical manipulations and approximations, we can get the following relationship to approximate the normalized open-circuit voltage waveform as a function of time in microseconds:

$$v_{SG,nor}(t) = 1.0203\left(e^{-4.16t} - e^{-0.0139t}\right) \tag{7.4}$$

Figure 7.8 indicates the MATLAB-generated waveform, which closely matches the shape of the normalized open-circuit voltage waveform together with its impact on an RC circuit comprising of 220 μF and 1 F capacitors in series with a 1 Ω resistor. Figure 7.8 clearly shows that due to very large time constants in supercapacitors, such a HV waveform which has a total duration of about 100 μs will not develop any adequate voltage across the capacitor. For example, if the peak voltage of the surge waveform is 1 kV, a 220 μF capacitor with 1 Ω resistor develops an approximate capacitor voltage of 600 V, while a 1 F capacitor's case does not show any measurable value.

Given the aforementioned situation, we simply predicted that an SC of farad-order value with 10- to 100-mΩ ESR may withstand HV transients of durations less than a few hundreds of microseconds. However, no publication indicated if this is the case, in the context of any other possible failure phenomenon which can occur due to a HV pulse. In analyzing the effect of a surge or a transient on an SC, similar to the case of a HV LSS output applied on the DUT, we can use the circuits in Figure 7.9. In this case, we have simplified the SC using the equivalent circuit in Figure 7.7(a). Based on the Thevenin equivalent circuit seen by the two points A and B in Figure 7.9(a), the case can be simplified to the circuit as shown in Figure 7.9(b).

In this case

$$V_{EQ} = V_{SG}\left[\frac{1}{1+\left[\left(R_{SG}+R_{S}+r_{path}\right)/R_{p}\right]}\right] \tag{7.5}$$

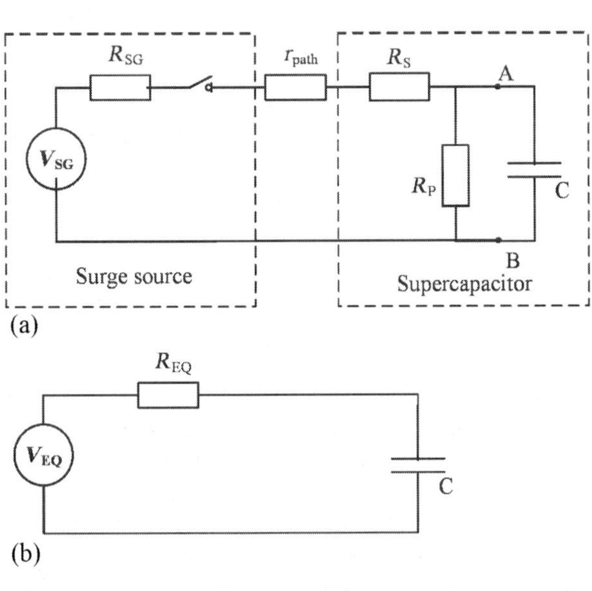

Figure 7.9 Surge source connected to a supercapacitor. (a) Case with simplified equivalent circuit of an SC and (b) Thevenin equivalent circuit with simplifications.

$$R_{EQ} = R_{Thevenin} = \frac{R_{SG} + R_S + r_{path}}{1 + \frac{R_{SG} + R_S + r_{path}}{R_p}} \quad (7.6)$$

where R_{SG} and r_{path} are the internal resistance of the surge source and the path resistance of interconnections between the SC and the surge source, respectively. It can be easily seen that the time constants applicable to the combined effect will be always significantly higher than those with the case for $\tau = R_S C$, which will be the lowest possible time constant for the SC charging. This is based on the assumption that the value of R_p is significantly higher than other series path resistances. Given the aforementioned scenario, if a capacitor of value C is subjected to a step voltage of duration T with maximum amplitude of V_{max}, the worst-case voltage developed across the capacitor at the end of the step voltage will be,

$$V_{C,max} = V_{max}\left(1 - e^{\frac{-T}{R_S C}}\right) \quad (7.7)$$

which could be further simplified to,

$$V_{C,max} = V_{max}\left[\frac{T}{\tau} - \frac{1}{2}\left(\frac{T}{\tau}\right)^2 + \frac{1}{3!}\left(\frac{T}{\tau}\right)^3 - \cdots\right] \quad (7.8)$$

This indicates us an approximate value of $\frac{T}{\tau}V_{max}$ when the time constant τ is much larger than the step surge voltage duration is T. When a step voltage source of 100 μs duration with a maximum value of V_{max} is applied to a supercapacitor of 10 mΩ ESR and a capacitance value of 100 F, the capacitor could be charged only up to $10^{-4}\,V_{max}$. This comfortably predicts that a supercapacitor will not acquire enough charge to exceed its DC voltage rating during the period of single-shot application of a combination surge waveform with even 10 kV. Kularatna (2012) provides details of subjecting three different commercial supercapacitor families to surge waveforms up to 6.6 kV peak values, and it provided adequate confidence to use supercapacitors as surge-absorbing components.

7.4 Design approaches to a supercapacitor-based surge protector

Previous discussions indicate us that supercapacitors could have continuous energy storage capabilities in the range of energy carried in a transient surge into an electrical circuit, provided the capacitor value and maximum rated voltage of the device combined to give the $1/2CV^2$ value. Also, this fact should be compared with the transient energy absorption capabilities of MOVs and BBD used in typical surge protectors as shown in Figure 7.1(a). However, commercially available supercapacitors have very low DC voltage ratings, such as less than 4 V for single-cell devices, which is far below the instantaneous voltages occurring on the AC mains. Given this problem, a designer of a surge protector cannot just substitute a supercapacitor for a MOV or a BBD.

Given this situation, completely different approach should be used in designing a supercapacitor-based surge absorber circuit.

Supercapacitor-assisted surge absorber (SCASA) is a patented new technique, developed by the research team led by the author Kularatna et al. (2014), based on the concept in the circuit of Figure 7.10(a). In this circuit, typical NLDs such as a MOV or a BBD is combined with a magnetic component and an SC-based subcircuit. However, compared to a typical surge arrestor without supercapacitors, where NLDs are placed directly across the pairs of wires such as the neutral and the live (or neutral or live and earth), the NLD is placed in between the load side end of a primary coil of a transformer and the return wire as in Figure 7.10(a). Given this configuration, when a surge occurs at the AC input, and when the instantaneous voltage due to surge exceeds the firing voltage of the NLD, high instantaneous current flows through the primary coil, developing a voltage across the primary turns. This in turn develops an induced voltage across the secondary coil, and by winding the secondary coil in the opposite direction, secondary induced voltage can be generated to oppose the surge voltage. Result is to create a voltage across the critical load to be protected, which could be less than the instantaneous surge voltage. By adjusting the turns ratio, we can adjust secondary voltage in such a way that the instantaneous voltage across the supercapacitor-based subcircuit can be varied.

Once a superimposed HV transient travels along the mains input, NLD fires and enters into conduction stage, developing a transient voltage across the connected winding (v_P). When the HV transient exceeds the firing voltage of the NLD, it conducts heavily creating a surge current through the primary coil. Due to induction, secondary coil also develops a voltage (v_S) and the two windings are configured to create this induced secondary voltage higher than that of the primary winding and to oppose the transient so that the critical load end sees the difference between these two voltages as shown in Figure 7.11. Graphs in Figure 7.11 are for a circuit with a MOV (type Q20K275) which is having a maximum clamping voltage of 710 V.

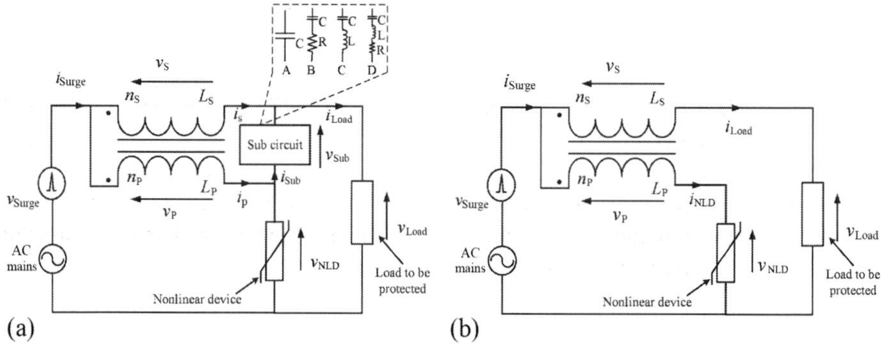

Figure 7.10 SCASA technique basics. (a) Overall configuration showing variations of the SC subcircuit and (b) SCASA circuit without SC subcircuit to explain the effect of magnetic component.

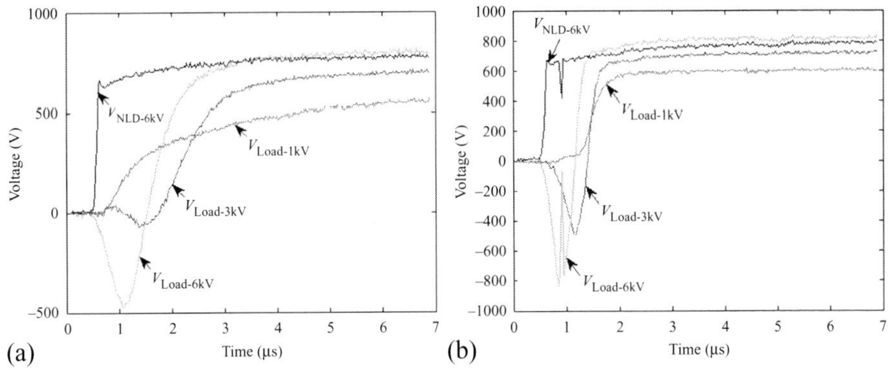

Figure 7.11 Load and varistor voltage variation for 1, 3, and 6 kV surge input for different toroid with 6:30 turn ratio. (a) Powdered core ($\mu_r = 60$) and (b) ferrite core ($\mu_r = 10{,}000$).

As indicated in Figure 7.10(a), a supercapacitor-based subcircuit configuration could have a few possible variations (A, B, C, and D). SC-based subcircuit receives the voltage difference between the two coils to create a circulating current through the subcircuit, absorbing part of the surge energy into the supercapacitor. With the ability of the supercapacitor to absorb part of the surge energy, which will be dissipated in the closed loops formed, transient surge energy burden on the NLD is significantly reduced. In addition, supercapacitor subcircuit will perform a useful filter function to reduce the ringing waveform created by the surge.

If generalized impedance Z_{Sub} is used to represent the subcircuit options A-D as in Figure 7.9(a), we can establish the following mathematical relationships:

$$v_p = L_p \frac{di_p}{dt} + M \frac{di_s}{dt} \tag{7.9}$$

$$v_s = L_s \frac{di_s}{dt} + M \frac{di_p}{dt} \tag{7.10}$$

$$v_{sub} = \left(v_s - v_p\right) = i_{sub} Z_{sub} \tag{7.11}$$

L_p and L_s are the self-inductances of the primary and secondary coils, respectively and M is the mutual inductance between the two windings. Voltages v_p, v_s, and v_{Load} are the instantaneous values at the transformer primary, secondary, and at the load, respectively, when the superimposed surge, v_{Surge}, is applied to the circuit. v_{NLD} and i_{Sub} are the voltage across the NLD and the transient related current flowing through the SC-based subcircuit.

7.4.1 Selection of the magnetic core

In the SCASA technique, overall performance is mainly governed by the capability of the magnetic component, where its leakage inductance combined with the

Table 7.4 Properties of different toroid cores used for testing (data sheet parameters)

Core type	Magnetics part no.	Relative permeability (μ_r)	A_L value (nH/T^2)
Powdered core	0077083A7	60	81 ± 8%
Ferrite core	ZW41605TC	10,000	0%

transformer action assists creating a lower effective clamping voltage across the critical load. Depending on the permeability of the selected core material, overall performance varies, since the secondary winding voltage due to the superimposed surge depends on the core's saturation behavior. Figure 7.10(b) shows the test circuit used to estimate the transformer action in suppressing the overall surge when no subcircuit is connected.

Two transformers with primary to secondary turns ratio (n_p/n_s) of 6:30 with approximately equal cross sections but different core materials were used for this test and some essential data for two cores are presented in Table 7.4. Figure 7.11 shows the voltage waveforms at the load when the overall circuit is subjected to peak surge voltages such as 1 kV ($V_{\text{Load-1 kV}}$), 3 kV ($V_{\text{Load-3 kV}}$), and 6 kV ($V_{\text{Load-6 kV}}$). The $V_{\text{NLD-6 kV}}$ line in each figure shows the voltage across the NLD for a 6-kV surge input. Results in Figure 7.11(a) are for a powdered-core transformer with relative permeability (μ_r) of 60. In Figure 7.11(b), results are for a ferrite core of relative permeability of 10,000. It can be seen from Figure 7.11(b) that the secondary voltages generated due to surge are much narrower than the case in the powdered core shown in Figure 7.11(a) for higher surge voltages. Hence, it stores less energy from the source, which will be absorbed and dissipated by the SC subcircuit. More details on these aspects are in Kularatna and Fernando (2014) and Fernando and Kularatna (2013 and 2014).

7.4.2 Impact of the supercapacitor subcircuit

In the SCASA technique, supercapacitor subcircuit plays the role of absorbing part of the surge energy superimposed on the incoming pair of wires. Since this is placed across the far end of the transformer-winding pair, difference of the instantaneous voltages across the two windings create a voltage difference charging the supercapacitor. Overall effect is that instantaneous voltage across the critical load becomes much less than the surge-related voltage across the NLD.

Figure 7.12 depicts the case of circuit arrangement used to measure the effect of different capacitance values used within the subcircuit used in the SCASA technique. For performance comparison purposes, using the configuration as shown in Figure 7.10(a), authors used an ordinary 50 nF ceramic capacitor as well as a 5 F supercapacitor as the subcircuit and applied 1-6 kV lightning surges generated by the LSS. This is the type B subcircuit with a resistor value of 1 Ω.

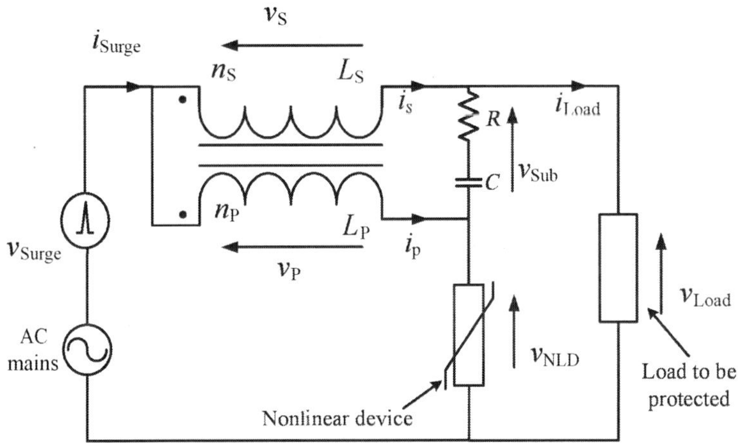

Figure 7.12 Circuit developed for final testing of SCASA.

When a ceramic capacitor of 50 nF is used as shown in Figure 7.13(a1-a3), voltage across the subcircuit starts oscillating at different frequencies varying from about 122 to 343 kHz. These waveforms are due to several situations occurring at the magnetic component such as

- Core parameter nonlinearity variation (due to core saturation)
- Transformer leakage inductance variation due to the ride of HV surge through the windings

Compared to this scenario, a supercapacitor of 5 F replacing the 50 nF ceramic shows very much lower voltages across the critical load. Figure 7.13(b1-b3) show the scenario. With the 5 F supercapacitor, it can be clearly seen that there are no oscillations and a voltage generated across the subcircuit due to the current i_{Sub}, which opposes the voltage across NLD (v_{NLD}), creates lower voltage across the load than v_{NLD}. Another important aspect of the SCASA technique is that the supercapacitor subcircuit placed does not get subjected to high 50 or 60 Hz voltage components since it is placed across the differential AC voltage across the two transformer windings, where AC mains input end is common. In this design, leakage impedance of the transformer plays a major role, and hence the selection of the right core for the transformer.

Mathematical analysis to detail this situation is beyond the scope of this chapter, but Eqns (7.9)–(7.11) form the basis. More details can be found in the research thesis related to the project (Fernando, 2015).

Results in Figure 7.14 show the overall effect of the surge on the critical load end with a B-type subcircuit as shown in Figure 7.12 and also depict the overall performance comparison of the SCASA configuration with a traditional commercial surge protector. A commercial version was recently built and was undergoing detailed laboratory tests at the University of Waikato.

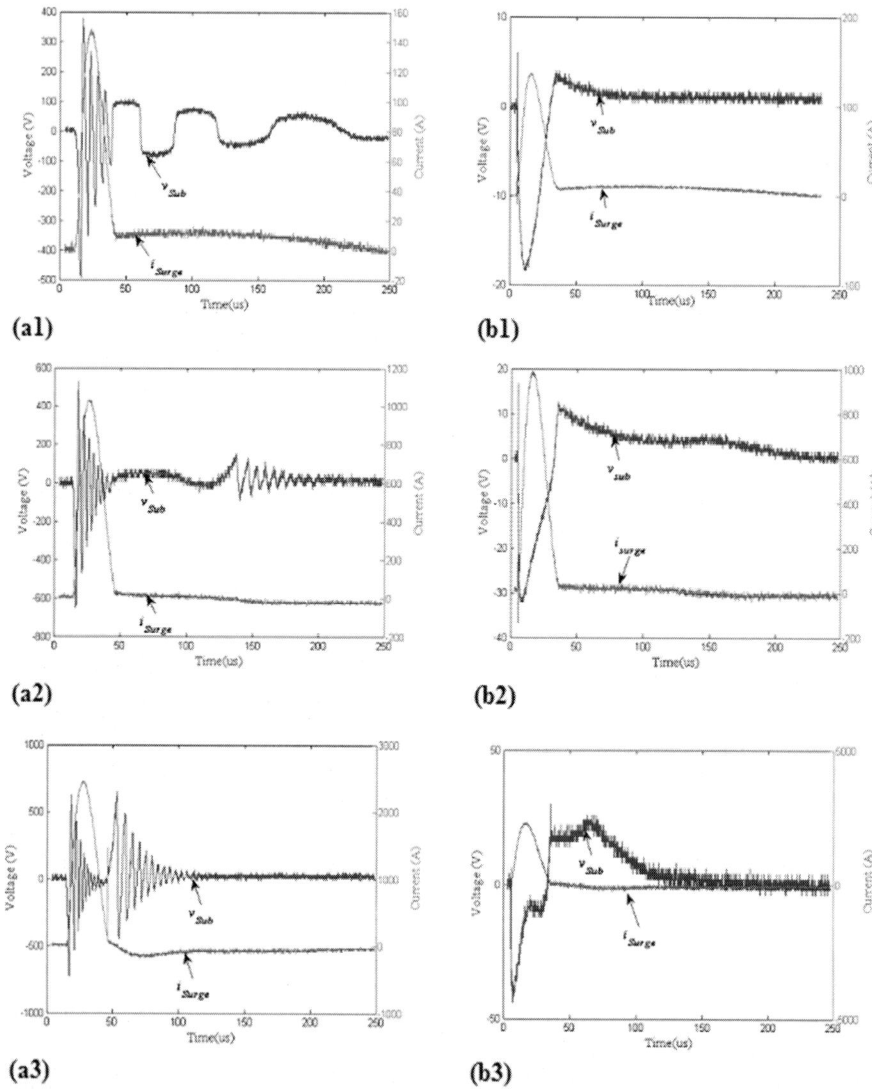

Figure 7.13 Impact of the supercapacitor subcircuit in the SCASA technique compared with a ceramic capacitor used in lieu of supercapacitor subcircuit. (a1) Effect of 50 nF ceramic capacitor as the subcircuit for a 1 kV surge. (a2) Effect of 50 nF ceramic capacitor as the subcircuit for a 3 kV surge. (a3) Effect of 50 nF ceramic capacitor as the subcircuit for a 6 kV surge. (b1) Effect of SCASA subcircuit with a 5 F supercapacitor for a 1 kV surge. (b2) Effect of SCASA subcircuit with a 5 F supercapacitor for a 3 kV surge. (b3) Effect of SCASA subcircuit with a 5 F supercapacitor for a 6 kV surge.

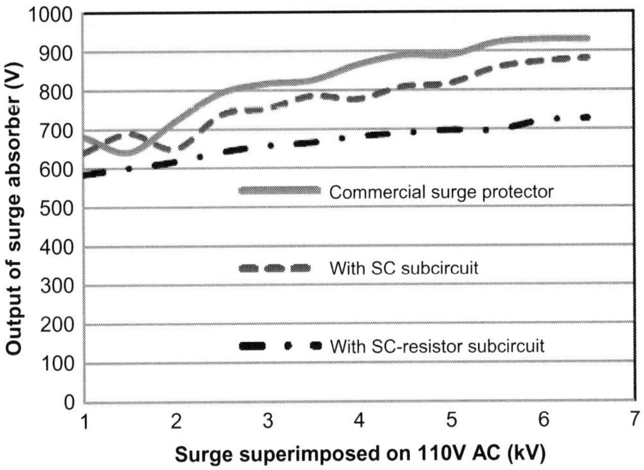

Figure 7.14 Comparison of SCASA performance with a common commercial surge protector circuit.

7.5 Conclusion

SCASA technique makes use of the ability of a supercapacitor to absorb transient HVs due to large time constants in a circuit loop, and forms the basis of an entirely new and nontraditional application of the ELDCs. This useful ability of supercapacitors can be extended to other major areas such as uninterruptible power supplies, and this is an entirely new applications research area at the University of Waikato, and early work under the concept of surge-resistant UPS is ongoing (Kularatna et al., 2011b; Kumaran, 2011). The subject area opens up another unique application set for EDLCs.

References

Fernando, J., 2015. Ph.D. Thesis, Characterization of supercapacitors for special applications. The University of Waikato.

Fernando, J., Kularatna, N., June 2013a. A supercapacitor based enhancement for standalone surge protector circuits. In: Proceedings of ISIE-2013 Conference, Taipei, Taiwan, pp. 1–6.

Fernando, J., Kularatna, N., September 2014. Supercapacitor assisted surge absorber (SCASA) technique: selection of supercapacitor and magnetic components. In: Proceedings of ECCE 2014.

IEEE recommended practice on characterization of surges in low-voltage (1000 V and less) AC power circuits, IEEE Standard C62.41.2, 2002a.

IEEE recommended practice on surge testing for equipment connected to low-voltage (1000 V and less) AC power circuit, IEEE Standard C62.45, 2002b.

James, S., Kularatna, N., Steyn-Ross, A., Kunnemeyer, R., Tantrigoda, D., November 2010. Investigation of failure patterns of desk top computer power supplies using a lightning surge simulator and the generation of a database of comprehensive surge propagation study. In: Proceedings of IEEE IECON 2010 Conference, AZ, USA, pp. 1275–1280.

Kularatna, N., 2012. DC Power Supplies, Power Management and Surge protection for Power Electronic Systems. CRC Press, USA, 390 p. (Chapter 9).

Kularatna, N., Fernando, L.J., March 21, 2014. Power and telecommunication surge protection apparatus. New Zealand Patent Number 604332.

Kularatna, N., Fernando, J., Pandey, A., James, S., 2011a. Surge capability testing of supercapacitor families using a lightning surge simulator. IEEE Trans. Ind. Electron. 58 (10), 4942–4949.

Kularatna, N., Tilakaratne, L., Kumaran, P.K., November 2011b. Design approaches to supercapacitor based surge resistant UPS techniques. In: Proceedings of IEEE-IECON 2011, pp. 4094–4099.

Kumaran, P.K., 2011. ME Thesis, Development of supercapacitor based SRUPS.

Littelfuse, May 2014a. Varistor products data sheet for TMOV and iTMOV series.

Littelfuse, January 2014b. Transient voltage suppressor data sheets, 1.5KE series.

Nelms, R.M., Cahela, D.R., Tatarchuk, B.J., 2003. Modeling double-layer capacitor behavior using ladder circuits. IEEE Trans. Aerosp. Electron. Syst. 39 (2), 430–438.

Spyker, R.L., Nelms, R.M., 2000. Classical equivalent circuit parameters for a double-layer capacitor. IEEE Trans. Aerosp. Electron. Syst. 30 (3), 829–836.

Supercapacitors in a rapid heat transfer application

8

8.1 Introduction

State-of-the-art supercapacitors (SCs) come in all sizes from about fractional farads to over 5000 F in single-cell devices. For smaller value devices their equivalent series resistance (ESR) values range from about 25 to 100 mΩ. The larger capacitance devices from about 300 to 5000 F have much lower ESR values such as 0.3 mΩ to about 5 mΩ. Given this summary, and the discussion we had in Chapter 1 about the maximum power transfer capability of a voltage source given by $\frac{V^2}{4r_{\text{int}}}$ where V is the voltage source value and r_{int} is the internal resistance of the voltage source, provided that we terminate the voltage source with a load resistance, R_L equal to r_{int}. Now if we apply this fact to a typical battery and a SC to estimate the maximum possible output power, we could see that the internal resistance device is the primary determining factor. In this chapter we discuss the case where we could use a bank of SCs to fast-heat water flowing in domestic water systems and to minimize the waste of treated water due to the stored cold water in domestic pipes between the central water heater and the individual faucet. The discussion will also demonstrate how a bank of SCs with an appropriate control system could circumvent the problem of developing a direct AC mains-based high-power fast-acting water heating solution, due to power delivery limits of an electrical subcircuit with a maximum limitation of about 2.3 kW.

8.2 Problem of wasted water in day-to-day situations at home

In domestic hot water supply systems and the central heating systems, whether it is gas or electrical, the hot water unit is usually placed in a central location and small diameter pipes carry hot water from the central unit to individual faucets. Table 8.1 provides a typical scenario in New Zealand residences.

Based on information in Table 8.1, one can estimate the wasted amount of water on a daily basis, in addition to the frustration due to initial cold water flushed through the hot water supply activation. Though fast-acting gas heater systems are becoming popular, the wasted water due to cold water stored in the buried pipes between the central gas unit and the individual faucet is unavoidable in domestic situations. In New Zealand with a population of approximately 4.5 million, more than 15 million liters of treated water is wasted annually due to this problem.

Table 8.1 **Typical hot water distribution scenario in new zealand residences**

Parameter	Average value
Average distance from central hot water unit to individual faucets	3-50 m
Water flow rate	4-8 l/min
Hot water temperature	35-55 °C
Time delay for hot water due to water storage in wall-buried pipes	3-40 s
Wasted amount of water due to delay in hot water	0.3-3 l/use
Average number of times for faucet usage	5-20/day

8.3 Problem of traditional heating from direct AC mains supply and heating system specifications

First requirement in this situation is to supply the energy to flowing water at an adequate rate. Following heat transfer relationship is the starting point to analyze the requirement.

$$P = \rho f_{\mathrm{w}} c \Delta\theta \tag{8.1}$$

where P is the power (W), ρ is the density of water (kg l^{-1}), f_{w} is the water flow rate (l s^{-1}), C is the specific heat of water (J kg^{-1} C^{-1}), and $\Delta\theta$ is the temperature rise (°C).

Based on the above relationship, if the water fallow rate is 8 l/min, the power delivery should happen at a rate of approximately 11.2 kW for an average temperature rise of nearly 20 °C. Given the 10 A maximum current limit from a domestic AC mains wall outlet of 230 V AC/50 Hz the maximum power delivery is now limited to 2.3 kW (RCCB, 2011), which is very much below the requirement. However, total energy requirement to heat the water with an average 20 °C temperature rise within 30 s at a flow rate of 8 l/min, is only about 150 Wh (540 kJ), which is compared with the energy required for an average weight electric car to drive about 1 km distance.

Considering the following key design criteria in Table 8.2, we can start working on a different solution based on prestored energy.

8.4 Commercial solutions for eliminating water wastage due to storage in buried plumbing

There are a few commercially available solutions such as tank-less systems (Laing, 2013), under the sink tanks (Permenter, 2004), and recirculation systems (Fazekas, 1984), to reduce or eliminate this delay, but costs range from 500 to 6000 NZ dollars. However, these solutions do not eliminate the problem completely and economically. The tank-less system heats water as it passes through the unit. This option saves

Table 8.2 Essential specifications for fast-heating solution based on energy storage

Specification	
Total maximum temperature rise	20 °C
Average water flow rate	6 l/min
Estimated time to receive central hot water from the gas heater	20 s
Total energy requirement (with losses neglected)	47 Wh
Power capability of the heating element	8.4 kW
Maximum voltage allowed for safety reasons	60 V DC or 42.4 V_{peak} (based on AS/NZS 60950.1:2003 Standard)

energy as only the water used is heated. However, this solution requires replacement of the existing faucets and heating system (Laing, 2013). Under-sink tanks are an alternative option, but this system contains a mini tank installed under the sink and a mixing faucet. This solution is expensive and requires regular maintenance (Permenter, 2004). Another solution is a recirculation system; this system is made up of a closed loop that connects to each hot water usage point and a pump for recirculating water. These systems cannot be easily retrofitted and running costs are expensive (Fazekas, 1984). More details are in Kulartna et al. (2014).

8.5 Practical requirements for a localized solution

The above power and energy requirement estimated as a typical worst case requirement indicated that the required energy can be stored in a suitable storage device and can be converted to a short-term high power delivery at the required rate into flowing water, without any additional water tanks, etc. However, several prime requirements as listed below are essential in such a system to be acceptable for regulatory authorities.

8.5.1 Electrical safety and isolation

Due to safety reasons, a power source over 60 V DC cannot be utilized based on safety extra low voltage requirement for electrical shock safety. If mains power is directly used, galvanic isolation via a transformer supplying less than 60 V (rms) may be required in case a high-power heater coil is placed inside pipe carrying domestic water, in addition to the overload safety provided via a residual type circuit breaker (Datasheet; Particular requirement for instantaneous water heaters, 2004).

8.5.2 Domestic installation regulations

Short-term high power delivery as discussed in an application such as the fast water heating requires special wiring and protection devices (Datasheet). It is impractical to use an existing electrical subcircuit to deliver this rate of power. In case high-power

30 A subcircuit is to be used, maximum power level can be increased to about 7.5 kW, but this is still beyond the power requirement of about 10-15 kW.

However, since the system required for inline water heating system such as this requires to be powered for a period of less than 30 s in general (due to this the central gas or electrical heating system can kick in within 2-5 s in general) and the statistical nature of water usage at a domestic faucet, prestored energy storage solution is a viable and cost-effective solution.

8.6 SC-based solution with prestored energy

Developing a SC-based energy storage system with an energy storage capacity in the range of 25-200 Wh is practically feasible based on large single-cell capacitors from multiple sources of manufacturers. For example, Maxwell Technologies, USA (Maxwell Technologies) and LS Mtron, Korea (LS Mtron) are representative companies who manufacture devices suitable for an application like this, and their important specifications for the application are summarized in Table 8.3.

As we see from Table 8.3, these large single devices, usually available up to about 5000 F in general, have extremely low ESR values of less than 1 mΩ and current delivery capabilities of 500-3000 A in general. Based on the simple theoretical considerations in Chapter 1 based on Eqn (1.9) and the related discussion, if a series capacitor bank of ten 3000 F from either manufacturer will allow an initial maximum power of 54 kW at a voltage of 25 V across the 10-element series bank. This is nearly impossible from a 230 V/50 Hz or 120 V/60 V single-phase domestic socket outlet. In fact, this is the maximum power rating of the single 3000 F SC cell charged to 2.5 V, suitably terminated by a load equal to its DC ESR.

In a 10-element bank charged up to 25 V will be able to hold about an energy content of 93 kJ or 26 Wh. To reach the energy stored level of 150 Wh, we require six parallel strings of 10-element 3000 F capacitor banks in general, but the designer is challenged with prohibitive cost of the energy storage bank for a domestic application such as inline water heating. Figure 8.1 shows the arrangement of the SC bank of 60 elements. Figure 8.1(b) shows the general case of Thevenin equivalent circuit for a

Table 8.3 Comparison of device specifications for fast energy delivery system such as inline water heating

Specification	Capacitance (F)	Rated voltage (V)	DC ESR (mΩ)	Maximum current (A)	Stored energy (Wh)
Maxwell Technologies	3000	2.7	0.29	2396	3.04
	1500	2.7	0.47	1426	1.52
	650	2.7	0.8	640	0.66
LS Mtron	3000	2.7	0.29	1900	3.04
	2000	2.7	0.35	1500	2.03
	1200	2.7	0.58	930	1.22

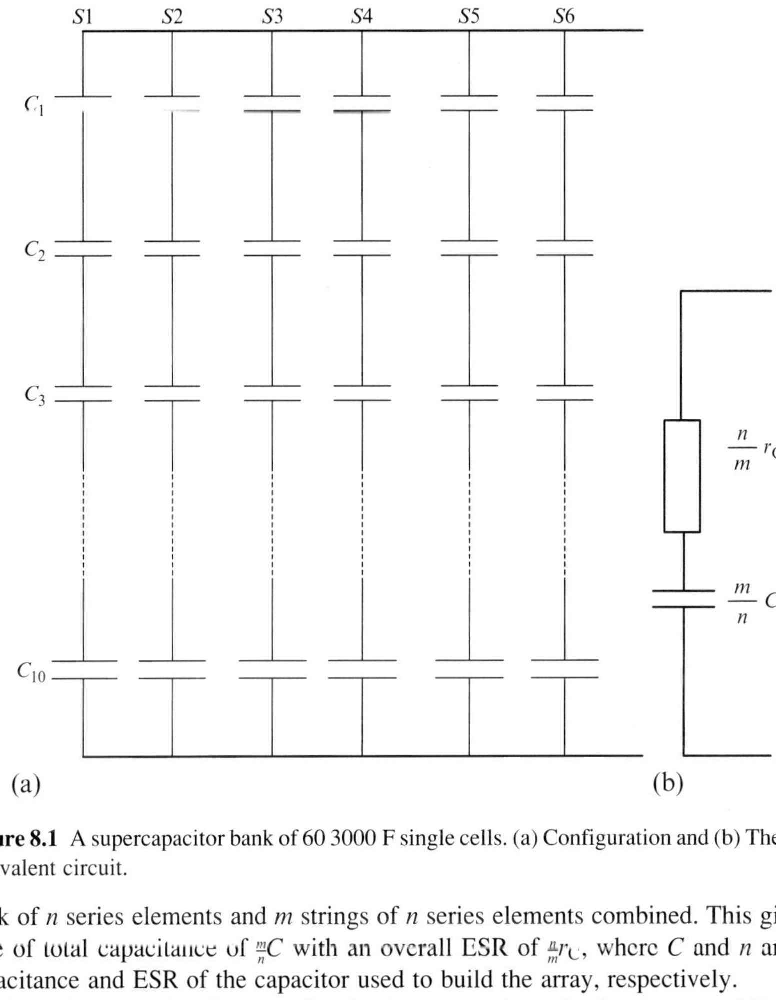

Figure 8.1 A supercapacitor bank of 60 3000 F single cells. (a) Configuration and (b) Thevenin equivalent circuit.

bank of n series elements and m strings of n series elements combined. This gives a case of total capacitance of $\frac{m}{n}C$ with an overall ESR of $\frac{n}{m}r_C$, where C and n are the capacitance and ESR of the capacitor used to build the array, respectively.

Given the case of such a capacitor bank connected to a load resistance of R_L via a switch as shown in Figure 8.2(a), we can show that the instantaneous voltage across the load is given by,

$$v_{RL}(t) = \frac{V_{C0}R_L}{R_L + r_{eq}} e^{-\frac{t}{(R_L + r_{eq})C_{eq}}} \tag{8.2}$$

and the energy transferred to the load and the ESR is given by

$$E(0) - E(t) = \frac{1}{2} C_{eq} V_{C0}^2 \left[1 - e^{-\frac{2t}{(R_L + r_{eq})C_{eq}}} \right] \tag{8.3}$$

where V_{C0} is the initial capacitor bank voltage, $C_{eq} = \frac{m}{n}C$, and $r_{eq} = \frac{n}{m}r_C$.

The above discussion gives rise to the graphs in Figure 8.2(b) and (c) for normalized capacitor voltage and the percentage energy discharged over time.

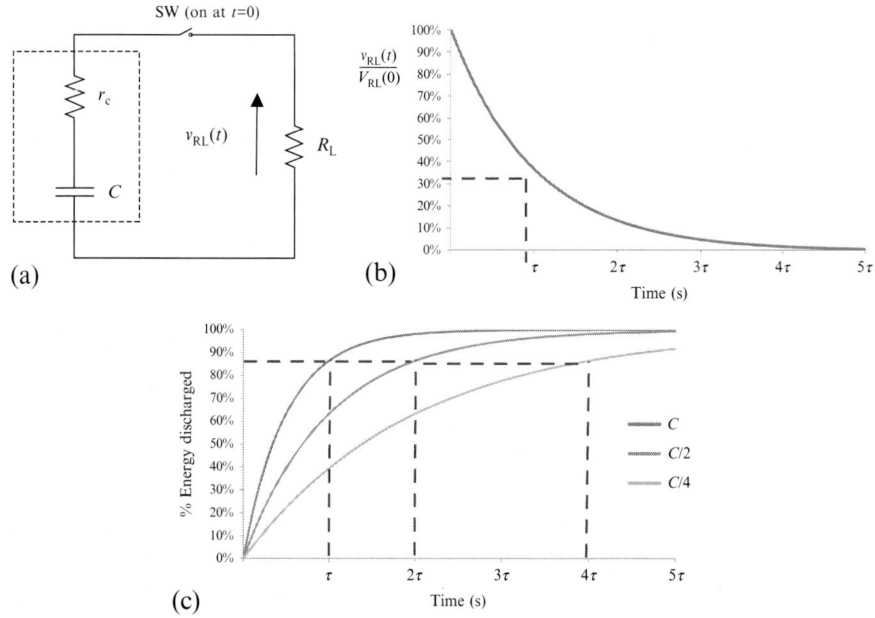

Figure 8.2 A supercapacitor bank discharging into a resistive load. (a) Equivalent circuit, (b) capacitor voltage versus time, and (c) percentage of delivered energy versus time.

In a case like this, the discharge time constant is given by,

$$\tau = \frac{mC}{n}\left(\frac{n}{m}r_C + R_L\right) \tag{8.4}$$

and the approximate average power, P_{ave}, delivered by the bank into the load is given by, (with individual capacitors with maximum voltage of V per capacitor)

$$P_{ave} = \frac{nV^2}{10\left(\frac{n}{m}r_C + R_L\right)} \tag{8.5}$$

In general, it is quite easy to show that a SC-based solution is practicable, compared to any other energy source such as a battery bank. In the case of a battery bank, since the life cycle for full depth of discharge (DOD) is in the range of 500-2000 for lead-acid or Li-ion chemistry, in addition to the gradual rise of the internal resistance with the DOD (Coleman et al., 2007), it does not become very much commercially attractive solution. However, in order to reduce the cost of the overall solution, it is possible to combine a suitable battery bank with a smaller SC bank. A discussion on this solution is beyond the scope of the chapter. In this discussion, we have used the most simplified model for a SC (Miller, 2010). Since the state-of-the-art SCs have relatively low leakage currents, an SC bank made of single cells without any balancing circuits can be fully charged and about 80-85% energy can be reused overnight, allowing the bank to recharge slowly based on a small capacity DC power supply suitable for an SC bank.

8.7 Results from an ongoing prototype development exercise

A complete system concept suitable for an application such as this is given in Figure 8.3. In order to maintain and control the temperature rise within the range of 20-30 °C, an overall system was built by combining three heating coils (in the range of 70-200 mΩ) within the last 50 cm length of the plumbing before the faucet. As shown in Figure 8.3, in order to develop the control system temperature sensors were placed at the input and the output ends of this short piece of retrofit plumbing, in addition to a flow rate sensor. Multiple energy storage banks were used in the early development stages, which can independently feed independent heating elements under the control of a processor-based system.

8.7.1 Early experimental results

Based on the above design approach, the research team developed a proof-of-concept laboratory system based on SC banks having different capacities and rated voltages. Thermocouples were inserted into both the inlet and outlet sides of the tube. The temperature versus time plots shown in Figure 8.4 are for various sizes of SC banks, switched with different delays.[1] Early experimental data indicated that it was a practically feasible solution.

Figure 8.4(b) shows one such SC bank switched into heater coils compared with the case of energy fed from the 230 V AC mains via an isolation transformer capable of

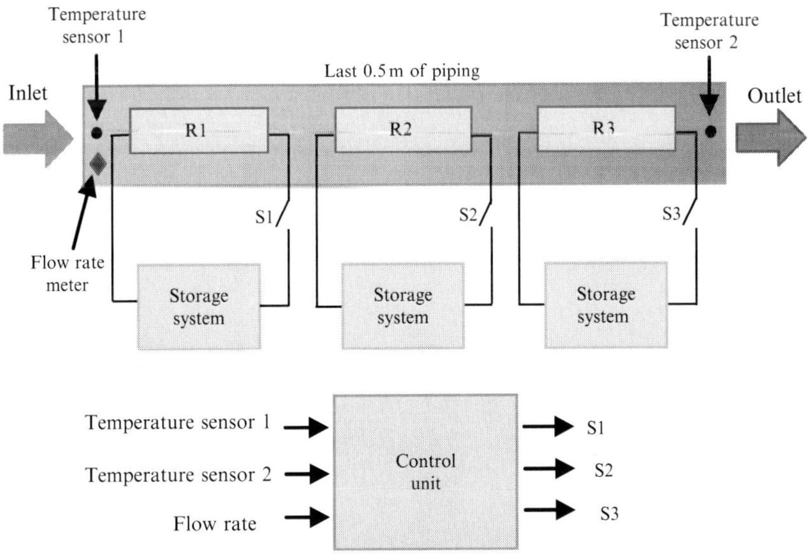

Figure 8.3 Complete system architecture of an SC-based inline water heating system.

[1]The details are commercially sensitive and hence not presented here.

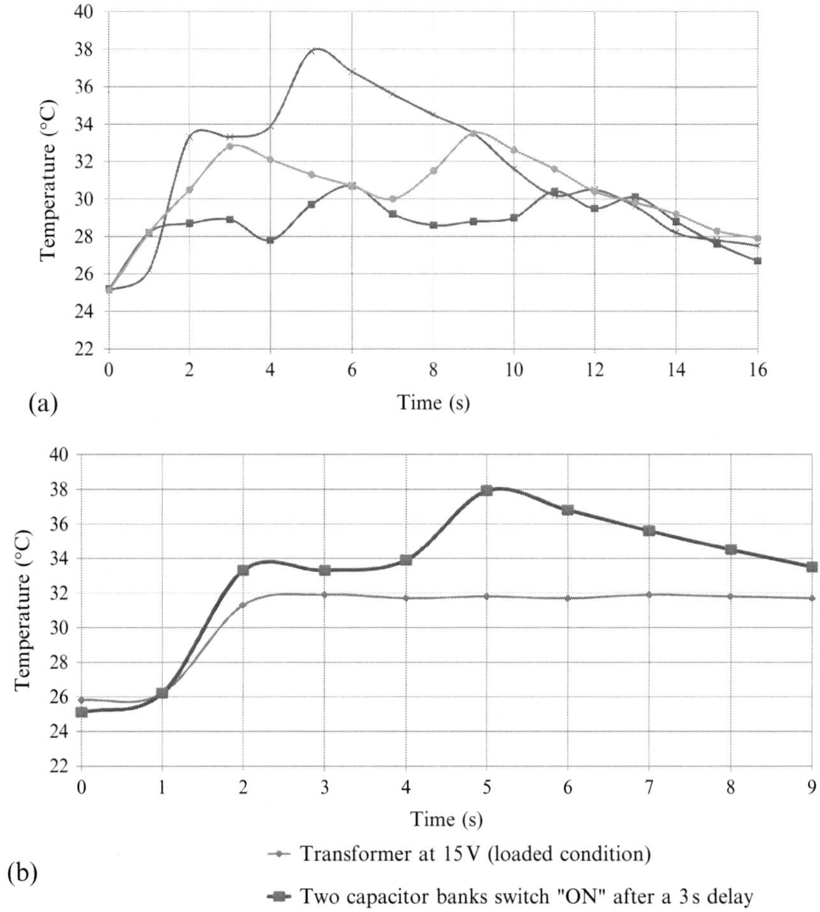

Figure 8.4 Temperature rise versus time in the inline water heating system configurations. (a) Switching 15 V capable SC banks with different change over delays and (b) Sc bank delivery compared with an isolation transformer with loaded terminal voltage of 15 V into a 100 mΩ heater coil.

maintaining approximately 15 V at the terminals of the heater coils, fed from a 2.3-kW rated socket outlet. Same terminal voltage capability SC bank pair was used on the same heater elements, with 3 s delay between the banks. As we clearly see, while the transformer can provide a steady state of temperature rise, its internal losses were prohibitive for a commercial solution.

Also, the transformer's initial start-up delay due to the LR circuit formed with the heating element was clearly observed as shown in Figure 8.4(b). SC banks can be upscaled to reach the required average power level (hence temperature rise) based on suitably prestored energy. In order to optimize the time constants (by modifying the resistance value of the inline heating element), different capacitor heater coil combinations were attempted.

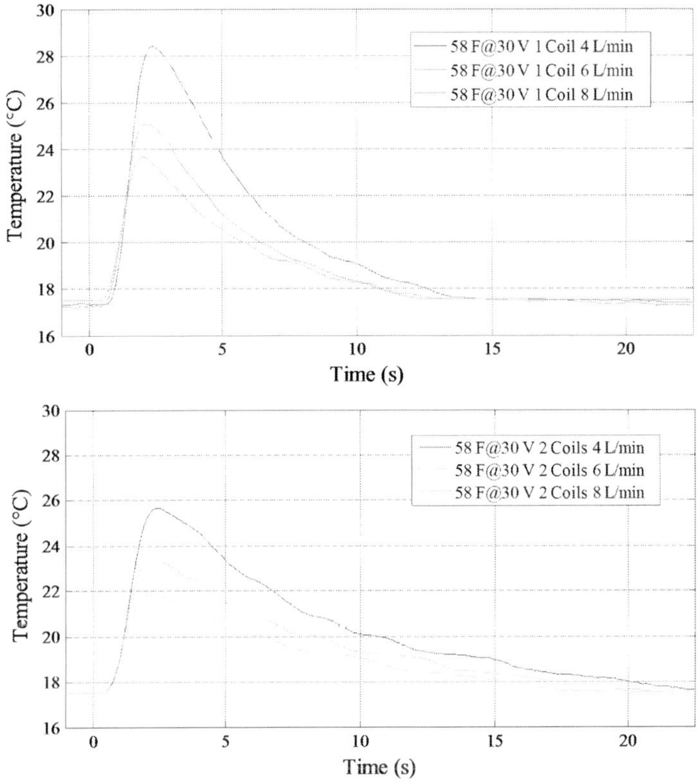

Figure 8.5 The outlet temperature versus time at different flow rates. (a) 90 mΩ heating element and (b) 180 mΩ heating element.

Figure 8.5 depicts the temperature of the outlet water versus time at different flow rates based on a series-connected SC bank energy storage with a terminal voltage rating of 30 V and an overall equivalent capacitance of 58 F. Given the examples with 90 and 180 mΩ heating coils at 4-8 l/min flow rates, we can see the possibility of tuning the circuit time constants to manage the heating process, which usually completes within less than 30 s, before the central gas heater kicks in.

8.8 Specific advantages of SC energy storage

Above proof-of-concept work clearly proves that the SC energy storage is an elegant solution due to the following reasons.

- **i.** With relatively low leakage commonly available SCs it is possible to precharge a bank and maintain almost 85% of the total energy at the end of a 8-h period. (i.e., night charging for early morning hot water usage).
- **ii.** Due to the statistical nature of the water usage, solution is practical with prestored energy.
- **iii.** Since the heater coil delivers the power continuous to the cold water only for a period of less than 30 s, overheating of system components due to excessive power levels does not occur.

iv. Practical time constants associated with typical SC banks and a heating coil (as per Figure 8.2(a)) are in the range of 1-10 s which is comparable with the delays associated with centrally heated hot water.

8.9 Implementation challenges

Several implementation challenges were identified during the course of the initial prototyping stage as given below:

i. Optimizing the SC bank with the required energy storage capability
ii. Limiting the energy transfer during the switching ON to prevent overheating or bubble forming in flowing water
iii. Developing an inline version of a high-power heater coil
iv. Sizing the SC bank with a cost consideration and current trends in SC prices (Brouji et al., 2009)
v. System integration aspects

An interesting research challenge set from this work is to develop a fast SC bank-charging system based on flowing options:

(a) Dividing the SC bank into three smaller subunits and to develop an energy circulating system
(b) Developing a fast charging system based on three concepts:

  i. Limited energy transfer into an SC bank at an extremely rapid charging rate
  ii. Investigating the possibility of extracting higher energy from AC mains for a short period of time, using the advantage of in-rush current allowance built into the residual current circuit breakers and other circuit breakers used in domestic systems
  iii. Developing special circuit topologies for fast but energy-limited charge transfers into SC banks

Work on these areas is currently the broader research interests of the SC applications research group at the University of Waikato. With the supercapacitor prices expected to drop heavily within the next 5 years (Ultracapacitor market forecast 2015-2020), developing this technique and other non-traditional applications further may open up new research and development areas in power lectronics.

References

Brouji, E., et al., 2009. Impact of calendar life and cycling ageing on supercapacitor performance. IEEE Trans. Veh. Technol. 58 (8), 3917–3929.

Coleman, M., et al., 2007. State-of-charge determination from EMF voltage estimation: using impedance, terminal voltage, and current for lead-acid and lithium-ion batteries. IEEE Trans. Ind. Electron. 54 (5), 2550–2557.

Datasheet, System pro M compact miniature circuit breaker S 200/S200M [online]. Available from: http://www05.abb.com/global/scot/scot209.nsf/veritydisplay/44c5b5ac208f3f25c1257ad7004ec86a/$file/2CDC002157D0202_view.pdf.

Fazekas, D.J., May 24, 1984. Control means and process for domestic hot water re-circulating system. US Patent 06613452.

Kularatna, N., Tilakaratne, L.N., Kumaran, P.K., November 2011. Design approaches to supercapacitor based surge resistant UPS techniques. In: Proceedings of IEEE-IECON 2011, pp. 4094–4099.

Kulartna, N., Gattuso, A., Gurusinghe, N., Jayasuriya, T., du Toit, J., November 2014. Prestored supercapacitor energy as a solution for burst energy requirements in domestic inline fast water heating systems. In: Proceedings of IECON 2014, USA.

Laing, O., April 11, 2013. Water delivery system and method for making hot water available in domestic hot water installation. US Patent 20130279891.

LS Mtron [online]. Available from: http://www.lsmtron.com/page/productMain.asp.

Maxwell Technologies [online]. Available from: http://www.maxwell.com/ultracapacitors/.

Miller, J.R., 2010. Introduction to electrochemical capacitor technology. IEEE Electr. Insul. Mag. 26 (4), 40–47.

Particular requirement for instantaneous water heaters. AS/NZS 60335.2.35, 2004.

Permenter, J.B., February 2, 2004. Insulated hot water storage tank for sink. US Patent 10788150.

Residual current operated circuit breakers without integral overcurrent protection for household and similar uses (RCCBs)—part 1: general rules. AS/NZS 61008.1:2011, 2011.

Ultracapacitor Market Forecast 2015-2020 [online]. Available from: http://www.marketresearchmedia.com/?p=912.

Appendix A: capacitors and AC line filtering

Co-author: Nicoloy Gurusinghe

One of the most common applications of capacitors is to apply them as AC line frequency filters in AC to DC converters. Usually a half-wave or a full-wave bridge rectifier is used together with an electrolytic capacitor of a sufficient value, to smoothen out the rectified waveform so that the output is an average DC voltage with a limited peak to peak 50/60 or 100/120 Hz ripple waveform superimposed on the DC average. In this short discussion we will compare the performance of electrolytic capacitors and supercapacitors in AC-DC converters to show why the present-day supercapacitors are not suitable for the specific application.

Figure A.1 depicts a simple full bridge rectifier and a capacitor connected to a load and the output voltage versus time. In a simplified case, we can show that,

$$V_{\mathrm{DC}} \approx V_{\mathrm{peak}} - 2V_{\mathrm{D}} - \frac{I_{\mathrm{L}}}{4fC} \tag{A.1}$$

where V_{DC} is the approximate DC output voltage, V_{D} is the diode forward voltage drop, I_{L} is the average DC load current, f is the AC line frequency, and C is the smoothing capacitor value. In this case we assume that the two diodes in series conduct for only a very short time within each half cycle of the waveform, and, the approximate time the capacitor discharges in each half cycle is very close to the half-cycle period of the AC input mains.

However, in practice, the capacitor used is not ideal and it comes with a fixed equivalent series resistance (ESR) in series with the capacitance value. Given this situation, we can analyze the situation using the case in Figure A.2, where the AC input source (and the diodes in series) presents an internal resistance of R_{S} and this resistor comes together with the capacitor which has an ESR of R_{C} in series with a capacitance value of C. Figure A.2(a) depicts the circuit, and Figure A.2(b) shows the Fourier series components of the rectified DC source (before smoothing by the capacitor), which presents a set of harmonic components of the rectified AC waveform, together with an average DC component.

Given the above nonideal situation, the capacitor is expected to filter out the ripple, and the output of the circuit is expected to provide a DC average voltage with minimized ripple frequency components. In this situation, the transfer function applied to the general case will be,

$$\frac{V_{\mathrm{o}}(s)}{V_{\mathrm{in}}(s)} = \frac{R_L(R_{\mathrm{C}}Cs+1)}{[(R_{\mathrm{L}}+R_{\mathrm{C}})R_{\mathrm{S}}+R_{\mathrm{L}}R_{\mathrm{C}}]Cs+(R_{\mathrm{L}}+R_{\mathrm{S}})} \tag{A.2}$$

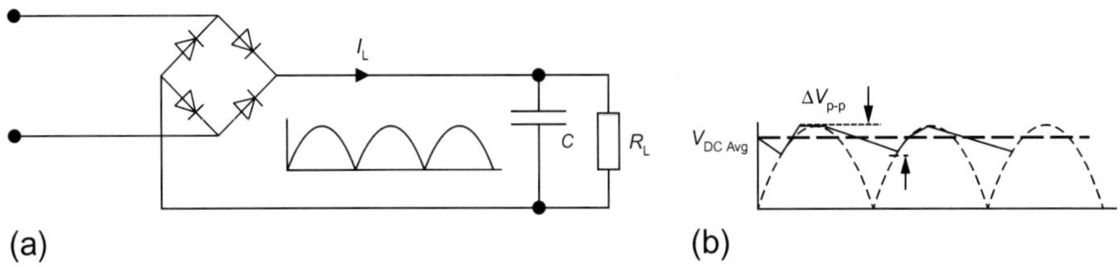

Figure A.1 Unregulated DC power supply derived from an AC source, using a bridge rectifier and a smoothing capacitor. (a) Simple case without a transformer and (b) approximate case of DC output with superimposed ripple.

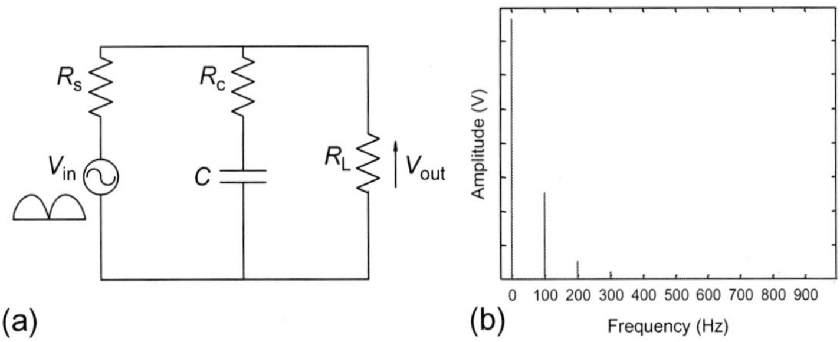

Figure A.2 Conceptual view of the rectification process considering nonideal case of capacitor and the AC source-rectifier combination. (a) Conceptual circuit with AC source rectifier and (b) spectrum of the full-wave rectified sine waveform.

The purpose of the capacitor is to act as a low-pass filter to minimize the harmonic components generated by the rectification process, and to achieve a higher DC average value. Figure A.3(a) depicts the rectified voltage waveform and the current drawn from the AC source. In this case we have used components with details as in Table A.1, with a bridge rectifier. Figure A.3(b) indicates fast Fourier transform (FFT) waveform of the rectified output, in this case without any capacitor. As we see from Figure A.3(c) with the electrolytic capacitor added to the circuit, DC output becomes smooth with only a small ripple superimposed. Figure A.3(d) indicates the FFT plot where DC component has significantly increased (note that FFT plot is in dB scale) compared to the case in Figure A.3(b), with most higher order harmonics are suppressed due to capacitor acting as an effective low-pass filter.

Figure A.3(e) indicates the case of a 25 F supercapacitors in series used instead of an electrolytic capacitor. As seen here input current drawn by the circuit is significantly increased, since at 100 Hz and higher order harmonics, the impedance shown by the capacitor is gradually decreased, compared its approximately constant ESR. Additional current drawn from the input creates heat in the two resistors (output resistance of the AC source, R_s and the capacitor ESR).

It is important to note here that the electrolytic capacitor acts as a ripple filter due to its low time constant compared with the time of the AC waveform's half period

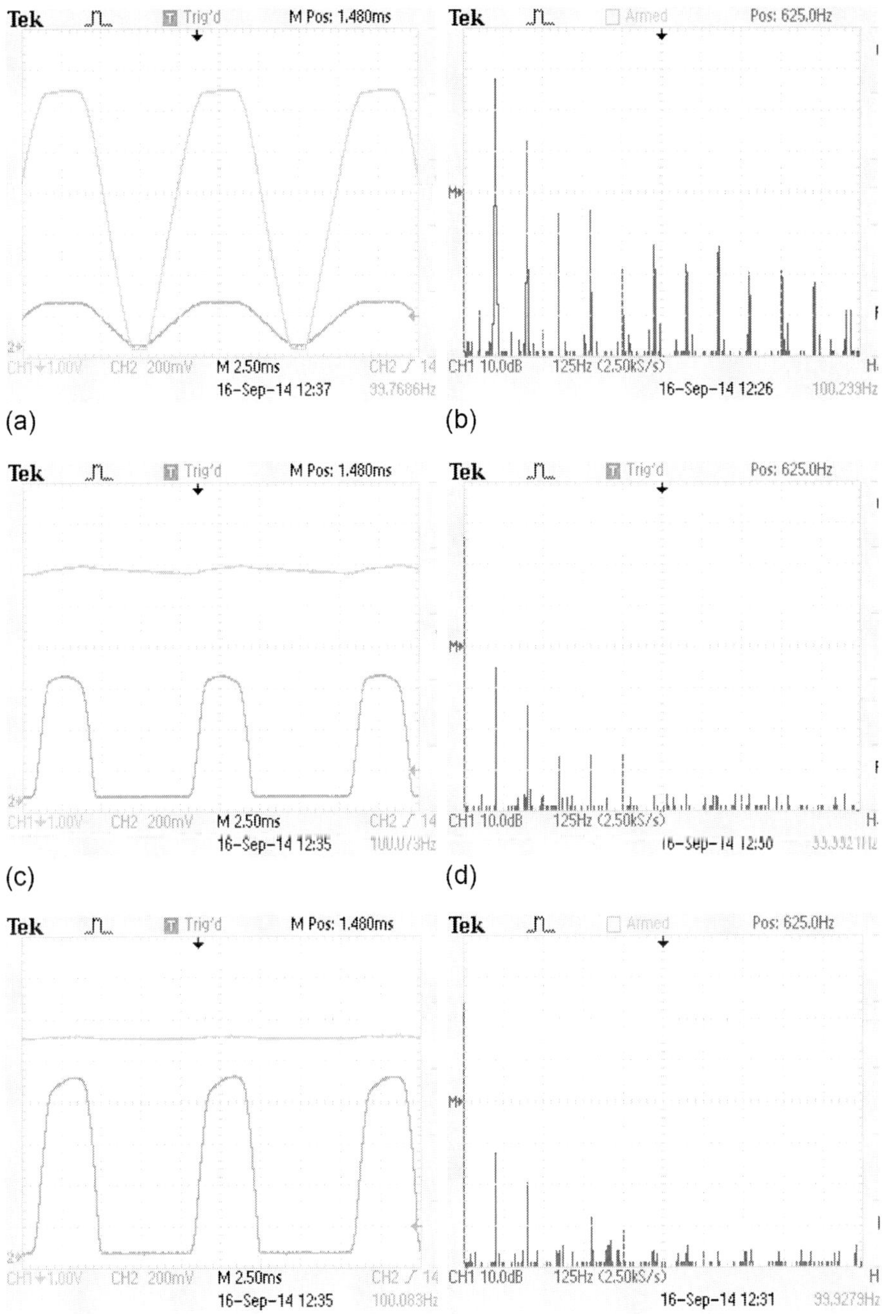

Figure A.3 Time-domain plots (oscillographs) and FFT plots of lightly loaded full-wave rectified case and filtered output. (a) Time-domain plots with no capacitor connected, (b) FFT plot with no capacitor, (c) time-domain plots with an electrolytic capacitor, (d) FFT plot with electrolytic capacitor, (e) time-domain plot with a supercapacitor, and (f) FFT plot with supercapacitor.

Table A.1 Test setup data used to show the impact of different capacitors as AC line filters

Parameter	Value	Remarks
AC output from the isolation transformer	5 V rms	Approx. 200 W
Load resistance	20 Ω	±10% tolerance
Electrolytic capacitor	3300 μF, ESR of 20 mΩ	Nichicon corporation
Supercapacitors (two in series)	25 F, 42 mΩ	Maxwell technologies

duration of 10 ms. If we look at the FFT plots carefully, we can see that the DC component at the output is increased by about 25 dB compared to the case of rectifier bridge only.

A careful look at the DC component in Figure A.3(f) for the supercapacitors indicates that it is about 2 dB lower than the case of electrolytic capacitor in Figure A.3(d). Bode plots in Figure A.4 indicates the two cases where the electrolytic capacitor providing a useful low-pass filtering action with about a phase angle of around −60°, whereas the supercapacitors showing a phase angle of almost zero degrees around 100 Hz frequency.

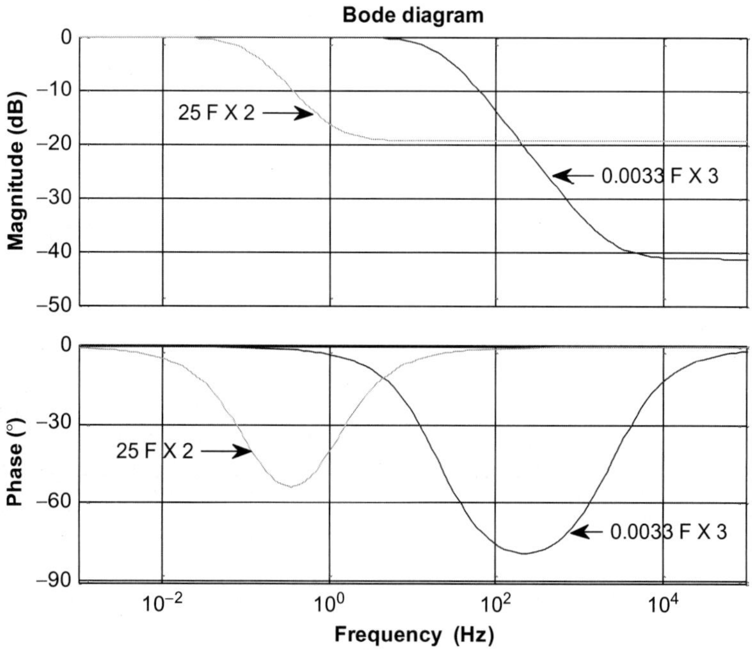

Figure A.4 Gain and phase plots for two cases of filters.

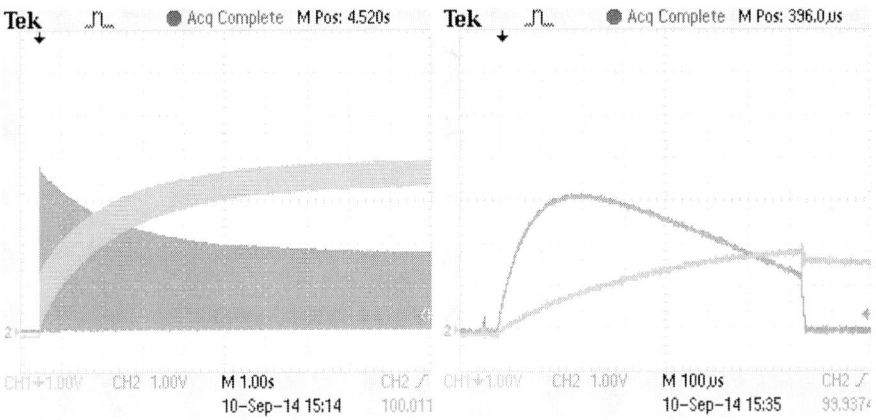

Figure A.5 Time delays to stabilize the DC output of capacitor filters. (a) 25 F supercapacitor and (b) 3300 μF electrolytic capacitor.

Due to supercapacitor impedance reduced at 100 Hz and above significantly, current drawn from the rectified AC source increases due to the simple voltage divider formed due to R_S (combined with R_L) and R_C and without the additional impedance of the capacitor. This means that the supercapacitor circuit wastes energy as heat inside these two components. This is the impact of large time constant of the supercapacitor in general compared to the electrolytic capacitor.

Another issue with supercapacitors is the initial time to stabilize the DC output. This is indicated in Figure A.5. In the case of supercapacitor in Figure A.5(a) it takes about 6 s to stabilize the DC output, whereas in Figure A.5(b) it takes about 750 μS only to stabilize the DC output.

In summary, this discussion details the case of present-day supercapacitors' inability to perform AC line filtering in AC-DC converters, due to their long-time constants compared to electrolytic capacitors. This issue is discussed in Chapter 5. Miller et al. (2011) and Burke and Miller (2008) provide more details and present approaches to develop better supercapacitors suitable for AC ripple filtering.

References

Burke, A.F., Miller, J.R., 2008. Electrochemical capacitors: challenges and opportunities for real-world applications. Electrochem. Soc. Interface 17, 53.

Miller, J.R., Outlaw, R.A., Holloway, B.C., 2011. Graphene electric double layer capacitor with ultra-high-power performance. Electrochim. Acta 56, 10443–10449.

Index

Note: Page numbers followed by *np* indicate footnote, *f* indicate figures and *t* indicate tables.

D

E

F

G

H

Edwards Brothers Malloy
Thorofare, NJ USA
December 1, 2014